C0-CDB-073

Lipid Analysis

The Practical Approach Series

SERIES EDITORS

D. RICKWOOD
Department of Biology, University of Essex
Wivenhoe Park, Colchester, Essex CO4 3SQ, UK

B. D. HAMES
Department of Biochemistry and Molecular Biology, University of Leeds
Leeds LS2 9JT, UK

Affinity Chromatography
Anaerobic Microbiology
Animal Cell Culture (2nd Edition)
Animal Virus Pathogenesis
Antibodies I and II
Biochemical Toxicology
Biological Membranes
Biomechanics—Materials
Biomechanics—Structures and Systems
Biosensors
Carbohydrate Analysis
Cell–Cell Interactions
Cell Growth and Division
Cellular Calcium
Cellular Neurobiology
Centrifugation (2nd Edition)
Clinical Immunology
Computers in Microbiology
Crystallization of Nucleic Acids and Proteins
Cytokines
The Cytoskeleton
Diagnostic Molecular Pathology I and II
Directed Mutagenesis
DNA Cloning I, II, and III
Drosophila
Electron Microscopy in Biology
Electron Microscopy in Molecular Biology
Electrophysiology
Enzyme Assays
Essential Molecular Biology I and II
Experimental Neuroanatomy

Fermentation

Flow Cytometry

Gel Electrophoresis of Nucleic Acids (2nd Edition)

Gel Electrophoresis of Proteins (2nd Edition)

Genome Analysis

Haemopoiesis

HPLC of Macromolecules

HPLC of Small Molecules

Human Cytogenetics I and II (2nd Edition)

Human Genetic Diseases

Immobilised Cells and Enzymes

In Situ Hybridization

Iodinated Density Gradient Media

Light Microscopy in Biology

Lipid Analysis

Lipid Modification of Proteins

Lipoprotein Analysis

Liposomes

Lymphocytes

Mammalian Cell Biotechnology

Mammalian Development

Medical Bacteriology

Medical Mycology

Microcomputers in Biochemistry

Microcomputers in Biology

Microcomputers in Physiology

Mitochondria

Molecular Genetic Analysis of Populations

Molecular Neurobiology

Molecular Plant Pathology I and II

Monitoring Neuronal Activity

Mutagenicity Testing

Neural Transplantation

Neurochemistry

Neuronal Cell Lines

Nucleic Acid and Protein Sequence Analysis

Nucleic Acids Hybridisation

Nucleic Acids Sequencing

Oligonucleotides and Analogues

Oligonucleotide Synthesis

PCR

Peptide Hormone Action

Peptide Hormone Secretion

Photosynthesis: Energy Transduction

Plant Cell Culture

Plant Molecular Biology

Plasmids

Pollination Ecology

Postimplantation Mammalian Embryos

Preparative Centrifugation

Prostaglandins and Related Substances

Protein Architecture

Protein Engineering

Protein Function

Protein Purification Applications

Protein Purification Methods

Protein Sequencing

Protein Structure

Protein Targeting

Proteolytic Enzymes

Radioisotopes in Biology

Receptor Biochemistry

Receptor–Effector Coupling

Receptor–Ligand Interactions

Ribosomes and Protein Synthesis

Signal Transduction

Solid Phase Peptide Synthesis

Spectrophotometry and Spectrofluorimetry

Steroid Hormones

Teratocarcinomas and Embryonic Stem Cells

Transcription and Translation

Virology

Yeast

Lipid Analysis

A Practical Approach

Edited by

RICHARD JOHN HAMILTON
Chemical and Physical Sciences,
Liverpool John Moores University
Liverpool

and

SHIELA HAMILTON
The Open University
Chorlton House
Manchester

OXFORD UNIVERSITY PRESS
Oxford New York Tokyo

Oxford University Press, Walton Street, Oxford OX2 6DP
Oxford New York Toronto
Delhi Bombay Calcutta Madras Karachi
Kuala Lumpur Singapore Hong Kong Tokyo
Nairobi Dar es Salaam Cape Town
Melbourne Auckland Madrid
and associated companies in
Berlin Ibadan

Oxford is a trade mark of Oxford University Press

A Practical Approach is a registered trade mark of the Chancellor, Masters, and Scholars of the University of Oxford trading as Oxford University Press*

Published in the United States
by Oxford University Press Inc., New York

A catalogue record for this book is available from the British Library

Library of Congress Cataloging in Publication Data
Lipid analysis: a practical approach/edited by Richard John Hamilton and Shiela Hamilton.—1st ed.
p. cm.—(Practical approach series)
Includes bibliographical references and index.
1. Lipids—Analysis—Technique. I. Hamilton, R. J. (Richard John) II. Hamilton, Shiela III. Series.
QP751.L545 1993 92-20910
ISBN 0–19–963098–4 (h/b)
ISBN 0–19–963099–2 (p/b)

Typeset by Cambrian Typesetters, Frimley, Surrey
Printed in Great Britain by
Information Press Ltd, Eynsham, Oxford

Preface

It is forty years since the invention of gas liquid chromatography by A. J. P. Martin and A. T. James revolutionized not only separation science in general but more particularly the study of lipids. Before GLC, lipid analysis depended on distillation and fractional crystallization—to achieve 'pure' materials for further study. The first molecules that Martin and James separated were fatty acids, and interest, stimulated by their discovery, revitalized the study of lipids.

On this ruby anniversary of the first GLC separation of fatty acids, it is appropriate that a new text should be available which explains how many additional techniques can be used for lipid analysis. It is especially apposite that the text is largely written by workers from Liverpool, which has been the centre of lipid studies for over seventy years since the father of British lipid chemistry, Professor T. P. Hilditch, worked at Liverpool University. Liverpool is still one of the three main centres of the oils and fats industry in the United Kingdom.

One illustrious non-Liverpudlian chapter author is Professor Frank Gunstone, who nevertheless has close links with Liverpool where he worked with Professor Hilditch. Professor Gunstone's introductory chapter outlines how lipids are grouped and named. In Chapter 7 Professor Gunstone explains how high resolution nuclear magnetic resonance spectroscopy (NMR) can be used to determine the nature of commercial fats and oils.

Chapter 2 explains how lipids can be extracted from natural sources, animal, vegetable, and mineral. Classical methods are described—as are methods based on new phase separation materials developed from high performance lipid chromatography (HPLC) stationary phases. This chapter also includes derivative formation for GLC, HPLC, and mass spectrometry (MS).

Chapter 3, by two authors from north of the border, describes one of the cheapest and more versatile techniques available to the lipid scientist, thin layer chromatography (TLC). The separations by GLC are mainly illustrated in Chapter 4, by new separations on capillary columns, with emphasis being given to the care of such columns.

HPLC is now available for lipid class separations, though it is less needed for separation of methyl esters. The availability of the 'mass' detector and a splitter will allow HPLC to be used as a preparative technique. The technique is outlined in Chapter 5.

Although ^{13}C NMR is making inroads into the study of lipids in biological systems, radio tracer work is still important and is well covered with illustrative experiments in Chapter 6.

NMR was initially less than valuable for lipid chemists when only ^{1}H

spectra were available. Methyl, methylene, and methine hydrogens attached to unusual functional groups (e.g. epoxide, alcohol, and ketone) were readily noted but the methylene 'envelope' for a C_{16} or C_{18} acid was less than helpful. ^{13}C NMR has opened up many of the secrets hidden by ^{1}H NMR and these are discussed in Chapter 7.

Chapter 8, on mass spectrometry, deals with the fragmentation patterns which confirm the presence of specific lipids in natural sources.

The book is aimed at the beginner in lipid analysis, whether chemist, ecologist, biologist, or microbiologist. It should be possible to follow the protocols with only a rudimentary appreciation of lipids. However, there are sufficient new examples in the text to make it of interest to the connoisseur of lipid analysis.

Liverpool R. J. HAMILTON
December 1991

Contents

3. Thin-layer chromatography 65

R. James Henderson and Douglas R. Tocher

Contributors

RICHARD P. EVERSHED
Department of Biochemistry, University of Liverpool, PO Box 147, Liverpool, L69 3BX, UK.

FRANK D. GUNSTONE
Chemistry Department, The University, St Andrews, Fife, Scotland, UK.

RICHARD J. HAMILTON
School of Chemical and Physical Sciences, Liverpool John Moores University, Byrom St., Liverpool L3 3AF, UK.

SHIELA HAMILTON
The Open University, Chorlton House, 70 Manchester Road, Chorlton-cum-Hardy, Manchester M21 1PQ, UK.

R. JAMES HENDERSON
NERC Unit of Aquatic Biochemistry, Department of Biological and Molecular Science, University of Stirling, Stirling FK9 4LA, Scotland, UK.

PETER A. SEWELL
School of Chemical and Physical Sciences, Liverpool John Moores University, Byrom Street, Liverpool L3 3AF, UK.

COLIN G. TAYLOR
School of Chemical and Physical Sciences, Liverpool John Moores University, Byrom Street, Liverpool L3 3AF, UK.

DOUGLAS R. TOCHER
NERC Unit of Aquatic Biochemistry, Department of Biological and Molecular Science, University of Stirling, Stirling FK9 4LA, Scotland, UK.

Abbreviations

ACTH	adrenocorticotropic hormone
ANS	1-analino-8-naphthalene sulfonate
BHA	butylated hydroxyanisole
BHT	butylated hydroxytoluene
BMP	2-bromophenacyl-1-methylpyridine
BSTFA	*N,O*-bis (trimethylsilyl) trifluoroacetamide
CI	chemical ionization
DCF	dichlorofluorescein
DCI	desorption chemical ionization
DEAE	diethylaminoethyl
DG	diglyceride
DMCS	dimethyldichlorosilane
DMDS	dimethyl disulfide
DMEQ	dimethoxy-1-methyl-2(1H)-quinoxalinone
DPH	diphenylhexatriene
ECL	equivalent chain-length
EI	electron ionization
ESCRs	external standard channel ratios
FAB	fast atom bombardment
FAMEs	fatty acid methyl esters
FFA	free fatty acids
FID	flame ionization detector
FSH	follicle-stimulating hormone
FTID	flame thermionic ionization detector
GC	gas chromatography
GLC	gas–liquid chromatography
HPLC	high-performance liquid chromatography
LPC	lysophosphatidylcholine
LPE	lysophosphatidylethanolamine
LSC	liquid scintillation counting
MG	monoglyceride
MS	mass spectrometry
MTBE	methyltertiarybutylether
NARP	non-aqueous reversed-phase chromatography
NICI	negative ion chemical ionization
NPB	*N*-phenyl-1-naphthylamine
ODS	octadecylsilane
PBPB	*p*-bromophenacyl bromide
PC	phosphatidylcholine
PE	phosphatidylethanolamine
PFB	pentafluorobenzyl
PI	phosphatidylinositol
PM	photomultiplier

PS	phosphatidylserine
PTV	programmed temperature vaporizing
QIP	quench index parameter
RI	refractive index
RIA	radioimmunoassay
RPTLC	reversed-phase thin-layer chromatography
SFC	supercritical fluid chromatography
SIDA	substoichiometric isotope dilution analysis
SIM	selected ion monitoring
SM	sphingomyelin
TBDMS	*tert.*-butyldimethylsilyl
TEAE	triethylaminoethyl
THF	tetrahydrofuran
TIC	total ion current
TLC	thin-layer chromatography
TMS	trimethylsilyl
TMSI	trimethylsilyl imidazole
UV	ultraviolet

1

Introduction

FRANK D. GUNSTONE

1. Introduction

The analysis of lipids can be undertaken to answer at least four questions at both the qualitative and quantitative level:

(a) what fatty acids (or alcohols or amines) are present,
(b) what different lipid classes are present,
(c) what is the fatty acid composition of each separate lipid class, and
(d) since lipids generally contain two or three acyl chains, how are these associated in individual lipid molecules?

This chapter is devoted to the structural nature of the acids (alcohols, amines) and the lipids so that these analytical questions and their answers can be understood.

Prior to the development of chromatographic methods of separation these questions were answered by measuring *average* properties such as unsaturation (iodine value), chain-length (saponification equivalent), or phosphorus content. These have now been replaced by ever-improving chromatographic separation procedures accompanied, when necessary, by spectroscopic methods of identification. These topics are covered in the following chapters of this book.

2. Fatty acids

2.1 Major structural features

Because fatty acids are made biosynthetically from a limited number of substrates by a limited number of pathways certain structural features recur frequently. More than 1000 acids have now been identified from natural sources but the number occurring frequently in the most common lipids is much fewer than this. Most lipid chemists and biochemists will probably encounter not more than 50 different acids.

On the basis of these many fatty acid structures it is possible to make four general statements. Although each of these generalizations is valid there are exceptions to each one and some of these are themselves important.

(a) Most fatty acids are straight-chain compounds with an even number of carbon atoms in each molecule. Chain-lengths range from C_2 to C_{80} but the C_{12}–C_{22} members are most common. Acids with an odd number of carbon atoms or branched or cyclic acids are also known.

(b) Mono-unsaturated acids have one double bond with *Z* (*cis*) configuration which usually occurs in one of a limited number of positions in the carbon chain. Acids with *E* (*trans*) olefinic unsaturation and acetylenic unsaturation have, however, been reported.

(c) Polyunsaturated acids have two or more *Z* (*cis*) double bonds each separated by one methylene group. This means that the pentadiene unit —$CH{=}CHCH_2CH{=}CH$— occurs once or more in most of these acids though other patterns of unsaturation are also known.

(d) Substituted acids are rare but natural hydroxy and epoxy acids have been recognized.

Examples of structures displaying these features will be given in the following sections.

All fatty acids have a systematic name which, in its full form, is a complete description of the structure but most also have a trivial name. These latter are much used because they are simpler despite the fact that they are less informative. Another common descriptor consists of numbers such as 18:1. The digits before the colon indicate the total number of carbon atoms in the chain (assumed to be unbranched) and the digit after the colon indicates the number of unsaturated centres (assumed to be *Z*-olefinic). If desired, further symbols may be added to indicate the position and configuration of the unsaturation.

When assigning a systematic name to a fatty acid numbering starts at the carboxyl group. There are, however, occasions when it is better to number from the end methyl group. Such numbers were prefixed by the Greek letter ω but *n*- is now preferred (*n* indicating the total number of carbon atoms in the chain). These points are illustrated in the descriptions of linoleic acid in *Figure 1*.

$$CH_3(CH_2)_4\overset{Z}{CH{=}CH}CH_2\overset{Z}{CH{=}CH}(CH_2)_7COOH$$

COOH

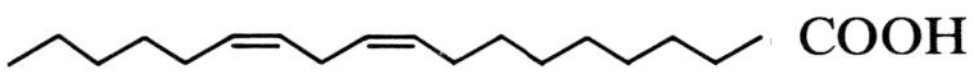

linoleic acid, 18:2 (9*Z*, 12*Z*), 18:2 (*n*-6)
cis-9, *cis*-12-octadecadienoic acid

Figure 1. Linoleic acid

2.2 Saturated acids

Saturated acids have the general formula $CH_3(CH_2)_nCOOH$ where n is usually an even number and the acid has $n + 2$ carbon atoms. Some of the more important members are listed and named in *Table 1* along with the melting point of the acid and the boiling point of the methyl ester.

Table 1. The more common natural saturated acids

Chain length	Systematic name	Trivial name	m.p. (°C) acid	b.p. (°C) methyl ester
2	ethanoic	acetic	16.6	57
4	butanoic	butyric	–5.3	103
6	hexanoic	caproic	–3.2	151
8	octanoic	caprylic	16.5	195
10	decanoic	capric	31.6	228
12	dodecanoic	lauric	44.8	262
14	tetradecanoic	myristic	54.4	114/1 mm
16	hexadecanoic	palmitic	62.9	136/1 mm
18	octadecanoic	stearic	70.1	156/1 mm
20	icosanoic (eicosanoic)	arachidic	76.1	188/2 mm

2.3 Monoene acids

Monoene acids have the general structure $CH_3(CH_2)_nCH{=}CH(CH_2)_mCOOH$. The best-known example is oleic acid in which $n = m = 7$. It is a C_{18} acid with unsaturation between C–9 and C–10. Some other common members of this class are listed in *Table 2*. They include C_{16} to C_{22} acids and four C_{18} isomers.

Table 2. The more common natural monoene acids[a]

Symbol	Name	Family
16:1 (9)	palmitoleic	*n*-7
18:1 (9)	oleic[b]	*n*-9
18:1 (6)	petroselinic	*n*-12
18:1 (11)	*cis*-vaccenic	*n*-7
20:1 (9)	gadoleic	*n*-11
20:1 (11)	cetoleic	*n*-9
22:1 (13)	erucic	*n*-9

[a] All are *Z* (*cis*) isomers
[b] *E* (*trans*) isomer is called elaidic

2.4 Polyene acids

The important polyene acids have two or more *Z* (*cis*) double bonds separated from each other by a single methylene group. Linoleic acid (*Figure 1*) is a typical member of this group.

The most important polyene acids fall into families the members of which are biosynthetically related. All the members of a family have a common structural feature *viz.* $CH_3(CH_2)_xCH{=}$. The two commonest families *n*-6 ($x = 4$) and *n*-3 ($x = 1$) are detailed in *Table 3*.

Table 3. The *n*-6 and *n*-3 families of polyene acids[a]

n-6 family		
$CH_3(CH_2)_4(CH{=}CHCH_2)_2(CH_2)_6COOH$	linoleic	18:2 (*n*-6)
$CH_3(CH_2)_4(CH{=}CHCH_2)_3(CH_2)_3COOH$	γ-linolenic	18:3 (*n*-6)
$CH_3(CH_2)_4(CH{=}CHCH_2)_3(CH_2)_5COOH$	dihomo-γ-linolenic	20:3 (*n*-6)
$CH_3(CH_2)_4(CH{=}CHCH_2)_4(CH_2)_2COOH$	arachidonic	20:4 (*n*-6)
$CH_3(CH_2)_4(CH{=}CHCH_2)_4(CH_2)_4COOH$	–	22:4 (*n*-6)
$CH_3(CH_2)_4(CH{=}CHCH_2)_5CH_2COOH$	–	22:5 (*n*-6)
n-3 family		
$CH_3CH_2(CH{=}CHCH_2)_3(CH_2)_6COOH$	α-linolenic	18:3 (*n*-3)
$CH_3CH_2(CH{=}CHCH_2)_4(CH_2)_3COOH$	stearidonic	18:4 (*n*-3)
$CH_3CH_2(CH{=}CHCH_2)_4(CH_2)_5COOH$	–	20:4 (*n*-3)
$CH_3CH_2(CH{=}CHCH_2)_5(CH_2)_2COOH$	EPA[b]	20:5 (*n*-3)
$CH_3CH_2(CH{=}CHCH_2)_5(CH_2)_4COOH$	–	22:5 (*n*-3)
$CH_3CH_2(CH{=}CHCH_2)_6CH_2COOH$	DHA[c]	22:6 (*n*-3)

[a] All unsaturated centres have *Z* (*cis*) configuration
[b] E icosapentaenoic acid
[c] Docosahexaenoic acid

2.5 Other acids (non-oxygenated)

Most of the common natural fatty acids belong to one of the three classes discussed in Sections 2.2–2.4: saturated, monoenes, and methylene-interrupted polyenes. In this section there is a brief indication of other types of natural unsaturated acids whilst the following Section 2.6 covers oxygenated acids.

(a) Acids with conjugated unsaturation. These are mainly C_{18} acids with 9,11,13 or 8,10,12 triene conjugation and are of seed origin. Two examples are:

18:3 (9*Z*,11*E*,13*E*) α-eleostearic acid

18:3 (8*E*,10*E*,12*Z*) calendic acid

(b) Acids with acetylenic or allenic unsaturation. An interesting example of the former type is crepenynic acid with the structure:

$$CH_3(CH_2)_4C{\equiv}CCH_2CH{=}CH(CH_2)_7COOH$$

Laballenic acid is an example of the very small group of acids with an allene unit. It has the structure:

$$CH_3(CH_2)_{10}CH{=}C{=}CH(CH_2)_3COOH$$

(c) Branched-chain acids are mainly of animal origin. The substituent group is most commonly methyl though more than one such group may be present. Examples include iso acids, ante-iso acids, and polybranched acids:

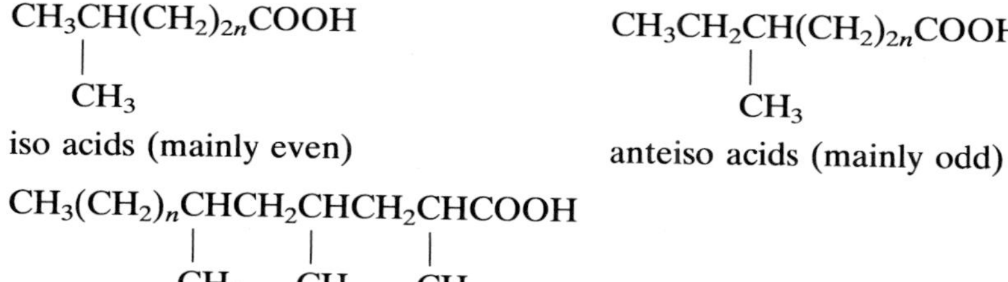

2,4,6-trimethylalkanoic acids (n = 15, 17, or 19)

(d) Acids containing a ring system most commonly have a cyclopropene ring (present in some seed oils), a cyclopropane ring (present in bacterial lipids), or a cyclopentene ring (present in some seed oils). Examples of these are as follows:

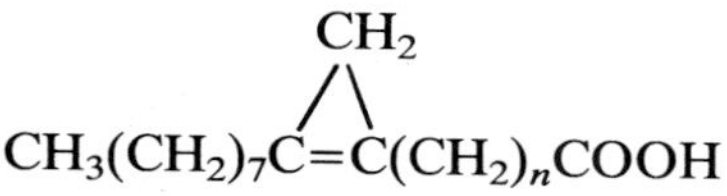

$$CH_3(CH_2)_7\overset{CH_2}{C{=}C}(CH_2)_nCOOH$$

n = 6 malvalic acid

n = 7 sterculic acid

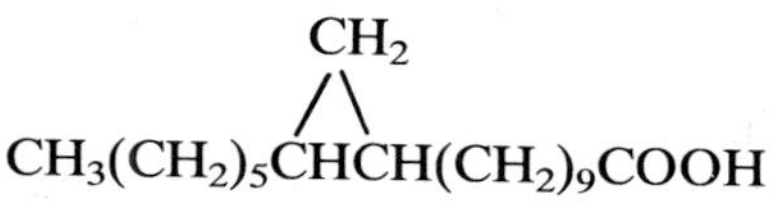

$$CH_3(CH_2)_5\overset{CH_2}{CHCH}(CH_2)_9COOH$$

lactobacillic acid

$$\text{(cyclopentenyl)}{-}(CH_2)_nCOOH$$

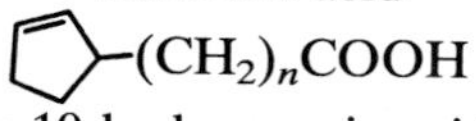

n = 10 hydnocarpic acid (C_{16})

n = 12 chaulmoogric (C_{18})

(e) Acids with *E* unsaturation in place of the more common *Z* configuration occur among the conjugated polyene acids [see (a) above] but also in acids such as vaccenic and hexadec-3-enoic.

$$CH_3(CH_2)_5CH\overset{E}{=}CH(CH_2)_9COOH$$

vaccenic acid (18:1)

$$CH_3(CH_2)_{11}CH\overset{E}{=}CHCH_2COOH$$

hexadec-3-enoic acid

(f) Some polyene acids are not completely methylene-interrupted in their patterns of unsaturation. For example, some acids have Δ^5 unsaturation as well as at the more common 9 and 9,12 positions.

e.g. $CH_3(CH_2)_4CH{=}CHCH_2CH{=}CH(CH_2)_2CH{=}CH(CH_2)_3COOH$
18:3 (5,9,12)

2.6 Other acids (oxygenated)

Acids with additional functional groups are uncommon but a number of hydroxy, epoxy, and furanoid acids have been identified. Some examples of these are given below:

$CH_3(CH_2)_5CH(OH)CH_2CH{=}CH(CH_2)_7COOH$ ricinoleic acid (12-hydroxyoleic)

$CH_3(CH_2)_4CH{=}CHCH_2CH_2CH(OH)(CH_2)_7COOH$ isoricinoleic acid

$CH_3(CH_2)_4\overset{O}{\widehat{CHCH}}CH_2CH{=}CH(CH_2)_7COOH$ vernolic acid

$CH_3(CH_2)_4CH{=}CHCH_2\overset{O}{\widehat{CHCH}}(CH_2)_7COOH$ coronaric acid

(furanoid acid: furan ring bearing H_3C and CH_3 at positions 3 and 4, $CH_3(CH_2)_4$ and $(CH_2)_{10}COOH$ at positions 2 and 5)

3. Other monofunctional compounds

Lipid chemistry is dominated by the long-chain acids produced by hydrolysis of acylglycerols, phospholipids, and sterol esters but some lipids furnish long-chain alcohols (wax esters) or aldehydes (present in waxes and produced by reaction of αβ-unsaturated ethers or sphingosine bases).

$RCOOH$	$RCHO$	RCH_2OH
carboxylic acids	aldehydes	alcohols

Apart from the change in functional group the aldehydes and alcohols structurally resemble the acids. For the most part, the natural aldehydes and alcohols are saturated or monoene compounds. Polyene aldehydes and alcohols are less common.

4. Lipids

Fatty acids (and alcohols and aldehydes) usually occur naturally in a combined form. They are most often present as esters though they form amides in the sphingolipids. Fatty acids are generally linked to glycerol either alone as mono-, di-, or tri-acylglycerols or in combination with an additional structural unit as in the phospholipids, glycosyldiacylglycerols, and ether lipids. Alternatively the fatty acids may be esterified with sterols (sterol esters) or with long-chain alcohols (wax esters) in place of glycerol.

Some general terms which may be used in this book are listed below. These lipid types contain the material indicated in parenthesis: glycerolipid (glycerol), phospholipid (phosphoric acid), glycolipid (carbohydrate), sulfolipid (sulfur-containing group), sphingolipid (sphingosine or other long-chain base), and ether lipid (long-chain alkyl group combined as an ether).

5. Acylglycerols (glycerides)

Reserve lipids are mainly triacylglycerols (triglycerides). These are usually complex mixtures produced from a pool of several fatty acids and occurring in isomeric forms. For example, even two acids, such as palmitic and oleic, can form eight possible triacylglycerols (*Figure 2*). This number rises rapidly with an increasing number of fatty acids (*Table 4*) and even when some possibilities of isomerism are ignored the mixture is still too complicated to be fully investigated by our present analytical facilities.

$\mathrm{CH_2O}P$	$\mathrm{CH_2O}P$	$\mathrm{CH_2O}P$	$\mathrm{CH_2O}Ol$	$\mathrm{CH_2O}P$	$\mathrm{CH_2O}Ol$	$\mathrm{CH_2O}Ol$	$\mathrm{CH_2O}Ol$
|	|	|	|	|	|	|	|
$\mathrm{CHO}P$	$\mathrm{CHO}P$	$\mathrm{CHO}Ol$	$\mathrm{CHO}P$	$\mathrm{CHO}Ol$	$\mathrm{CHO}Ol$	$\mathrm{CHO}P$	$\mathrm{CHO}Ol$
|	|	|	|	|	|	|	|
$\mathrm{CH_2O}P$	$\mathrm{CH_2O}Ol$	$\mathrm{CH_2O}P$	$\mathrm{CH_2O}P$	$\mathrm{CH_2O}Ol$	$\mathrm{CH_2O}P$	$\mathrm{CH_2O}Ol$	$\mathrm{CH_2O}Ol$

Figure 2. Triacylglycerols containing only palmitic (*P*) and/or oleic acids (*Ol*)

Table 4. The number of triacylglycerols containing 5, 10, or 20 fatty acids

Number of fatty acids	5	10	20
all isomers	125	1000	8000
excluding optical isomers	75	550	4200
no isomers distinguished	35	220	1540

The physical state of a triglyceride mixture depends mainly on the constituent fatty acids. When these are largely saturated the lipid is solid at ambient temperature and is called a fat or butter (for example, cocoa butter). With more unsaturated acids the triacylglycerol mixture is liquid at room temperature and the product is an oil.

Diacylglycerols and monoacylglycerols occur naturally only as minor components. The possibilities of isomerism in these esters is set out in *Figure 3*.

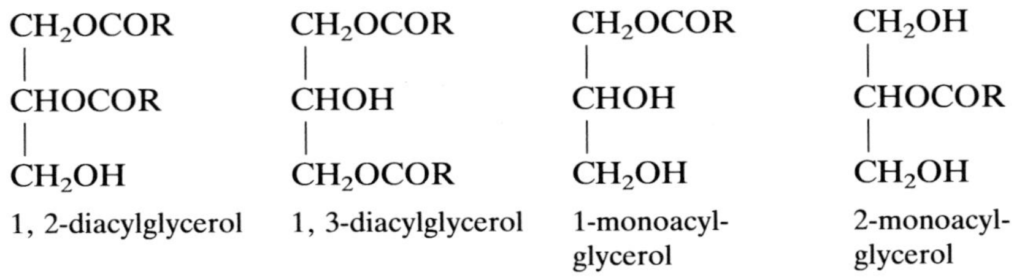

Figure 3. Isomeric di- and mono-acylglycerols

Triacylglycerols can be examined analytically at various levels. The simplest study indicates the composition of the total fatty acids both qualitatively and quantitatively, but this reveals nothing about the way these acids are distributed among the glycerides. Enzymic processes have been exploited to find out which acids are attached to each of the three glycerol carbon atoms. The recognition and determination of individual molecular species is a difficult challenge but, as later chapters will show, progress has been made toward this end.

Acylglycerols can be hydrolysed by water in the presence of acid or base to liberate the fatty acids and glycerol. If the water is replaced by an alcohol then alcoholysis occurs and the glycerol esters are converted to esters of the reactant alcohol. Reaction with methanol, for example, provides a convenient route to the methyl esters which are much used in gas chromatography.

$$\mathrm{RCOOMe} \xleftarrow[\text{or MeOH—H}^+]{\text{NaOMe—MeOH}} \begin{array}{l}\mathrm{CH_2OCOR}\\ |\\ \mathrm{CHOCOR}\\ |\\ \mathrm{CH_2OCOR}\end{array} \xrightarrow{\mathrm{H_2O,\ HO^-\ or\ H^+}} \mathrm{RCOOH}$$

In these reactions the acyl cleavage is complete. This can also be achieved using a lipase (enzyme) as catalyst. Of greater interest to the analyst is the observation that certain enzymes (such as pancreatic lipase) act specifically to catalyse reactions at the 1 and 3 positions of glycerol only and furnish the 2-monoacylglycerols which can then be analysed.

$$\begin{array}{l}\mathrm{CH_2OCOR}\\ |\\ \mathrm{CHOCOR}\\ |\\ \mathrm{CH_2OCOR}\end{array} \xrightarrow[\text{hydrolysis}]{\text{enzymic}} \underbrace{\begin{array}{l}\mathrm{CH_2OCOR}\\ |\\ \mathrm{CHOCOR}\\ |\\ \mathrm{CH_2OH}\end{array} + \begin{array}{l}\mathrm{CH_2OH}\\ |\\ \mathrm{CHOCOR}\\ |\\ \mathrm{CH_2OCOR}\end{array}}_{\text{diacylglycerols}} \longrightarrow \underbrace{\begin{array}{l}\mathrm{CH_2OH}\\ |\\ \mathrm{CHOCOR}\\ |\\ \mathrm{CH_2OH}\end{array}}_{\text{2-monoacylglycerol}}$$

6. Phospholipids

Phospholipids are important components of lipid membranes. When hydrolysed each molecule furnishes fatty acids, glycerol, phosphoric acid, and a low-molecular-weight alcohol represented by XOH in the following equation. This component is most commonly one of the alcohols listed in *Table 5*. When X = H the compounds are called phosphatidic acids.

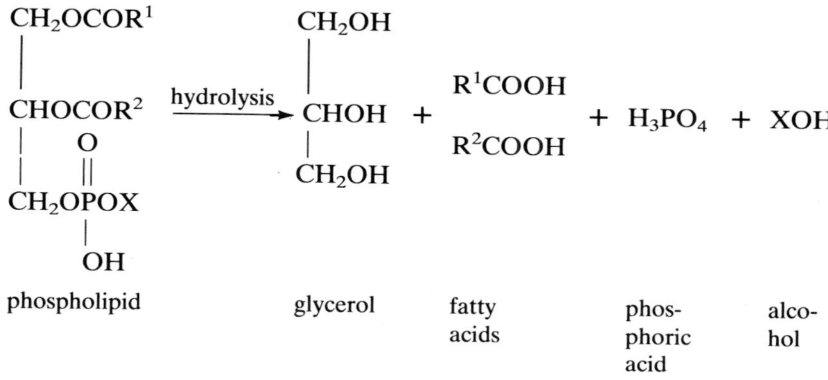

Table 5. The more common phospholipids

Structure of alcohol	Name of alcohol	Name of lipid type	
$HOCH_2CH_2\overset{+}{N}Me_3\bar{X}$	choline	phosphatidylcholine	(PC)
$HOCH_2CH_2NH_2$	ethanolamine	phosphatidylethanolamine	(PE)
$HOCH_2CH(NH_2)COOH$	serine	phosphatidylserine	(PS)
$(HO)_6C_6H_6$	inositol[a]	phosphatidylinositol	(PI)
$HOCH_2CH(OH)CH_2OH$	glycerol	phosphatidylglycerol	(PG)

[a] 1,2,3,4,5,6-hexahydroxycyclohexane

Complete hydrolysis of a phosphatidyl ester is usually effected in acidic media. With alkali only the acyl groups are split off readily leaving intermediates such as glycerophosphorylcholine (etc.) which is hydrolysed slowly to choline (etc.) and the isomeric mixture of α- and *β*-glycerophosphoric acids.

$$\text{PC} \xrightarrow[\text{hydrolysis}]{\text{mild}} \begin{cases} \text{fatty acids} \\ \text{GPC} \end{cases} \xrightarrow[\text{hydrolysis}]{\text{strong}} \text{choline} + \alpha\text{- and } \beta\text{-GPA}$$

Enzymes operate more specifically and effect selective hydrolysis of one or other of the four ester bonds. Each of the enzymes phospholipase-A^1, phospholipase-A^2, phospholipase-C, and phospholipase-D cleave the phosphatidyl ester to two products as indicated below. Phosphatidyl esters which have been deacylated once and retain one acyl group are called lysophosphatidyl esters.

$$\begin{array}{l} CH_2OCOR^1 \quad \leftarrow A^1 \\ R^2COO\ CH \quad (\leftarrow A^2 \text{ at } R^2COO) \\ CH_2OP(=O)(OH)OX \quad (C \rightarrow \text{ at } CH_2O\text{–}P;\ D \rightarrow \text{ at } P\text{–}OX) \end{array}$$

$$\xrightarrow{A^1} R^1COOH + \text{lysophosphatidyl ester}$$

$$\xrightarrow{A^2} R^2COOH + \text{lysophosphatidyl ester}$$

$$\xrightarrow{C} \text{1, 2-diacylglycerol} + \text{phosphoryl ester}$$

$$\xrightarrow{D} \text{phosphatidic acid} + \text{XOH (choline etc.)}$$

7. Waxes

The term wax is used in a variety of ways. It sometimes describes material with characteristic physical appearance which is generally a mixture of several types of long-chain compounds including hydrocarbons, acids, aldehydes, alcohols, and esters. The term wax ester is used more specifically to describe the ester formed between long-chain alcohols and long-chain acids. The structure below is the ester of such an alcohol (22:1) and acid (20:1).

$$CH_3(CH_2)_7CH{=}CH(CH_2)_9COO(CH_2)_{12}CH{=}CH(CH_2)_7CH_3$$

8. Ether lipids

Ether lipids are of four types depending on whether they are based on triacylglycerols or phospholipids and on whether or not they are vinyl ethers.

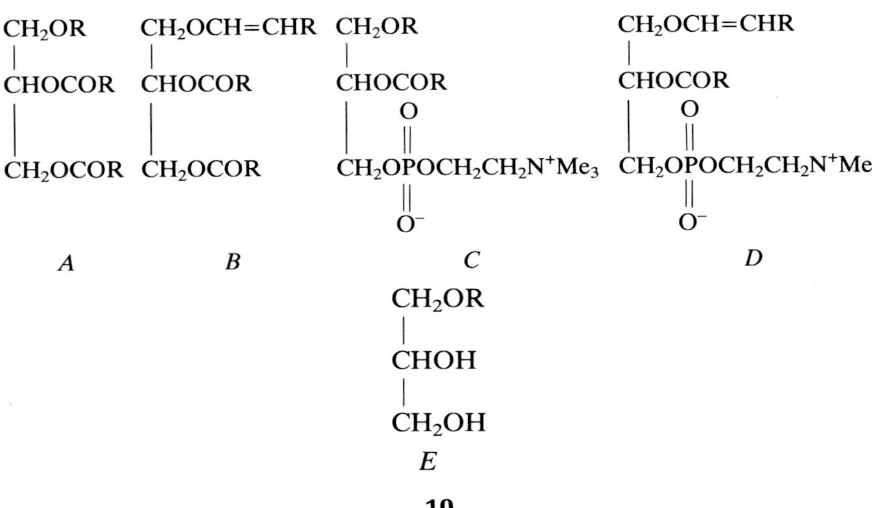

In the four structures shown above, R stands for any appropriate alkyl group. *A* and *B* are based on triacylglycerols and *C* and *D* on phospholipids (the compounds are shown as choline derivatives but other amino alcohols can replace the choline. *B* and *D* are vinyl ethers whilst *A* and *C* have ether groups which if unsaturated have unsaturation in more conventional positions (such as Δ^9) away from the ether function.

Hydrolysis of *A* and *C* will yield the usual hydrolysis products but instead of glycerol the alkoxy derivative *E* is formed. The vinyl ethers *B* and *D* behave similarly when submitted to alkaline hydrolysis but under acidic conditions the more reactive vinyl ether produces an aldehyde via the expected enol. Because such compounds were found in blood and gave aldehydes on hydrolysis they were called plasmalogens.

Glycerolipids containing two ether linked groups are known but are rare.

$$\mathrm{R'OCH{=}CHR} \longrightarrow \mathrm{R'OH} + \underset{\text{enol}}{[\mathrm{RCH{=}CHOH}]} \longrightarrow \underset{\text{aldehyde}}{\mathrm{RCH_2CHO}}$$

9. Glycosyldiacylglycerols

These glycerolipids contain two fatty acyl chains but the C-3 glycerol hydroxyl is associated with a sugar unit such as galactose. Lipids of this type are important components of the chloroplast.

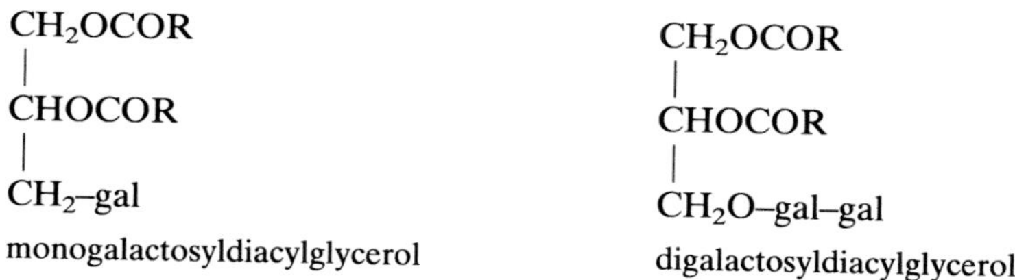

10. Sterol esters

Most fats contain minor quantities of sterols which are present both as free sterols and as esters with the usual range of long-chain fatty acids. The sterol esters undergo the usual hydrolysis or methanolysis reactions but react more slowly than the more common glycerolipids.

11. Sphingolipids

There are several subgroups of sphingolipids and some of them have fairly complex structures. However, they all share one structural feature: they contain long-chain amino alcohols acylated at the amino group to form amides. The free primary alcohol group is also derivatized in some way.

$$\begin{array}{l} R^1CHCH_2OH \\ \quad | \\ \quad NHCOR \end{array}$$

The long-chain bases (sphingosines) are C_{12}–C_{22} compounds of two types:

E	
$RCH{=}CHCH(OH)CH(NH_2)CH_2OH$	sphingosines
$RCH_2CH(OH)CH(OH)CH(NH_2)CH_2OH$	phytosphingosines

The corresponding amides are known as ceramides.

The primary alcohol group in the ceramides is further associated with sugars (glucose or galactose) to form cerebrosides and the more complex gangliosides or with phosphoryl esters such as phosphorylcholine or phosphorylethanolamine to form sphingomyelins.

References

Most books on fatty acids and lipids contain information on the chemical nature of these compounds. Readers are referred to any of these or to *The lipid handbook* (ed. F. D. Gunstone, J. L. Harwood, and F. B. Padley). Chapman & Hall, London, 1986.

2

Extraction of lipids and derivative formation

SHIELA HAMILTON, RICHARD J. HAMILTON, and PETER A. SEWELL

1. Introduction

The extraction of most lipids involves the use of solvents. The handling and use of solvents are the first, most important stage in the consideration of means of extraction.

1.1 Safe handling

All practising chemists and biochemists recognize the necessity to handle *any* chemical carefully. Recently, there has been the need to assess the hazards of each operation in the science laboratory and to assess the use of every chemical. In the experimental sections which follow, it is assumed that the proper handling of chemicals will be given due care and attention.

It is becoming more usual in research laboratories to perform most chemical reactions in fume cupboards or other similar vented spaces. Some of the small-scale reactions covered in the subsequent sections of the book are carried out in screw-capped vials or centrifuge tubes. Provided that the operations of mixing the reagents and the placing them in the tubes are performed in the fume cupboard, the subsequent warming or leaving to stand can take place on the laboratory bench. Some of the experimental procedures are quoted exactly as given in the original literature. Thus, there are mentions of chloroform and benzene, two solvents which have been all-but-banned from many laboratories. Alternatives to these two which are currently being used are dichloromethane for chloroform and toluene for benzene.

1.2 Hazard assessment

Hazard assessment should be considered under the guidance given by the Health and Safety legislation for each individual country. Many of the chemicals used in the extraction of lipids are toxic and/or flammable.

Diethyl ether, in particular, and ethers, in general, will develop peroxides when stored for a long time. Such peroxides are explosive, and ethers should not be distilled to dryness.

Chloroform is usually sold with 2% ethanol added as a stabilizer. To obtain chloroform for use in chromatography, it is best to remove the ethanol which would otherwise make the chloroform significantly more polar than pure chloroform.

Removal of ethanol from chloroform can be achieved by placing 30 g silica gel in a chromatography column with a tap and a sintered filter or a plug of glass wool. If 500 ml of chloroform is allowed to flow through the column at the rate of 50 ml per minute the ethanol will bind to the silica gel, leaving alcohol-free chloroform. This purified chloroform may be an added hazard in that, on storing, it will decompose in a photochemical reaction to produce the highly toxic phosgene. It is advisable therefore to prepare only sufficient of the alcohol-free chloroform for use at one time. Chloroform:methanol mixtures will rapidly leach the skin lipids from hands. Further contact with the solvent will give rise to irritation. It is therefore advisable to handle all solvents with care.

1.2.1 Autoxidation

All lipids if left for a time in air and in the daylight will go rancid. This rancidity is caused either by hydrolysis or by oxidation. The oxidation reaction has such a low activation energy that it is impossible to avoid completely the attack between oxygen and the unsaturated lipid. If some precautions are not taken the lipids extracted from tissues will contain the oxidation product, which may be taken for an original naturally occurring lipid. The autoxidation reactions are catalysed by the presence of metal ions and by light. There are therefore a number of instructions later to 'keep the extracted lipid in the dark'. Since free radicals are implicated in the autoxidation, it is valuable to introduce synthetic antioxidants into the extracted lipid. Butylated hydroxytoluene BHT should be added at the level of 50 to 100 mg per litre. Many vegetable sources of lipids contain tocopherols, natural antioxidants.

Ensuring that the air content of the vapour above the extracted lipid is kept to a minimum is an alternative way to minimize autoxidation. This is done by flushing containers and reaction vessels with nitrogen and by evaporating solvents in a rotary evaporator at a temperature not exceeding 40 °C, with the flasks being of as small a volume as possible. It is also important that the lipid extract is not evaporated to dryness. It is best to leave 3–5 ml of solvent with the lipid in the evaporator flask. It is then possible to transfer the last 3–5 ml of solvent containing the extracted lipid into a vial with a Pasteur pipette and to remove the solvent by blowing a stream of nitrogen on to the surface of the solution. It may be necessary to suspend the vial over an electrical heating plate so that the higher boiling solvents such as petroleum ether 60–80 °C and water can be removed in a reasonable time.

1.3 Purification of solvents

Most solvents are purified by distillation with or without some pre-treatment.

When large quantities of lipid are to be extracted, it may not be important that the quality of the extraction solvent is poor and that there are minor quantities of contaminants in the solvent. This is not the case when small quantities of lipid are to be extracted. In such instances, it is vital that the highest quality of solvent is purchased. Analar quality solvents should be used and it may be necessary to purify Analar solvents.

It is important to ensure that all glassware used for distillation is kept free from grease of any kind whether hydrocarbon in nature (for example, Apiezons and Vaseline) or silicone.

It must be recognized that if the solvent or the solvent vapour comes in contact with rubber or plastic, some contaminants will be extracted into the solvent.

All solvents contain some proportion of lipid contaminants. So, if a large quantity of solvent is to be used for the extraction or for the preparation of derivatives, when that solvent is evaporated this proportion of lipid will be incorporated into the extracted lipid or derivative. Distillation of the solvent in the first place will remove these contaminants.

The following is given for hydrocarbon solvents; for example, hexane, petroleum ether of various boiling point ranges. It may need to be amended for other solvents, especially with regard to the drying agent.

Protocol 1. Distillation of solvents

Apparatus

- 2-litre round-bottomed flask
- Vigreux distillation column
- double-surface water condenser
- steam bath, or electrical heater
- thermometer

Reagents

- Linde 5A sieve
- light petroleum (40–60 °C)

1. Place 300 g of Linde 5A sieve into a Winchester of light petroleum (b.p. 40–60).
2. Allow to stand for 24 h.
3. Filter off the drying agent.
4. Place 1400 ml of dried solvent in the 2-litre round-bottomed flask.

Protocol 1. *Continued*

5. Attach the Vigreux distillation column, thermometer, double-surface condenser, and receiver adaptor with drying tube.
6. Arrange for the adaptor to fit into a 2-litre conical flask.
7. Heat the solvent on the steam bath.
8. Discard the first 100 ml of distillate.
9. Allow the distillation to proceed collecting the solvent which boils between 40 to 60 °C.
10. Stop the distillation when 50 to 100 ml of the solvent remain in the 2-litre round-bottomed flask.

The Vigreux column should prevent any high-boiling lipid contaminants from being distilled. As a precaution against the possibility of the peroxides exploding on heating to dryness it is always important to leave some liquid in the flask which is being heated.

Diethyl ether is often contaminated by peroxides, which may be tested for in the following manner:

Shake 2 ml of the ether with 1 ml of a 10% potassium iodide solution. If there is any peroxide present it will react with the iodide to form iodine, which can be monitored by adding a drop of starch indicator solution. A blue colour indicates the presence of iodine which in turn represents some peroxide (see *peroxide value*, *Protocol 27*).

Protocol 2. Removal of peroxides and the drying of diethyl ether

Apparatus

- 2-litre separatory funnel
- sodium die press
- 2-litre conical flask
- drying tube containing anhydrous calcium chloride

Reagents

Ferrous sulfate solution: dissolve 60 g of ferrous sulfate in a mixture of 6 ml of concentrated sulfuric acid and 110 ml of water to prepare a concentrated solution of ferrous salts.

Dilute 20 ml of this concentrated ferrous solution with 100 ml of water

1. Place 1 litre of diethyl ether which has been shown to contain peroxides, into a 2-litre separatory funnel.
2. Add the 120 ml of ferrous solution prepared above.

3. Shake the two solutions vigorously.
4. Allow to settle, and remove the lower aqueous layer which contains the ferrous salts.
5. Transfer the upper layer which contains the diethyl ether to a 2-litre conical flask.
6. Add 200 g of anhydrous calcium chloride and allow to stand for 24 h.
7. Filter the ether through a large, fluted filter paper, which will remove the drying agent, into a dry 2-litre conical flask.
8. Fill the die of a sodium press with lumps of sodium.
9. Place the conical flask containing the ether on to the stand of the press.
10. Screw the plunger down and ensure that the sodium wire goes into the diethyl ether in the conical flask.
11. When all the sodium has been forced into the conical flask, remove the conical flask from the press.
12. Insert into the ground-glass joint of the conical flask, a drying tube containing calcium chloride.
13. Allow the ether to sit overnight and then examine the surface of a sodium wire.
14. If the surface of the sodium is still bright, remove the drying tube from the conical flask and replace it with a ground-glass stopper.
15. If the surface of the sodium wire is badly pitted, it is likely that the calcium chloride drying was not successful, so repeat step 6.

When the ether has been dried it is necessary to distil the solvent.

Protocol 3. Distillation of diethyl ether

Apparatus

- 2-litre round-bottomed flask
- double-surface water condenser
- heating mantle

Reagents

- dry diethyl ether

1. Set up a distillation apparatus consisting of a 2-litre round-bottomed flask attached to a double-surface water condenser.
2. Ensure that the flask can be heated with a heating mantle.
3. Lag the surface of the flask and the condenser with aluminium foil or insulating tape.

Protocol 3. *Continued*

4. Pour the dried ether into the conical flask.
5. Adjust the heating of the mantle so that the ether distils at the rate of 1 ml per minute.
6. Allow the distillation to proceed.
7. Discard the first 50 to 100 ml of condensate.
8. Collect the bulk of the liquid which boils at 35 °C.
9. Do not distil the last 100 ml from the distillation flask.

It may be that the commercial sample of ether which you have bought does not need to be dried nor does it appear to contain any peroxides. In such an event it is possible to use only *Protocol 3*. The last step 9 in this protocol is there as a double security to ensure that no peroxides wil become concentrated in the distillation flask. Each solvent has its own characteristics and this is true for its purification.

Ethanol or ethyl alcohol is available as *rectified spirit*, which is a constant boiling mixture of 95.6% ethyl alcohol and 4.4% water, or as *absolute alcohol*, which is 99% pure.

Rectified spirit can be dehydrated with quicklime to produce absolute alcohol, though the price differential makes it unlikely that lipid laboratories will consider it worthwhile carrying out the procedure in *Protocol 4*.

Protocol 4. Preparation of absolute alcohol (ethanol)

Apparatus

- Muffle furnace
- 5-litre round-bottomed flask
- double-surface water condenser
- steam bath
- desiccator with calcium chloride drying-agent

1. Heat 500 g of quicklime in the Muffle furnace at 220 °C for 2 h.
2. Remove the quicklime from the furnace and place in a desiccator to cool.
3. Arrange the 5-litre flask and condenser for reflux.
4. Place the contents of a Winchester bottle of rectified spirits (2.4 litres) into the flask and add the quicklime.
5. Connect the condenser and fit a drying-tube containing anhydrous calcium chloride.

6. Reflux the mixture on the water bath for 6 h.
7. Rearrange the apparatus for distillation inserting a splash head connector between the flask and the double-surface condenser.
8. Distil off the first 50 to 100 ml of distillate and discard this fraction.
9. Collect the bulk of the absolute alcohol in a round-bottomed flask.
10. Ensure that the side-arm of the receiver adaptor is attached to a drying-tube containing anhydrous calcium chloride.
11. Transfer the absolute alcohol carefully to a stoppered bottle, since the solvent is very hygroscopic.

To produce extemely dry ethanol, absolute ethanol is reacted with magnesium in the presence of iodine to form magnesium ethoxide, which subsequently reacts with water.

$$Mg + 2\ C_2H_5OH \rightarrow H_2 + Mg\ (OC_2H_5)_2$$

$$Mg\ (OC_2H_5)_2 + 2\ H_2O \rightarrow Mg\ (OH)_2 + 2\ C_2H_5OH$$

Protocol 5. Preparation of extremely dry ethyl alcohol

Apparatus

- 2-litre round-bottomed flask
- double-surface condenser
- steam bath
- drying-tube containing anhydrous calcium chloride

Reagents

- magnesium turnings
- 95% ethanol
- Iodine

1. Arrange the round-bottomed flask, condenser, and drying-tube for reflux.
2. Place 5 g of clean, dry magnesium turnings and 0.5 g of iodine in the flask.
3. Add 75 ml of 99% alcohol.
4. Warm the mixture over the steam bath until the iodine has disappeared. If hydrogen is not generated, a further 0.5 g of iodine may need to be added.
5. Continue the heating until all the magnesium is converted to magnesium ethoxide.

Protocol 5. *Continued*

6. Add 900 ml of absolute alcohol and reflux for 30 min.
7. Rearrange the apparatus for distillation.
8. Distil the dry ethanol which should be 99.95% pure.

Methanol can be produced in the same fashion from commercial absolute methanol.

Chloroform contains up to 2% of ethanol, which prevents the formation of phosgene; see Section 1.2. After removal of the ethanol, it is not advisable to store chloroform for any length of time.

2. Preliminary treatment of tissues

The use of live animals is strictly controlled in most countries and may require a licence. Even with dead animals, the animal should be dissected immediately after death. It is advisable to extract all tissue as soon as possible to minimize degradation of lipid components. Even with dead tissue, changes may occur; for example, the enzymes (lipases, phospholipases A, C, and D) will hydrolyse diacylphospholipids to lysophospholipids and free fatty acids, to diacylglycerols and to phosphatidic acid, respectively. The presence of these latter components in large amounts may indicate poor sample handling, since they are normally present in very minor proportions. Some of the enzymes may be inactivated by macerating green plant material in propan-2-ol (isopropyl alcohol) before the extraction proper is begun. An alternative is to place the green tissue in boiling water for 1 min.

Any tissue which is not to be extracted immediately should be frozen rapidly on dry ice and stored in sealed glass containers at −20 °C under nitrogen. They should be allowed to thaw out before being homogenized with the extraction solvent.

Tissues should be stored in glass, because if stored in plastic the tissues will absorb plasticizers (such as dibutyl phthalate) from the plastic.

2.1 Solvent extraction from tissues

There are two approaches to the extraction of lipids from tissues: namely, to carry out an exhaustive extraction of all lipids present and subsequently to separate out the classes of lipid in which you are interested, *or* to be more selective in the extraction process in the first place.

The simplest example of this is to contrast the methods which were formerly used to isolate the hydrocarbons from the surfaces of leaves. In one method (1) the leaf was macerated and extracted with benzene in a Soxhlet extractor, and the lipid extract was hydrolysed with sodium hydroxide. The unsaponifiable matter was then chromatographed over alumina to give the hydrocarbons.

Currently, the recommended method (2) is to dip the leaf in chloroform at room temperature, which will remove the surface constituents only. This extract can then be separated by thin-layer chromatography (TLC) on silica gel. The selectivity is introduced by only contacting that part of the plant in which you are interested with solvent and by using a solvent which will remove from the leaf only those components which are lipid. There is no need to use a Folch procedure.

Lipids have been extracted from tissues by solvent extraction (liquid–liquid). Neutral lipids may be extracted by relatively non-polar solvents such as hexane, diethyl ether, chloroform, or benzene. Alternatives to the latter two are dichloromethane and toluene. Diethyl ether and chloroform may stimulate phospholipase D to action if used on plant tissues. It is necessary to use chloroform:methanol to extract phospholipids.

Membrane-associated lipids and lipoproteins require polar solvents such as methanol or ethanol to disrupt the hydrogen bonding between lipid and protein. Glycolipids are soluble in acetone.

Where the lipid is covalently linked to other carbohydrate or protein molecules, it is necessary to break the covalent bond by hydrolysis.

Although reagent grade solvents may be used for extraction, when small quantities of lipids are to be recovered it is important to re-distil the solvents. Failure to do so will result in substantial quantities, up to 100 mg per 2 litres of solvent, of impurities.

Solvents to be used for extraction or for chromatography may need to be de-aerated by bubbling nitrogen through them. Highly unsaturated lipids will be oxidized in the presence of such dissolved air. Three methods for the liquid–liquid extraction of lipids widely cited in the literature are those of Folch, Lees, and Stanley (3), Bligh and Dyer (4), and Ways and Hanahan (5). All three methods use chloroform:methanol as the extracting solvent.

Protocol 6. The Folch method

Reagents

- chloroform:methanol mixture (2:1 v/v) This is the extracting solvent
- chloroform:methanol:water (3:48:47 v/v) 'upper phase'; this is one of the washing solvents.
- chloroform:methanol:water (86:14:1 v/v) 'lower phase'; this is one of the washing solvents.
- 'upper phase' containing 0.02% $CaCl_2$, 0.017% $MgCl_2$, 0.29% NaCl or 0.37% KCl.

Protocol 6. *Continued*

It is assumed that the tissue sample has a density of 1.0 g per litre.

1. Homogenize the tissue with chloroform:methanol (2:1) to a final dilution 20 times the volume of the tissue sample, i.e. the homogenate from 1 g of tissue is diluted to a volume of 20 ml. The time of homogenization will vary with the sample but a minimum of 3 min is usually required.
2. Filter the homogenate through a suitable paper into a glass-stoppered bottle. (Centrifugation may be used instead of filtration.) For the purpose of computation, this extract corresponds to 0.05 its volume of tissue, i.e. 1 ml of extract corresponds to 0.05 g of tissue.
3. Wash the crude extract with 0.2 of its volume of either water or salt solution.
4. Allow the solution to separate into two phases. The volumes of the upper and lower phases are 40 and 60% of the total volume respectively.
5. Remove the upper phase by siphoning.
6. Rinse the interface three times with pure 'upper phase', i.e. the chloroform:methanol:water 3:48:47 so that the lower phase is not disturbed. This has the effect of removing any 'fluff' at the interface.
7. Finally, add methanol so that the lower phase and the rinsing liquid form one phase.
8. Dilute the resulting solution to any desired volume by the addition of chloroform:methanol (2:1).

Steps 7 and 8 may be omitted if it is intended to remove the solvent under vacuum to yield a dry extract for weighing.

It is possible to depart from the procedure in *Protocol 6*, but it is essential that, whilst washing the extracts, the ratio of chloroform:methanol:water is 8:4:3 v/v. In calculating these proportions it is important to remember that the extract contains all the water from the tissue. This Folch procedure is the most widely quoted extraction procedure, with special reference given to the washing stages. Where difficulties arise, the 'modified' Folch method due to Ways and Hanahan allows 95 to 99% recover of lipids.

In the modified Folch method, the tissue sample is assumed to have a density of 1.0 g per millilitre.

Protocol 7. Modified Folch method

Apparatus

- Waring blender

Reagents

- chloroform
- methanol

1. Homogenize 1 g of tissue for 1 min with 10 ml of methanol.
2. Add 20 ml of chloroform.
3. Continue the homogenization for a further 2 min.
4. Filter the homogenate through a suitable filter paper into a glass-stoppered bottle.
5. Suspend the solid residue in 30 ml of chloroform:methanol (2:1 v/v).
6. Homogenize for a further 3 min.
7. Filter once more.
8. Wash the solid residue on the filter paper with 20 ml of chloroform.
9. Further wash the solid residue with 10 ml of methanol.
10. Combine the filtrates in a measuring cylinder from the above extractions.
11. Add an aqueous solution containing 0.88% potassium chloride whose volume corresponds to one-quarter that of the combined filtrates.
12. Shake the mixture thoroughly.
13. Allow to settle.
14. Remove the upper layer, which is the aqueous layer, by aspiration. (This minimizes the disturbance to the lower layer which contains the lipid.)
15. Add water:methanol (1:1 v/v) to the lower layer. The volume of this water:methanol solution is one-quarter of the lower layer.
16. Shake this mixture thoroughly.
17. Allow to settle.
18. Remove the upper layer by aspiration.
19. Recover the lipid from the lower layer, which consists of chloroform: methanol:water, by evaporation of the solvent in a rotary film evaporator.

    This evaporation should be accomplished at as low a temperature as possible. If the water does not azeotrope off in the first evaporation it may be necessary to add further portions of chloroform to the evaporator flask where the extract is held.
20. Re-dissolve the lipid after weighing in a small volume of chloroform and store at −20 °C.

In dealing with very unsaturated lipids it may be necessary to modify step 20 to ensure that the lipid is not heated in air. In step 19, recover the unsaturated lipid but ensure that the dried lipid stays at water-bath temperature for the minimum time possible. When the evaporator is opened, it is possible to fill the vacuum with nitrogen, not air. Step 20 may be modified so that the lipid is transferred to a small vial always with an atmosphere of nitrogen. Then the chloroform can be removed in a stream of nitrogen.

There is another extraction method due to Bligh and Dyer. This method again uses chloroform:water for the extraction, but the quantities are such that when mixed with water in the tissue, a single phase solution is formed. Alterations to the procedure may be made but the ratios of chloroform: methanol:water before and after dilution must be 1:2:0.8 and 2:2:1.8 respectively. This procedure is suitable for a tissue containing 80 ± 1% water and about 1% lipid.

Protocol 8. Bligh and Dyer method

Apparatus

- Waring blender
- Buchner flask and funnel

Reagents

- chloroform
- methanol

1. Homogenize 100 g sample of fresh or frozen tissue in a Waring blender for 2 min with a mixture of 100 ml of chloroform and 200 ml of methanol.
2. Add to this mixture 100 ml chloroform.
3. Blend for 30 sec.
4. Add 100 ml distilled water.
5. Blend for 30 sec.
6. Filter the homogenate through a Whatman No. 1 filter paper on a No. 3 Buchner funnel with a slight suction.
7. Compress the residue with the end of a spatula or a glass-bottle top to ensure maximum recovery of filtrate.
8. Transfer the filtrate to a 500 ml measuring cylinder.
9. Allow to settle for a few minutes to complete the separation and clarification.
10. Record the volume of the chloroform layer (at least 150 ml).
11. Remove the alcoholic layer by aspiration.

12. A small volume of the chloroform layer is removed to ensure complete removal of the top layer. The lower chloroform layer contains the lipid.

For quantitative lipid extraction, the lipid held in the tissue residue is recovered by blending the residue and filter paper with a further portion of 100 ml chloroform. The mixture is filtered through the original Buchner funnel. The Waring blender, jar, and residue are rinsed with a total of 50 ml chloroform. The filtrate is mixed with the original filtrate from step 12 prior to the removal of the alcoholic layer.

Highly polar lipids such as gangliosides may partition into the methanol: water phase.

Enzymes may be deactivated with isopropanol (4) as follows. The plant tissues are homogenized with a hundred-fold excess by weight of isopropanol. The mixture is filtered and the residue is shaken overnight with isopropanol: chloroform (1:1 v/v). The filtrates are then combined, and most of the solvent is removed on a rotary film evaporator. The lipid residue is taken up in a chloroform:methanol (2:1 v/v) and given a Folch wash.

The mass of fresh tissue extracted, together with the mass of lipid obtained from it should always be recorded. It may also be desirable to determine the amount of dry matter in the tissue so that the mass of lipid relative to a mass of dry matter may be calculated. For the Bligh and Dyer method it is always necessary to know the water content of the tissue.

Later in this chapter, specific examples of the extractions will be seen to illustrate how individual researchers have adapted the basic extraction procedures to their own needs.

Whilst these extraction procedures have been used for decades, it has only more recently been possible to adopt adsorption (liquid–solid) methods. The tissue matrix in which lipids occur is very complex and the physical forms in which they occur is such that no one method of extraction will be satisfactory for all possible situations.

Simple lipids are often part of large aggregates in storage tissues and are relatively easy to extract. Complex lipids are often constituents of membrane where they may be associated with proteins and polysaccharides. In which case, disruption of the cell walls is necessary for extraction. Water then assists extraction by causing swelling of the bio-polymers and it is usually a constituent of the extracting solvents.

2.2 Liquid–solid extraction

Liquid–liquid extraction needs large volumes of solvent. It is time-consuming and not very good for routine extraction. It often results in the formation of an emulsion which prevents phase separation with a resultant loss of material.

Liquid–solid extraction depends on the adsorption of the lipid on to a solid adsorbent from a liquid medium and its subsequent desorption. Examples of the adsorbents are silica gel or octadecylsilane bonded silica or Amberlite resins. It is possible to desorb different classes of lipid from the adsorbents by varying the eluting or desorbing solvent.

Now that it is possible to obtain chemically bonded silicas with different functional groups (for example, octadecylsilane C–18, C–8, C–CN, and C–NH2), it has been shown that very sophisticated extractions and subsequent group fractionations are possible.

Protocol 9. Separation of lipids in human milk

Apparatus

- chromatography column (30 cm × 2.2 cm) fitted with a sintered disc
- 100-ml beaker
- three 250-ml conical flasks

Reagents

- anhydrous sodium sulfate
- Celite 545
- calcium phosphate
- diethyl ether
- dichloromethane
- methanol

1. Mix thoroughly in a glass beaker a 1-ml sample of milk with 4 g of granular anhydrous sodium sulfate.
2. Add 3 g of Celite 545.
3. Grind to a uniform powder.
4. Pack 2 g of calcium phosphate/Celite 545 (1:9 w/w) into the glass chromatography column.
5. Place the mixture from step 3 on to the column.
6. Elute the lipids from the column with 100 ml distilled dichloromethane into a tared conical flask.
7. When the top of this portion of the solvent has reached the top of the column bed, remove the first receiving conical flask.
8. Remove the solvent from this first fraction to obtain the neutral lipids—for example, cholesterol and triglycerides.
9. Elute the second fraction with 50 ml of distilled dichloromethane: methanol (9:1 v/v) into a second tared conical flask.
10. Allow this solvent to elute until the column is dry (30 to 45 min).
11. Remove the solvent to yield the polar lipids (such as phospholipids).

The total lipids of milk also contain α-tocopherol. Since there may be impurities in the packaging materials which might absorb in the ultraviolet at the same place as the α-tocopherol, it is suggested that the packing material is washed with 100 ml of dichloromethane:chloroform (9:1 v/v) and then allowed to dry before packing the column.

This solid–liquid method has been shown to give good agreement with the results obtained by the modified Folch method. It has an added advantage that it allows a preliminary separation into lipid classes at the same time.

Another procedure which uses octadecylsilane-treated silica to absorb the lipids is given in *Protocol 10*. (6)

Protocol 10. Use of ODS silica

Apparatus

- 1-litre separatory funnel
- 500-ml conical flask
- 1-litre conical flask

Reagents

- octadecylsilane silica (ODS)
- ethanol
- methyl formate

1. Plant material was extracted into an acidified water.
2. Place 500 ml of the acidified plant extract in a 1-litre conical flask.
3. Add 150 g of pre-treated ODS silica.
4. Swirl the flask until the coloured materials are absorbed.
5. Filter off the ODS silica.
6. Place the ODS silica in a 500-ml conical flask.
7. Add 150 ml of an ethanol:water (15:85 v/v) and swirl the two phases together. Decant off the solvent.
8. Repeat this process with four further portions of ethanol:water.
9. Discard the ethanol:water extracts.
10. Add 150 ml of methyl formate.
11. Swirl the two phases for 2 min.
12. Decant off the methyl formate.
13. Repeat steps 11, 12, and 13 with four further portions of 150 ml of methyl formate.
14. Combine the portions of methyl formate.
15. Place this methyl formate in a separatory funnel.

Protocol 10. *Continued*

16. Add ethanol:water to the funnel.

17. Allow the two layers to separate out.

18. Discard the lower layer (ethanol:water).

19. Recover the prostaglandin lipids by removing the solvent by evaporation.

Lipid extracts should be stored under an apolar solvent such as chloroform at −20 °C in a vial with a screw-cap. It is important to displace the air from the vial with nitrogen. Where the sample is going to be stored for long periods, it may be a good idea to place the sample into a glass ampoule from which the air may be removed by evacuation and the ampoule then sealed.

With highly unsaturated lipids, it may be necessary to add 0.005% butylated hydroxyanisole (BHA) or butylated hydroxytoluene (BHT) antioxidant to minimize the risk of autoxidation. This reagent may be added before concentration of the lipid or it can be added to the glass vial before storage.

3. Treatment of samples

The need to modify the basic extraction procedures has already been mentioned. The following examples illustrate the nature of these modifications.

Protocol 11. Extraction of amniotic fluid (7)

Apparatus

- centrifuge
- centrifuge tubes

1. Centrifuge amniotic fluid for 20 min at 1500 *g*.

2. Store supernatant at −20 °C until extraction.

3. Extract the lipids with 20 vol. of chloroform:ethanol (2:1 v/v) for 2 h at 40 °C with shaking.

4. Remove the non-lipid contaminants by a Folch procedure to give lipids in concentration corresponding to 10–20 g/litre.

Another extraction which involves a pre-treatment of the biological fluid is given in *Protocol 12*. Blood samples should be processed immediately after withdrawal from the patient.

Protocol 12. Extraction from blood

- centrifuge
- centrifuge tubes

1. Add a 3.8% citrate solution to a plasma sample at a level of 1 ml per 10 ml of blood which prevents clotting.
2. Allow a serum sample to clot by standing at 37 °C for 1 h in the absence of the citrate solution.
3. Remove red cells by centrifugation at 1500 *g* for 15 min.
4. Extract the lipids into chloroform:methanol in a modividation of the Folch procedure.

By contrast, it may be necessary to extract from vegetable sources. Glucosinolates are found in the seeds of the plant families Cruciferae, Capparaceae, Tovariaceae, Resedaceae, and Moringaceae. They are hydrophilic compounds and can be obtained best by removing the seed oils with petroleum ether (40–60 °C). This extraction can be achieved by a Soxhlet extraction.

However, the Swedish tube method of extraction (*Protocol 13*) is preferred (8).

Protocol 13. Extraction of glucosinolates

Swedish tube method

Apparatus

- Stainless-steel tube
- Stainless-steel balls

Reagents:

- light petroleum (40–60)
- methanol

1. Place seeds in a stainless-steel tube with light petroleum (30–60).
2. Add stainless-steel balls and seal the tube with a screw-cap.
3. Shake tube to promote grinding of seeds as well as extraction of triglycerides.
4. Filter off the miscella (the oil-laden light petroleum).
5. Extract the glucosinolates with methanol.

Protocol 13. *Continued*

For vegetative samples

1. Freeze the vegetative sample to −40 °C.
2. Grind the sample in a frozen state at −20 °C.
3. Extract with boiling methanol.

The extraction from unicellular organisms is complicated by the large mass of liquid which is often present. In *Protocol 14*, the need to treat first with methanol to inactivate the enzymes (9).

Protocol 14. Extraction from algae

Apparatus

- 250-ml separatory funnel

Reagents

- chloroform
- methanol
- anhydrous sodium sulfate
- silica gel G

1. Grow *Anabaena variabilis* at 35 °C.
2. Treat the wet paste of cell suspension under nitrogen and in the dark with 50 ml of methanol at 65 °C.
3. Treat the cell suspension twice with 50 ml of chloroform:methanol (2:1 v/v) at 55 °C.
4. Combine the extracts in a separatory funnel.
5. Mix in an equal volume of a saturated aqueous solution of sodium chloride.
6. Remove the lower layer.
7. Dry this lower layer over anhydrous sodium sulfate.
8. Remove the solvent under reduced pressure.
9. Chromatograph this residue on silica gel G with a chloroform:methanol: water (65:25:4 v/v)

Isolation of lipids is rarely achieved by using steam distillation but short chain acids and antioxidants can be obtained in this way.

Protocol 15. Isolation of BHA and BHT from potato granules

Apparatus

- Kuderna, Danish evaporator concentration device
- 4-ml concentrator tube
- distillation column
- 500-ml separatory funnels
- 100-ml and 250-ml volumetric flasks
- 2-ml and 10-ml pipettes
- 25-ml, 100-ml, and 250-ml graduated cylinders

Reagents

1. Weigh accurately 20.00 g of potato granules into the distillation apparatus.
2. Add 125 ml of distilled water.
3. Add 25 ml of magnesium oxide suspension to scrubber trap at top of distillation apparatus.
4. Allow steam to pass through the distillation vessel.
5. Adjust the rate of distillation to 20 ml condensate per minute.
6. Collect 200 ml of distillate into a 250-ml volumetric flask.
7. Stop the distillation.
8. Add 50 ml of 95% alcohol through the condensing column.
9. Transfer the distillate to separatory funnel.
10. Add 2 g of sodium chloride.
11. Add 25 ml of dichloromethane.
12. Shake the flask vigorously for 1 min.
13. Allow the two phases to separate.
14. Filter the dichloromethane layer through 20 g of anhydrous sodium sulfate into 125-ml Kuderna, Danish evaporator.
15. Repeat steps 11 to 14 twice more with 10 ml of dichloromethane.
16. Rinse the sodium sulfate twice more with 10 ml of dichloromethane.
17. Evaporate on steam bath to less than 1 ml.
18. Adjust the volume to 2 ml.

An aliquot of this solution can be injected into the gas–liquid chromatography (GLC) to monitor the level of butylated hydroxy toluene BHT and BHA.

In order to obtain phospholipids from a large sample of oil it may be necessary to use a solvent partition procedure as shown in *Protocol 16.*

Protocol 16. Solvent partition

Apparatus

- 1-litre separatory funnels
- rotary film evaporator

Materials

- 87% aqueous ethanol
  This solution is first saturated with respect to hexane by placing 500 ml of the 87% aqueous ethanol in a 1-litre separatory funnel. 500 ml of hexane is now added to the separatory funnel and the mixture shaken for 5 min.

  The mixture is allowed to settle when two layers are produced, the upper layer to be called solvent A in the following contains hexane which is saturated with respect to water and ethanol, and the lower layer contains 87% aqueous ethanol which has been saturated with respect to hexane to be called solvent B.
- hexane

1. Dissolve the 100 g of the lipid sample in the 500 ml of the solvent B.
2. Add this solution to a separatory funnel containing 500 ml of the solvent A.
3. Shake this mixture for 2 min.
4. Allow to stand so that the two layers will separate out.
5. Run the lower layer into a second separatory funnel containing a fresh sample of the solvent A.
6. Shake this mixture for 2 min.
7. Place the lower layer in a rotary evaporator.
8. Remove the solvent under reduced pressure.
   This extract contains the phospholipids and polar lipids.
9. The upper layers from these separations are equilibrated with a fresh portion of 500 ml of the solvent B.
10. Allow to stand so that the two layers can be separated.
11. The hexane can now be removed under reduced pressure.
    This extract contains the bulk of the triglycerides and other non-polar lipids.

It is possible to replace the hexane by light petroleum ether (b.p. 40–60 °C)

It is also important to ensure that the temperature of the extract is kept as low as possible when the solvent is being removed. It may be that at the end a small amount of water remains in the rotary flask. In which case it is advisable to add a small amount (50 ml) of a solvent such as acetone or chloroform to the extract. The whole can now be distilled under reduced pressure so as to azeotrope off the water with the chloroform.

Cholesterol can be determined quantitatively by GLC and by high-performance liquid chromatography (HPLC), and semi-quantitatively by TLC. However, it is sometimes necessary to extract the cholesterol from the lipid matrix. In which case, it is possible to make the digitonin:cholesterol complex which precipitates out in the presence of other lipids.

Digitonin is very expensive, but the solution mentioned in *Protocol 17* can be stored in the dark for up to 6 months.

Protocol 17. Digitonin:cholesterol complex

Apparatus

- vortex mixer
- 15–ml centrifuge tubes
- centrifuge

Reagents

- 1% digitonin

  Dissolve 1 g of digitonin in 50 ml of 95% ethanol. Dilute to 100 ml with water.
- Acetone:95% ethanol (1:1)

1. Place an aliquot of the solution of the lipid mixture in chloroform in the 15-ml centrifuge tube. This solution may contain up to 3 mg of cholesterol.
2. Evaporate to dryness under a stream of nitrogen.
3. Add 1 ml of the digitonin solution.
4. Allow the two to mix and leave for 10 min.
5. Centrifuge at 1000 *g* for 5 min.
6. Discard the supernatant.
7. Allow the precipitate to dry.
8. Suspend the residue in 4 ml of acetone.
9. Mix on the Vortex blender.
10. Centrifuge at 300 r.p.m. for 5 min.
11. Discard the supernatant.

Protocol 17. *Continued*

12. Remove as much of the solvent with a Pasteur pipette as possible.

13. Remove the rest of the solvent by evaporation under a stream of nitrogen.

The cholesterol can then be silylated as in *Protocol 38* and further analysed by GLC or HPLC.

Extraction with isolation of the desired portion of the lipid is highly desirable. In *Protocol 18*, it possible to isolate an aflatoxin mixture from corn meal by using the new *solid phase* extraction materials which are available. They can be purchased from the following, amongst others:

- Supelchem UK Ltd., UK
- Alltech UK Ltd., UK
- Waters Associates, USA

These materials have a number of advantages; namely, they are easy to use, convenient, can be used as a single tube or can be operated for many samples with multiple tubes, and they are reproducible.

They can be obtained in various forms, such as silica-, octyl-bonded, octadecyl-bonded, and diol-bonded. These forms correspond to the stationary phases which are used in HPLC (see Chapter 5).

In this example, the corn meal is extracted with a methanol:water mixture. This extract is placed on a solid phase from which the aflatoxin fraction can be eluted. The solutions and solvents can be pushed through the solid phase either with a syringe or by compressed air or by suction as in *Figure 1*. Other examples are found (10).

Protocol 18. Isolation of aflatoxins

Apparatus

- Supelclean LC-CN SPE tube

Reagents

- methanol:water 8:2
- 0.5% glacial acetic acid in water
- 20% tetrahydrofuran in 0.5% glacial acetic acid in water
- 25% tetrahydrofuran in hexane
- 1% tetrahydrofuran in dichloromethane

1. Condition the Supelclean LC-CN tube with 2 ml of 0.5% glacial acetic acid in water.

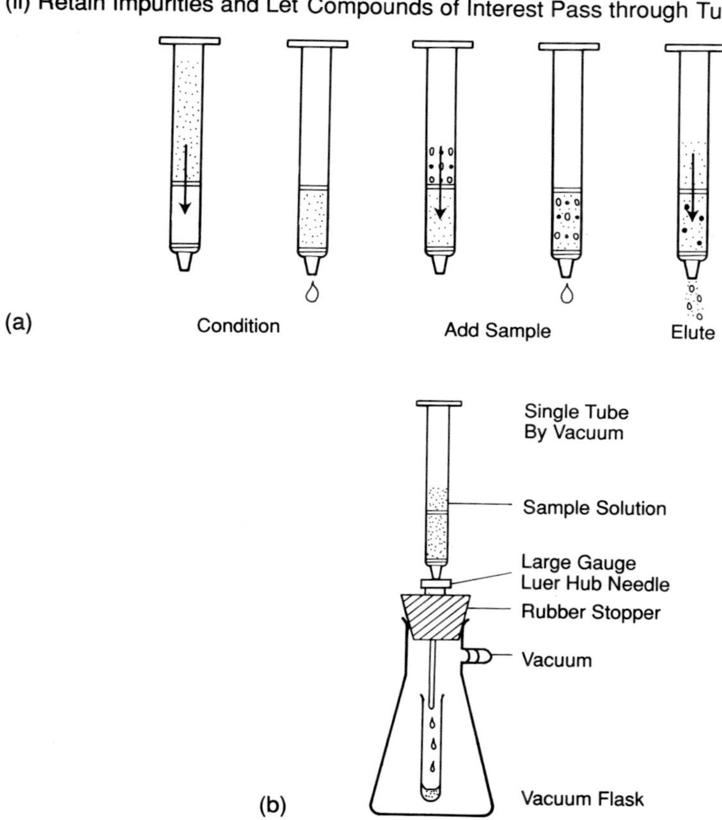

Figure 1. Solid phase extraction apparatus (after Supelco)
(a) Two systems for concentration
(i) Solid phase retains compounds of interest and impurities pass through tube
(ii) Solid phase retains impurities and compounds of interest pass through tube
(b) Use of vacuum to pull solvent through one Solid Phase tube

Protocol 18. *Continued*

2. Blend 50 g of corn meal with 100 ml of methanol:water (8:2) for 1 min.
3. Filter extract through Whatman No. 1 filter paper.
4. Mix 1 ml of this filtrate with 4 ml of 0.5% glacial acetic acid in water.
5. Add 1 ml of the sample solution from step 4 above to the tube.
6. Wash with 500 ml 20% tetrahydrofuran in 0.5% glacial acetic acid in water.
7. Wash with 2 ml of hexane.
8. Dry packing under nitrogen.
9. Add 3 ml of 25% tetrahydrofuran in hexane.
10. Dry packing under nitrogen.
11. Elute the aflatoxins with two portions of 2-ml aliquots of 1% tetrahydrofuran in dichloromethane.

It is claimed that this procedure allows samples of the aflatoxins as low as 3 p.p.b. to be determined.

There are a number of tests and procedures which are used by industrial concerns for the characterization of oils and fats. The *acid value* is important in commerce as oils and fats are commonly traded on a very small number of characteristics.

Protocol 19. Acid value

Apparatus

- 50 ml burette
- 250-ml wide-mouthed conical flask (iodine flask)

Reagents

- 0.5 M alcoholic potassium hydroxide

Weigh out about 30 g of potassium hydroxide into a conical flask. Add 1 litre of ethanol. The heat given off will help the potassium hydroxide to dissolve. This solution should be straw-coloured. If it is much darker than this it should be discarded.

Standardize against 1 M aq. sulfuric acid.

1. Weigh accurately approx. 2 g of fat or oil into a wide-mouthed conical flask.
2. Add 20 ml of hot neutralized ethanol.

3. Bring to the boil on a water bath.
4. Titrate whilst the solution is still hot to neutrality with 0.5 M potassium hydroxide, using phenolphthalein as an indicator.

The acid value is determined as the number of milligrams of potassium hydroxide per gram of fat.

Refined oils have such a low free acidity that it is necessary to take 10–20 g of the oil in step 1 of this protocol.

Saponification value

The *saponification value* is defined as the number of milligrams of potassium hydroxide required to react completely to saponify 1 g of the fat. It is inversely proportional to the molecular weight. Thus, an oil such as high erucic rape, which has high-molecular-weight fatty acids, will have a lower *saponification value*, 173, than an oil with lower-molecular-weight acids, such as palm kernel oil with a saponification value of 244.

Protocol 20. Saponification value

Apparatus

- two 250-ml round-bottomed flasks
- water condensers
- steam bath
- burette

Reagents

- 0.5 M alcoholic potassium hydroxide
- 1 M aqueous sulfuric acid

1. Weigh accurately approx. 2 g of the oil or fat into a 250-ml round-bottomed flask.
2. Add a few anti-bumping granules and add 25 ml of 0.5 M alcoholic potassium hydroxide from a pipette.
3. As a blank, add a few anti-bumping granules and 25 ml of 0.5 M alcoholic potassium hydroxide to a second 250-ml round-bottomed flask.
4. Place both flasks, fitted with condensers, on a steam bath.
5. Reflux both solutions for 30 min.
6. Remove the condenser and back titrate the excess alkali with 1 M aq. sulfuric acid to phenolphthalein, as quickly as possible, without cooling.

Protocol 20. *Continued*

The difference between the blank and the test gives the amount of potassium hydroxide absorbed by the oil.

$$\text{Saponification value} = \frac{28.05\ (\text{titration blank} - \text{titration of sample})}{\text{mass of sample}}$$

Unsaponifiable matter

The unsaponifiable matter is defined as that material which is extracted by diethyl ether from an aqueous solution of soap produced by saponification of the fat and which is non-volatile at 100 °C. There are a number of different protocols for the determination of the unsaponifiable matter depending on the nature of the solvent needed to extract the particular compounds in that oil.

The value of the unsaponifiable should be very low for most oils (for example, 0.2%) but for some oils (whale oil, for example) it is high at 17%.

Protocol 21. Unsaponifiable matter

Apparatus

- two 250-ml round-bottomed flasks
- 500-ml separatory flask
- steam bath

Reagents

- diethyl ether
- 0.5 M alcoholic potassium hydroxide
- anhydrous sodium sulfate
- phenolphthalein

1. Weigh accurately approx. 2 g of oil or fat into a 250-ml round-bottomed flask.
2. Add a few anti-bumping granules, and add 25 ml of 0.5 M alcoholic potassium hydroxide.
3. Place the flask, fitted with a condenser, on a steam bath.
4. Reflux for 30 min.
5. Remove the condenser and add 50 ml of water.
6. Transfer this aqueous ethanolic solution to a separatory funnel.
7. Extract three times with diethyl ether with 100-ml, then 50-ml, and finally 50-ml portions.

8. Wash the combined ether extracts with three 25-ml portions of water.
9. Wash the ether extract with 25 ml of 0.5 M aqueous potassium hydroxide.
10. Wash with water until the washings are neutral to phenolphthalein.
11. Dry the ether layer over anhydrous sodium sulfate.
12. Place the diethyl ether extract in a weighed, round-bottomed flask.
13. Remove the solvent in a stream of nitrogen.
14. Weigh the flask plus the residue.
15. Determine the unsaponifiable matter by difference.

If there are any free fatty acids present as a result of some conversion of the sodium salts to free acids, they may be titrated. Dissolve the unsaponifiable matter in a 10-ml portion of hot neutral ethanol. Titrate with 0.1 M sodium hydroxide to phenolphthalein.

Much of the classical work was performed on a substantial scale, i.e. 5 to 10 g of lipid was needed. *Protocols 22–24* illustrate how Ackman has adapted the classical procedures to work with the much smaller quantities of living tissue which the more sensitive analytical procedures have permitted (11).

Protocol 22. Extraction from an *Amphipod corophium* volutator (Pallas)

Apparatus

- centrifuge
- four 10.0-ml screw-capped centrifuge tubes

Reagents

- methanol
- 50% sodium hydroxide
- petroleum ether
- 12 M hydrochloric acid
- 14% boron trifluoride in methanol
- dichloromethane

1. Place approx. 0.13 g, i.e. eight individuals of the species of amphipod in a centrifuge tube (Teflon-lined screw-cap).
2. Add 1.0 ml of distilled water, 1.0 ml of methanol and 0.2 ml of 50% sodium hydroxide.
3. Flush the tube with nitrogen.
4. Cap the tube and heat for 1 h at 100 °C.

Protocol 22. *Continued*

5. Cool the reactants and add 1.5 ml of water.
6. Extract this aqueous alcohol phase by shaking it with 1.5 ml of petroleum ether.
7. If an emulsion forms, centrifuge the tube and its contents to produce two layers.
8. Transfer the upper layer by means of a Pasteur pipette.
9. Extract the bottom layer of aqueous ethanol a second time with 1.5 ml of petroleum ether.

   Repeat steps 7 and 8.
10. Sterols and other unsaponifiable material are in the petroleum ether extract.
11. Acidify the aqueous ethanol phase from 9 with 12 M HCl.
12. Extract this acidified phase with 2.0 ml of petroleum ether (twice).
13. Transfer this petroleum ether extract to a screw-capped tube.
14. Remove the solvent with a stream of nitrogen.
15. Add 1 ml of methanol and 12 ml of 14% BF_3 in methanol reagent.
16. Flush the tube with nitrogen and cap the tube.
17. Heat the reaction mixture for 1 h at 100 °C.
18. Cool the tube and add 2 ml of water.
19. Extract this aqueous methanol solution twice with 1.5 ml of petroleum ether.
20. Combine the petroleum ether extracts and add 1 ml of water.
21. Wash out the acidic material with water.
22. Remove the petroleum ether layer by Pasteur pipette and place it in a clean tube.
23. Remove the solvent with a jet of nitrogen.
24. Dissolve the methyl esters which have been formed in dichloromethane and inject an aliquot into the GLC.

Protocol 22 allows the unsaponifiable material to be obtained and also to measure the total fatty acid profile.

A second portion of the amphipod can be extracted as in *Protocol 24*.

Protocol 23. Extraction of total lipid

Apparatus

- centrifuge

- four 10.0-ml screw-capped centrifuge tubes

Reagents

- methanol
- chloroform
- petroleum ether

1. Place 0.61 g of the amphipod in a screw-capped centrifuge tube with 2 ml of chloroform and 1.0 ml of methanol.
2. Allow this extraction to proceed for 3 h with intermittent agitation.
3. Add 2 ml of water.
4. Remove the chloroform layer (the lower layer) with a Pasteur pipette and place in another tube.
5. Re-extract the residue with 1.0 ml of chloroform and 0.5 ml of methanol for 1½ h.
6. Remove the chloroform layer and combine with the one obtained earlier.
7. Centrifuge the chloroform solution and remove the chloroform layer. Any solid residue should remain at the bottom of the tube.
8. Remove the solvent with a stream of nitrogen and then complete the removal of solvent by applying high vacuum to the tube.

The lipid removed in this procedure corresponded to 1.3% and represented the lipid obtainable by a Folch extraction.

Protocol 24. Extraction of an amphipod

Apparatus

- Sorvall Omni-Mixer

Reagents

- methanol
- chloroform
- petroleum ether

1. Place a third portion of the amphipod which had been frozen corresponding to 2.8 g in a Sorvall Omni-Mixer (50-ml cup) with 70 ml of chloroform and 30 ml of methanol.
2. Homogenize for 1 min.
3. Filter the blend through filter paper.
4. Wash the filter paper with 10 ml of chloroform.
5. Add 20 ml of water to the filtrate.

Protocol 24. *Continued*

6. Agitate this mixture.
7. Separate the chloroform layer.
8. Evaporate the solvent under vacuum to obtain the lipid corresponding to 0.7% by mass of the sample.

In commercial practice it is usual to measure the degree of unsaturation in oil which is reported as the iodine value, which is expressed in grams of iodine absorbed by 100 g of fat.

The active species in the reagent is ICl so that the reaction is as follows:

$$\mathrm{CH{=}CH + ICl \rightarrow CHI\ CHCl.}$$

Protocol 25. Determination of iodine value

Apparatus

- glass weighing vessel
- two 250-ml wide-necked conical flasks (iodine flasks)
- burette

Reagents

- 17 M acetic acid and tetrachloromethane must be checked as free of oxidizable matter. To test for oxidizable material see later.
- potassium iodide solution with a concentration of 100 g per litre containing no free iodine or iodate.
- sodium thiosulphate solution 0.1 M accurately standardized.
- starch indicator solution with a concentration of 1 g per 100 ml freshly prepared.
- tetrachloromethane

1. Weigh the recommended mass of fat into a glass weighing vessel.
2. Place the weighing vessel and the fat into a 250-ml iodine flask.
3. Add 15 ml of tetrachloromethane.
4. Pipette into the flask 25 ml of the Wijs solution.
5. Place the stopper in the iodine flask and shake the flask gently.
6. Place the flask in a cupboard in the dark.
   If the fat has an iodine value below 150, the flask is left in the dark for 1 h. If the iodine value is above 150, allow the flask to stand for 2 h.
7. Add 20 ml of the potassium iodide solution and 150 ml of water.

8. Shake the contents of the flask gently.
9. Titrate the liberated iodine with the sodium thiosulfate solution.
10. Add the starch indicator towards the end of the titration until the blue colour disappears even after vigorous shaking.

Calculation

$$\text{Iodine value} = \frac{12.69\, M\,(V - V')}{m}$$

where M is the exact molarity of the 0.1 M sodium thiosulfate.
m is the mass of the test fat in grams.
V is the volume of 0.1 M sodium thiosulfate solution used for the blank determination in millilitres.
V' is the volume of 0.1 M sodium thiosulfate solution used for the determination in millilitres.

Test for oxidizable material
Shake 10 ml of the reagent with 1 ml of saturated aqueous potassium dichromate solution and 2 ml of concentrated sulfuric acid. A green colour indicates that there is some oxidizable material.

- Potassium iodide solution with a concentration of 100 g per litre containing no free iodine or iodate.

Protocol 26. Preparation of Wijs reagent

Apparatus

- brown Winchester bottle

Reagents

- iodine trichloride
- 17 M acetic acid
- tetrachloromethane
- potassium iodide
- sodium thiosulfate

1. Weigh 9.0 g of iodine trichloride (ICl_3).
2. Place in a brown glass bottle.
3. Add 700 ml of 17 M acetic acid and 300 ml of tetrachloromethane.

Protocol 26. *Continued*

Determination of halogen content

(a) Mix together 5.0 ml of Wijs solution, 5.0 ml of potassium iodide solution and 30 ml of water.

(b) Titrate with 0.1 M sodium thiosulfate solution in the presence of a few drops of the starch indicator solution.

(c) Add 10 g of powdered iodine to the bulk reagent.

Repeat first two steps.

The titration should equal 1½ times that of the first determination.

4. Allow the bulk solution to stand.

5. Decant the clear liquid into a brown glass bottle.

This solution can be kept for several months provided it is kept out of the light.

The peroxide value is used to indicate the degree to which a lipid has been oxidized either intentionally or by mistake. It is a measure of an oil's acceptability to the consumer.

Protocol 27. Peroxide value

Reagents

- glacial acetic acid:chloroform solution (3:2 v/v)
- saturated potassium iodide solution
- 0.01 M sodium thiosulfate
- a vegetable starch solution (1 g in 100 ml)

Apparatus

- 1-ml pipette with 0.01-ml divisions.
- 250-ml conical flask with a ground-glass joint
- 500-ml volumetric flask
- 100-ml volumetric flask
- 25-ml burette with 0.05-ml divisions

1. Weigh to the nearest 0.1 mg, 5.00 ± 0.05 g of the sample into the conical flask.

2. Add 30 ml of the acetic acid: chloroform (3:2) solution.

3. Allow the sample to dissolve.

4. Add 0.5 ml of the saturated potassium iodide solution from the pipette.

5. Swirl the solution for 1 min.
6. Add 30 ml of distilled water.
7. Titrate with 0.01 M sodium thiosulfate, adding it gradually with constant and vigorous shaking.
8. Continue the titration until almost all the yellow colour has almost disappeared.
9. Add approx. 0.5 ml of starch indicator solution.
10. Continue the titration until all the blue colour has disappeared.
11. Shake the flask near the end-point to ensure that all the iodine has been liberated from the chloroform layer.

Carry out a blank in which no oil is added to the flask in step 1. This blank should not exceed 0.1 ml of the thiosulfate solution.

$$\text{Peroxide value} = \frac{(V_s - V_b)\, M}{W} \times 1000$$

where V_s is the volume in millilitres of the sodium thiosulfate
V_b is the volume of the blank
M is the molarity
W is the mass of the oil sample

$$ROOH + 2KI + H_2SO_4 \rightarrow I_2 + K_2SO_4 + ROH + H_2O.$$

Whilst the peroxide value measures the oxidation as it is proceeding, the *anisidine value* is a measure of the oxidation products which have been formed from the degradation of the peroxides. Thus the anisidine value is measuring something that happened in the past. It may not be possible to stop the autoxidation of an oil if there is evidence from a large anisidine value.

Anisidine value is defined as 100 times the optical density measured in a 1-cm cell resulting from the reaction of 1 g of oil with 100 ml of the anisidine reagent.

Protocol 28. Anisidine value

Apparatus

- spectrophotometer suitable for use at 350 nm
- two glass cells of 1-cm light path
- 10-ml test-tubes, with ground-glass stoppers
- 25-ml volumetric flasks

Protocol 28. *Continued*

- 5-ml and 1-ml pipettes
- 1-ml pipette with rubber bulb

Reagents

- glacial acetic acid
- 0.25% (w/v) solution of *p*-anisidine in glacial acetic acid. *p*-Anisidine bought commercially is darkened by oxidation. It must be purified either by molecular distillation with a cold finger, at a pressure of 1 mmHg, or by recrystallization of the commercial product, i.e.:
  Dissolve 40 g of *p*-anisidine in 1 litre of water at 75 °C. Add 2 g of sodium sulfite and 20 g activated carbon. Stir for 5 min and then filter the solution through filter paper. Cool the filtrate to about 0 °C and allow to stand for 4 h. Filter off the crystals and wash with a small amount of water. Dry the crystals in air. These crystals can be stored in the dark and at a low temperature.
- spectroscopic quality hexane or iso-octane.

1. Melt the sample at 60–70 °C.
2. Homogenize thoroughly.
3. Provided the sample is clear and dry, place approx. 0.3 g of the sample (weighed accurately) in a 25-ml volumetric flask.
4. Dissolve the sample in hexane and make up to the 25-ml mark. This acts as the sample solution for the remaining steps.

   To determine the sample blank:
5. Pipette accurately 5 ml of the oil in hexane into a test-tube.
6. Add 1 ml of glacial acetic acid.
7. Into another test-tube, add 5 ml of hexane and 1 ml of glacial acetic acid. This acts as the blank solvent.
8. Fill one clean and dry matched cell with the sample solution.
9. Fill the second matched cell with the blank solvent.
10. Measure the absorbance at 350 nm (A_{sb})
    To determine the reagent blank:
11. Pipette 5 ml of hexane into a test-tube.
12. Add 1 ml of the anisidine solution.
13. Homogenize completely with minimum shaking and allow to react for 10 ml.
14. Into another test-tube, add 5 ml of hexane and 1 ml of glacial acetic acid.
15. After 10 min, fill one cell with the solution from steps 11, 12, and 13.
16. Fill the second cell with the solution from step 14.

17. After 10 min, measure the absorbance at 350 nm, which corresponds to the reagent blank (A_{rb}).

18. Pipette 5 ml of the sample from step 4 into a test-tube.

19. Add 1 ml of the anisidine reagent.

20. Homogenize completely with minimum shaking.

21. Prepare a solvent blank made up of 5 ml of solvent and 1 ml anisidine.

22. Just before 10 min have elapsed fill one cell with the solution from step 20.

23. Fill a second cell with the blank solution from step 21.

24. Measure the absorbance at 350 nm after exactly 10 min. (A_p)

The reagent product is called R_p and determined by difference equal to $A_p - A_{rb}$.

The anisidine value (AV) is given by the equation

$$AV = 25 \times \frac{1.2\,(R_p - A_{sb})}{m}$$

where m is the mass of the oil.

4. Derivative formation

Since GLC is the most popular separation technique for lipids, methyl esters are by far the most widely used derivative.

Metcalfe and Schmitz (13) showed that BF_3/methanol was a suitable reagent for the methylation of free fatty acids. By 1969, it was accepted by the American Oil Chemists' Society as an official method. It has one major advantage, i.e. it can be purchased from commercial chemical companies already prepared at the desired concentration (12). *Protocol 29* shows how the triglycerides of an oil are converted to free fatty acids which can then be esterified under acidic conditions.

Protocol 29. Methylation using BF_3–methanol

Apparatus

- 50-ml round-bottomed flask
- water condenser
- test-tube

Reagents

- sodium hydroxide approx. 0.5 M in methanol
  2 g of sodium hydroxide is dissolved in 100 ml of methanol containing not more than 0.5% of water

Protocol 29. *Continued*

- boron trifluoride in methanol (12–15%)
- heptane and light petroleum are redistilled before use

1. Place 250 mg of fat in a 50-ml round-bottomed flask.
2. Add 4 ml of the sodium hydroxide solution.
3. Heat the reaction mixture under reflux until the droplets of fat disappear, which normally takes 5 to 10 min.
4. Add 5 ml of methanolic boron trifluoride solution with a graduated pipette through the top of the condenser to the boiling liquid.
5. Reflux this mixture for 2 min.
6. Add 3 ml of hexane through the top of the condenser.
7. Boil the reaction mixture for a further 1 min.
8. Cool the mixture to room temperature.
9. Add a small portion of saturated sodium chloride solution and swirl the flask gently.
10. Add more sodium chloride solution to the flask so that the top of the liquid is in the neck of the flask.
11. Transfer 1 ml of the upper hexane layer to a test-tube.
12. Add anhydrous sulfate to the hexane layer to remove any water.
13. Inject this solution which corresponds to 100 mg/ml of methyl esters into the GLC.

Boron trifluoride:methanol is a reagent which is toxic and is relatively unstable on storage. There have been reports that artefacts can form (14) if the reagent is old, especially with cyclopropyl, cyclopropenyl, hydroxy, and epoxy acids. It is suggested that use of an old reagent can lead to the decomposition of polyunsaturated fatty acids (15). Boron trifluoride:methanol can also be used as an inter-esterification catalyst for the conversion of triglycerides to methyl esters.

Another acidic catalyst is formed from anhydrous hydrochloric acid which some workers claim is the best reagent (16).

Protocol 30. HCl:methanol

Apparatus

- 25-ml round-bottomed flask

Reagents

- bubble 2.5 g of hydrochloric acid gas from a cylinder into 100 ml of anhydrous methanol
- petroleum ether

1. Place 30 mg of lipids in a 25-ml round-bottomed flask.
2. Add 4.5 ml of methanolic hydrochloric acid.
3. Heat this mixture under reflux for 1 to 2 h.
4. Allow to cool and add 0.5 ml of water.
5. Extract the methyl esters with three 5-ml portions of petroleum ether.

Epoxy groups are unstable in acid, so it is not advisable to methylate epoxy fatty acids with any acidic reagent.

Hitchcock and Hammond claim that sulfuric acid:methanol is a very satisfactory reagent for methylation (17). It is easy to prepare and can be stored easily. It can be used to methylate free fatty acids and to inter-esterify triglycerides.

Protocol 31. H_2SO_4:methanol

Apparatus

- 100-ml measuring cylinder
- 20-ml screw-capped vial

Reagents

- toluene
- prepare the reagent by pouring 5 ml of conc. sulfuric acid carefully down the side of a measuring cylinder containing 100 ml of anhydrous methanol. This solution should be prepared with stirring. Add 50 ml of toluene which acts as a solubilizing agent for the lipid.
- diethyl ether

1. Place 50 mg of lipids in a screw-capped vial.
2. Add 5 ml of the sulfuric acid:methanol:toluene reagent.
3. Heat the sealed vial on a water bath for 60 min.
4. Allow the vial to cool.
5. Add 5 ml of water and 5 ml of diethyl ether.
6. Shake this mixture.
7. Remove the lower layer by Pasteur pipette.

Protocol 31. *Continued*

8. Decant off the upper layer, which is the organic layer, into anhydrous sodium sulfate.
9. Allow the drying to continue for 5 min.
10. Inject a 3-μl aliquot directly on to the GLC.

There is an alternative procedure for the methylation of free acids, i.e. diazomethane. This is a toxic gas which is always used in diethyl ether. It is best prepared in such a way that it can be distilled into the vessel containing the free fatty acids. The gas may decompose on contact with quartz ground-glass joints. It is preferable to prepare the gas in a specially designed piece of glassware as in *Figure 2*.

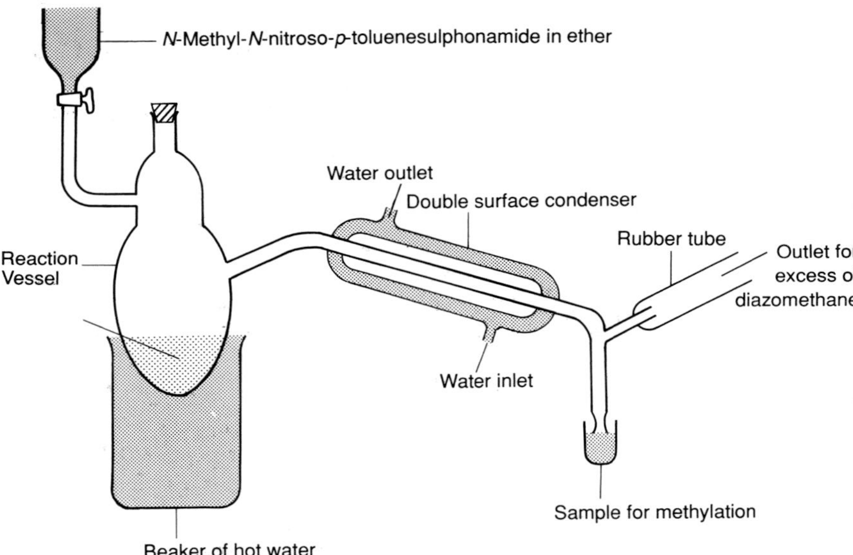

Figure 2. All-glass apparatus for generation of diazomethane. No quick-fit joints.

Protocol 32. Diazomethane preparation

Apparatus

- specially designed as in *Figure 2*.

Reagents

- *N*-methyl-*N*-nitroso-*p*-toluenesulfonamide (Diazald trade name)
- diethyl ether

- potassium hydroxide
- ethanol

1. Place 85.6 mg of Diazald dissolved in 50 ml of diethyl ether in the dropping funnel of the distillation vessel.
2. Add a solution of 16 mg of potassium hydroxide in 3 ml of water and 3.0 ml of methanol in the distillation bulb of the apparatus.
3. Arrange that the end of the condenser dips below the surface of an ethereal solution of the free fatty acids held in a conical flask.
4. Place the distillation bulb in a water bath.
5. Allow the alkali solution to drop into the Diazald solution slowly.
6. Control the distillation of the ether and the generated diazomethane so that the level of the solution in the conical flask rises steadily.
7. The distillation is complete when the solution of the acids becomes yellow due to the presence of excess diazomethane.

In this, it is especially important that the ether is not allowed to distil to dryness. The Diazald solution should not be allowed to go to dryness, and in cleaning out the apparatus after use the flask should be washed in ethanol and the washings treated with alkali in the fume cupboard. These quantities are sufficient for 4–5 mg of free fatty acids. This reagent is not suitable for the inter-esterification of triglycerides (18).

Base catalysed methanolysis was used earlier than the BF_3/methanol method, i.e. in 1956 (19). However, it was not until 1978 that the International Organization for Standardization (ISO) recognized it officially. This reagent is a very effective route to the inter-esterification of triglycerides and phospholipids. It cannot be used to methylate free fatty acids. It is possible to carry out the methanolysis on a small scale such that the reaction is complete in 5 to 10 min at room temperature (20).

Protocol 33. Methylation with sodium methoxide

Apparatus

- 50-ml volumetric flask fitted with a B14 ground-glass joint
- heating mantle or water bath

Reagents

- 0.25 M sodium methoxide
- iso-octane
- saturated sodium chloride

Protocol 33. *Continued*

1. Transfer approximately 150 mg of pre-molten fat or oil to a dry volumetric flask.
2. Add 5 ml of 0.25 M sodium methoxide in methanol:diethyl ether.
3. Reflux this mixture for 30 sec.
4. Cool the sample.
5. Add 3 ml of *iso*-octane.
6. Add 15 ml of saturated sodium chloride.
7. Shake this mixture for 15 sec.
8. Add sufficient further sodium chloride solution to bring the level of the liquid up into the neck of the volumetric flask.
9. Allow the layers to separate out.
10. Inject 2.5 μl of the upper layer directly into the GLC.

It is evident that the reaction can be carried out in a screw-capped vial without the need for the special volumetric flask.

Sodium methoxide can be made very simply by dissolving clean, freshly cut, sodium in dry methanol. It can be made up freshly for each batch of samples, or alternatively the reagent can be stored in an oxygen-free atmosphere for several months.

Phospholipids, glycosyl diglycerides and steryl esters can be transesterified by this reagent.

It is reported that sodium methoxide in methanol cannot be used with polyunsaturated fatty acids. However, wax esters and steryl esters react very slowly, so *Protocol 34* has been developed, which involves hydrolysis with base first, followed by methylation of the sodium salts of the free fatty acids.

Protocol 34. Methylation of jojoba esters (21)

Apparatus

- 10-ml screw-capped vials
- 100-ml separatory funnel

Reagents

- 5% sodium hydroxide in methanol
- 10% boron trifluoride in methanol
- diethyl ether

1. Dissolve 5 mg of jojoba oil in a minimum volume of toluene.
2. Add 1.2 ml of 5% sodium hydroxide in methanol.

3. Reflux this mixture for 1.5 to 2 h.
4. Cool the reaction mixture.
5. Add 1 to 2 ml of 10% BF_3 in methanol.
6. Reflux for half an hour.
7. Cool the reaction mixture.
8. Transfer the reaction mixture to a separating funnel by rinsing with 5 ml of diethyl ether and 20 ml of water.
9. Extract this mixture with four portions of 5 ml of diethyl ether.
10. Remove the ether on a steam bath under nitrogen.
11. Separate the methyl esters and the long-chain alcohols which this procedure generates by TLC.
12. Convert the alcohols to silyl ethers with bis(trimethyl silyl) trifluoroacetamide.

On a large scale, it is possible to extract from whole seeds and to convert the oil into methyl esters as in *Protocol 35*.

Protocol 35. Extraction and methylation in one operation (22)

Apparatus

- Waring blender
- 500-ml round-bottomed flask

Reagents

- conc. sulfuric acid
- methanol
- petroleum ether

1. Macerate 20.6 g of whole seeds in 50 ml of methanol in a laboratory grinder, to give a product with a particle size of coarse sand.
2. Transfer the slurry quantitatively to a 500-ml round-bottomed flask.
3. Add methanol to bring the total volume to 150 ml.
4. Heat this mixture under reflux for 4 h with a catalyst of 6 ml of conc. sulfuric acid.
5. Allow to cool.
6. Filter the reaction mixture through sintered glass.
7. Add 150 ml of petroleum ether to the filtrate.
8. Shake this mixture for 2 min.

Protocol 35. *Continued*

9. Allow to settle.
10. Separate off the upper layer, the petroleum ether.
11. Wash this layer with four portions of 150 ml of water; the last water wash should be neutral to litmus paper. If not, a further washing with water is needed.
12. Dry this petroleum ether layer over sodium sulfate.
13. Filter off the drying agent.
14. Remove the solvent by evaporation which yielded 7.7 g from sunflower seeds.
15. Distil under reduced pressure to give the pure sample of the methyl esters.

As an alternative to acid methanolysis in *Protocol 35*, methyl esters can be made by saponification followed by acid-catalysed methylation.

Protocol 36. Saponification of primrose oil (23)

Apparatus

- 16-ml Teflon-lined screw-capped test-tube

Reagents

- 0.5 M sodium hydroxide in methanol
- 1 M hydrochloric acid
- hexane
- 14% boron trifluoride in methanol
- diethyl ether

1. Mix 2 ml of 0.5 M sodium hydroxide in methanol with 100 μl of primrose oil sample in a 16-ml Teflon-lined screw-cap test tube.
2. Reflux this mixture at 70 °C for 20 to 30 min.
3. Cool the sample.
4. Acidify with M HCl to pH 3.
5. Extract with 5 ml of hexane.
6. Separate off the hexane layer by Pasteur pipette.
7. Extract twice more with 5 ml of hexane.
8. Evaporate off the combined hexane with a jet of nitrogen in a screw-capped test-tube.
9. Add 2 ml of 14% BF_3–methanol.

10. Flush out the test-tube with nitrogen.
11. Reflux for 45 min at 70 °C.
12. Allow the contents to cool.
13. Add 6 ml of water.
14. Add 3 ml of diethyl ether.
15. Extract the methyl esters into the ether.
16. Remove the ether under a jet of nitrogen.
17. Dissolve the methyl esters into 1 ml of hexane and inject into the GLC.

There is a need to make other derivatives of lipids.

Protocol 37. Preparation of acetates (24)

1. Place 0.2 g of lipid containing an alcohol group in a screw-capped vial.
2. Add 1 ml of ethanoyl chloride (acetyl chloride) dropwise.
3. Reflux the solution gently on a water bath for 1 min.
4. Remove the excess ethanoyl chloride and the by-product ethanoic acid in a stream of nitrogen.

$$ROH + CH_3COCl \rightarrow ROCOCH_3$$

Sterols and steryl esters can be determined after conversion of the free sterol to a silyl ether.

Protocol 38. Silylation procedure (25)

1. Dissolve 100 μg of cholesterol in 50 μl of dichloromethane in a screw-capped vial. This solvent prevents the precipitation of the ether derivative.
2. Add 50 μl of a proprietary silylating agent (for example, *tert.*-butyl dimethylchlorosilane in imidazole).
3. Allow to stand at room temperature (22 to 28 °C) for 1 h.
4. Add 1.0 ml of distilled water.
5. Add 1.0 ml of dichloromethane to the vial.
6. Shake the contents of the vial thoroughly.
7. Allow to settle.
8. Remove the lower layer (the dichloromethane).
9. Place this extract in a fresh vial.

Protocol 38. *Continued*

10. Repeat steps 5 to 9 with two further portions of dichloromethane on the aqueous layer form step 7.

11. Combine the dichloromethane extracts.

12. Dry the dichloromethane extracts with anhydrous sodium sulfate.

13. Filter of the drying agent or pipette off the dry dichloromethane extract with Pasteur pipette from the surface of the drying agent.

14. Remove the solvent with a stream of nitrogen.

15. Dissolve the silyl ethers in dichloromethane to give a final concentration of each component of 30 ng per litre.

$$ROH + CH_3\text{—}\underset{\underset{Cl}{|}}{\overset{\overset{CH_3}{|}}{Si}}\text{—}C(CH_3)_3 \rightarrow RO\text{—}\underset{\underset{C(CH_3)_3}{|}}{\overset{\overset{CH_3}{|}}{Si}}\text{—}CH_3$$

The preparation of derivatives which have suitable cleavage patterns for identification in the mass spectrometer is now an important part of the extraction and derivatization of lipids.

Protocol 39. Preparation of a pyrrolidide derivative for mass spectrometry (26)

1. Dissolve 10 mg of fatty acid methyl ester in 1 ml of freshly distilled pyrrolidine in a screw-capped vial.

2. Add 0.1 ml of acetic acid.

3. Heat the reaction mixture for 1 h at 100 °C.

4. Allow to cool and add 8 ml of dichloromethane.

5. Wash this solution with 4-ml portions of 2 M hydrochloric acid (several times).

6. Wash the dichloromethane solution with 4-ml portions of water (several times).

7. Dry over anhydrous sodium sulfate.

8. Remove the solvent with a stream of nitrogen.

$$\text{pyrrolidine (NH)} + RCO\,OCH_3 \longrightarrow RCO-N\text{(pyrrolidine)}$$

Methyl thio alkanes can be used to determine the position of unsaturation in alkenes. A similar procedure to that in *Protocol 40* can be used for mono and di unsaturated fatty acids.

Protocol 40. Preparation of α,β-bis(methylthio)alkanes (27)

1. Place 6 mmol dimethyl disulfide and 1 mmol of an alkene in a flat-bottomed flask equipped with a magnetic stirrer.
2. Dissolve 0.005 mmol of iodine in the mixture.
3. Purge the flask with a stream of nitrogen.
4. Seal the flask with a stopper.
5. Stir the reactants for 24 h at room temperature.
6. Take an aliquot and inject directly into the GLC.

$$\mathrm{RCH=CHR'} \xrightarrow[\mathrm{I_2}]{(\mathrm{CH_3{-}S})_2} \mathrm{RCH(SCH_3){-}CHR'(SCH_3)}$$

This product, the methyl thioalkane may decompose on standing for several days. It is possible to minimize the decomposition by dissolving the product in diethyl ether and washing with dil aqueous sodium hydroxide.

Another method of making silyl ethers for the determination of diacylglycerols is given in *Protocol 41*.

Protocol 41. Microdetermination by GC-MS of diacylglycerols (28)

Apparatus

- 3-ml screw-capped vial

Reagents

- *tert*.-butyl dimethyl chlorosilane in imidazole
- light petroleum
- anhydrous sodium sulfate

1. Heat 0.5 mg of diacylglycerol and 150 μl of *tert*.-butyl dimethyl chlorosilane/imidazole at 80 °C for 20 min in a screw-capped vial.
2. Allow the reaction mixture to cool.
3. Add 5 ml of light petroleum to the reaction mixture.

Protocol 41. *Continued*

4. Wash this petroleum ether solution with three portions of 0.5 ml water.
5. Dry the petroleum ether extract over sodium sulfate for 15 min.
6. Filter off the drying agent.
7. Remove the solvent by evaporation with a stream of nitrogen.
8. Analyse the *tert.*-butyldimethylsilyl ethers by GC-MS directly.

The silyl ethers can be stored in light petroleum at −20 °C. Derivative formation for HPLC requires that the derivative contains some functional groups which will make the detector more sensitive, i.e. an aromatic ring to permit the use of the ultraviolet detector.

Protocol 42. Succinimidyl 2-naphthoxyacetate as a fluorescent derivative (29)

Apparatus

- 12-mm or 32-mm capped vial

Reagents

- triethylamine
- succinimidyl naphthoxy acetate
- chloroform

1. Place an aliquot of lipid solution containing 3 μg of lipid containing phosphorus in a 12–32-mm vial.
2. Remove the solvent by heating the vial at 50 °C under a stream of nitrogen.
3. Add 5 μl of triethylamine to the dried lipid.
4. Prepare a solution of succinimidyl naphthoxy acetate at a concentration 1 mg per millilitre in chloroform.
5. Add 45 μl of this solution of reagent to the lipid.
6. Cap the vial with an aluminium seal crimped in place.
7. Swirl the vial vigorously for 10 sec.
8. Shake the reaction mixture in the dark at room temperature for 2 h.
9. Analyse the derivatized lipid immediately.

It is possible to store the derivative after the triethylamine has been removed by evaporation under nitrogen at 50 °C. The dry residue can then be re-dissolved in chloroform. This derivative is used to test for ethanolamine-

and serine-containing phospholipids which fluoresce at 228 nm excitation and 342 nm emission in the presence of the succinimidyl naphthoxy acetate.

Fatty acids can be converted into a derivative which makes them visible in the ultraviolet.

Protocol 43. Preparation of 2-naphthacyl esters (30)

Apparatus

- 10-ml conical flask

Reagents

- 2-naphthacyl bromide
- *N,N*-diisopropylethylamine
- dimethylformamide
- ethyl acetate
- anhydrous sodium sulfate

1. Dissolve 100 μm of fatty acid, 92 μm of 2-naphthacylybromide and 184 μm of *N,N*-diisopropyleth lamine in 10 ml of dimethylformamide.
2. Stir the reaction mixture at room temperature for 20 h.
3. Remove the solvent at 50 °C under vacuum.
4. Dissolve the residue in 5 ml of ethyl acetate.
5. Wash this solution with saturated sodium bicarbonate solution.
6. Wash the ethyl acetate solution further with water.
7. Dry the solution over anhydrous sodium sulfate.
8. Filter off the drying agent.
9. Remove the solvent at room temperature under nitrogen.

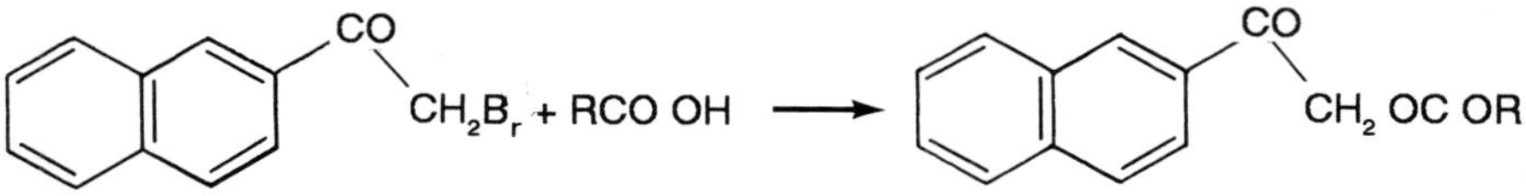

CO

CH_2B_r + RCO OH ⟶

CO

CH_2 OC OR

These esters absorb at 247–248 nm and allow a lower level of detection 4–90 ng.

The most frequently formed ultraviolet derivatives are phenacyl esters (31), 2-naphthacyl esters (31). The fluorescent derivatives can be made from 4-bromo-methyl-7-methoxy coumarin (32) and *p*-(9-anthryloxy) phenacyl bromide (33).

Protocol 44. Preparation of phenacyl esters (34)

Apparatus

- 3-ml Reacti-Vial (Pierce Chemical Co.)
- Stirrer hot plate
- 50 ml pear shaped flask

Reagents

- α-*p*-dibromoacetophenone
- 18-Crown-6 ether (1,4,7,13,16-hexaoxacyclooctadecane) (Aldritch Chemical Co.)
- 85% potassium hydroxide in methanol

1. Place a sample of organic acids (0.5mM) in methanol in a 50-ml pear-shaped flask.
2. Neutralize the acids with a solution of potassium hydroxide in methanol using phenolphthalein as an indicator.
3. Remove the solvent under vacuum.
4. Prepare a solution of α-*p*-dibromoacetophenone:18-Crown-6 ether (10:1) dissolved in acetonitrile.
5. Mix the potassium salt of the organic acids with an excess of α-*p*-dibromoacetophenone:18-Crown-6 ether (10:1) with enough acetonitrile added to bring the total volume to 1.5 ml in the Reacti-Vial.
6. Insert the stirrer bar.
7. Cap the Reacti-Vial with a septum disk.
8. Heat the mixture with stirring at 80 °C for 15 minutes.
9. Cool the vial and remove an aliquot for direct injection into the HPLC.

$K^+\ {}^-O_2CR$ + [18-crown-6] → [18-crown-6·K^+] ${}^-O_2CR$ ≡ (K^+) ${}^-O_2CR$

RCO_2^- (K^+) + Br–C_6H_4–C(=O)CH_2Br → Br–C_6H_4–C(=O)CH_2O_2CR + (K^+) Br^-

(K^+) Br^- + RCO_2K → RCO_2^- (K^+) + KBr

5. Isolation (extraction) of volatiles

The level of total aroma volatiles in food is often very low; for example, 10 to 100 p.p.m.

The large quantity of volatile components which do not contribute to the aroma (such as water), must be removed before an analysis of the aroma constituents can be performed. The flavour volatile fraction can contain up to 500 constituents.

Where the flavour components are not damaged by heat, a combined steam distillation and solvent partition system is recommended. The apparatus is shown in *Figure 3* was designed by Likens and Nikerson (35).

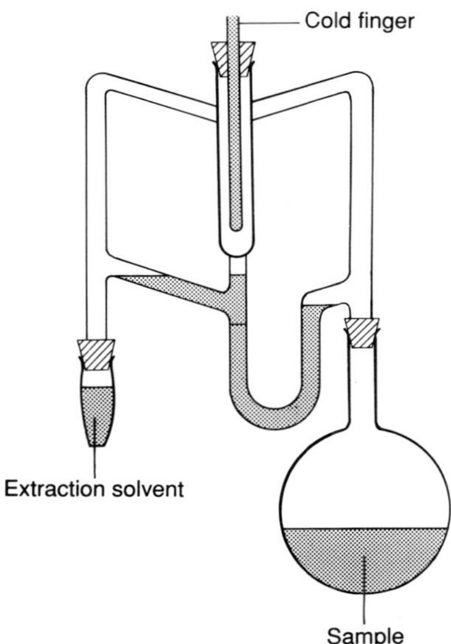

Figure 3. Nickerson/Likens apparatus for low-temperature solvent extraction of volatile components.

Protocol 45. Isolation of flavour volatiles

Apparatus

- 1-litre three-necked flask
- heating mantle
- magnetic stirrer
- trap cooled with solid carbon dioxide in acetone

Protocol 45. *Continued*

Reagents

- 2-methylbutane

1. Place the 100 g of the foodstuff in water held in the three-necked bolt-necked flask (1 litre).
2. Fit the flask with a stirrer.
3. Heat the mixture to 100 °C.
4. Allow the steam-distilled volatiles to be condensed in the double-surface condenser.
5. Allow this portion to be extracted into the solvent 2-methylbutane which is also condensing through the the condenser.
6. Let the volatiles flow back into the solvent flask through arm X. The water being denser is returned to the steam distillation flask via arm Y.
7. Allow the escaping low-boiling volatiles to pass through a solid carbon dioxide:acetone trap.

An alternative procedure involves the use of a porous polymer material, such as Porapak Q or Tenax. The volatiles in the head space from above the surface of a foodstuff are blown through the polymeric material from which they can be removed by heat (36).

Protocol 46. Isolation of headspace volatiles

Reagents

- Porapak Q

1. Place 100 mg of Porapak Q in a glass tube.
2. Heat the tube to 180–225 °C and pass a purge of oxygen-free nitrogen through the polymer.
3. Allow the tube to cool to room temperature.
4. Attach the tube containing the Porapak to the flask containing the food.
5. Transfer the volatiles from the head space to the tube with a purge of purified nitrogen flowing at 50 to 300 ml per minute.
6. Remove the tube containing the Porapak Q.
7. Connect the tube to a capillary tube chilled in solid carbon dioxide.

8. Heat the tube containing the adsorbed volatiles and collect them in the cooled capillary tube.

9. Seal the capillary tube at both ends for storage.

References

1. Martin, J. T. and Batt, R. F. (1958). *Ann. Appl. Biol.*, **46**, 375.
2. Hamilton, R. J. and Hamilton, S. (1972). In *Topics in lipid chemistry*, Vol. 3. (ed. F.D. Gunstone). J. Wiley and Sons, New York.
3. Folch, J. Lees, M., and Stanley, G. H. S. (1957). *J. Biol. Chem.*, **226**, 497.
4. Bligh, E. G. and Dyer, W. J. (1959). *Can. J. Biochem. Physiol.*, **37**, 911.
5. Ways, P. and Hanahan, D. J. (1964). *J. Lipid Res.*, **5**, 318.
6. Rogers, Van A., Van Aller, R. T., Pessoney, G. F., Watkins, E. J., and Leggett, H. G. (1984). *Lipids*, **19**, 304.
7. Biezenski, J. J., Pomerance, W., and Goodman, J. (1968). *J. Chromat.*, **38**, 148.
8. Troeng, S. (1955). *J. Am. Oil. Chem. Soc.*, **32**, 124.
9. Sandmann, G. and Boger, P., (1982). *Lipids*, **17**, 35.
10. Supelco literature.
11. Ackman, R. G. (1979). *Proc. Nova Scotia Institute of Science*, **29**, 501.
12. Brockerhoff, H. and Jensen, R. G. (1974). In *Lipolytic enzymes*. Academic Press, New York.
13. Metcalfe, L. D. and Schmitz, A. A. (1961). *Anal. Chem.*, **33**, 363.
14. Shepherd, A. J. and Iverson, J. L. (1975). *J. Chromat. Sci.*, **13**, 448.
15. Lough, A. K. (1964). *Biochem. J.*, **90**, 4C.
16. Christie, W. W. (1989). *Gas chromatography of lipids*. Oily Press Ltd., Scotland.
17. Hitchcock, C. and Hammond, E. W. (1980). In *Developments in food analysis*, (2nd edn) (ed. R. D. King). Applied Science Publishers, London.
18. Raie, M. Y. (1972). Ph.D. Thesis, Liverpool Polytechnic.
19. James, A. T. and Martin, A. J. P. (1956). *Biochem. J.*, **144**, 63.
20. Bannon, C. D., Breen, G. J., Craske, J. D., Ngo Trong Hai, Harper, N. L., and O'Rourke, K. L. (1982). *J. Chromat.*, **247**, 71.
21. Spencer, G. F., Plattner, R. D., and Miwa, T. (1976). *J. Am. Oil Chem. Soc.*, **54**, 187.
22. Harrington, K. J. and D'Arcy-Evans, C. (1985). *J. Am. Oil Chem. Soc.*, **62**, 1009.
23. Manku, M. S. (1983). *J. Chromat. Sci.*, **21**, 367.
24. Jamieson, G. R. and Reid, E. H. (1967). *J. Chromat.*, **26**, 8.
25. Smith, N. B. (1982). *Lipids*, **17**, 464.
26. Anderson, B. A. and Holman, R. T. (1975). *Lipids*, **10**, 716.
27. Francis, G. W. and Veland, K. (1981). *J. Chromat.*, **219**, 379.
28. Myher, J, J., Kuksis, A., Marai, L., and Yeung, S. K. F. (1978). *Anal. Chem.*, **50**, 551.
29. Shi-Hua Chen, S., Kou, A. Y., and Chen, H. H. Y. (1983). *J. Chromat.*, **276**, 37.
30. Cooper, M. J. and Anders, M. W. (1974). *Anal. Chem.*, **46**, 1849.
31. Wood, R. and Lee, T. (1983). *J. Chromat.*, 254, 237.

32. Bussell, N. E., Miller, R. A., Setterstrom, J. A., and Gross, A. (1979). In *Biology/biomedical applications of liquid chromatography*. Chromatography Science Series, Vol. 10 (ed. G. L. Hawk). Marcel Dekker, New York.
33. Korte, W. D. (1982). *J. Chromat.*, **243**, 153.
34. Durst, H. D., Milano, M., Kikta, E. J., Connelly, S. A., and Giushka, E. (1975). *Anal. Chem.*, **47**, 1797.
35. MacLeod, A. J. and Cave, S. J. (1975). *J. Sci. Food Agric.*, **26**, 351.
36. Jennings, W. G., Wohleb, R., and Lewis, M. J. (1972). *J. Food Sci.*, **37**, 69.

3

Thin-layer chromatography

R. JAMES HENDERSON and DOUGLAS R. TOCHER

1. Introduction

Thin-layer chromatography (TLC) has long been used for the separation of lipid mixtures into their component lipid classes. Despite the fact that the more modern technique of high performance liquid chromatography (HPLC) is finding applications in lipid analysis, several features still make TLC particularly useful for this purpose. For example, although sophisticated and expensive apparatus can be purchased for TLC, good results on a qualitative basis can be obtained relatively simply using the minimum of equipment. Analyses are performed rapidly and many samples can be analysed simultaneously alongside standards. The separated classes can be visualized and the chromatograms kept as a permanent record, or, alternatively, samples can be recovered for further analysis.

The present chapter considers the practical aspects of employing TLC for the analysis of lipids. It concentrates mainly on TLC in the ascending mode using glass plates coated with adsorbent because this is the most common form used in lipid analysis. Detailed descriptions of the historical and theoretical aspects of the technique can be found elsewhere (1). TLC with flame ionization detection (TLC-FID) is also considered since this technique is becoming increasingly popular for the quantitative analysis of lipid classes.

The stages of TLC are outlined separately in detail that is sufficient to allow the novice to undertake the analysis of lipids from any source.

1.1 General TLC procedure

Although lipid analysis by TLC may be carried out by various techniques, one underlying procedure is common to all. Basically, a lipid mixture is applied to an adsorbent which is coated in a thin layer on a support. The point of sample application is termed the origin. The mixture is then resolved into its components by differential migration, as a stream of solvent passes through the layer of adsorbent by capillary action. In a given solvent system each lipid component has a characteristic mobility which can be described as its R_f value. The R_f value is defined as the distance travelled by the component divided by the distance travelled by the solvent front, both distances being measured from the origin. Since lipids are generally colourless, the separated lipid components have to be rendered visible by chemical reagents. For quantitative analysis, the proportions of the individual components are then determined by various techniques available.

2. Materials and apparatus

In this section only the materials and apparatus employed for TLC using ascending chromatography are described (see *Figure 1*). Other forms of TLC such as overhead pressure chromatography, radial chromatography, gravity-fed chromatography, and short-bed TLC have not yet been used extensively for lipid analysis and all require specialized apparatus. Suppliers of the apparatus used for TLC include: (UK) Alltech Applied Science Ltd., Anachem, BDH Ltd., Camlab Ltd. (now suppliers of Desaga apparatus), Supelchem, Whatman Labsales Ltd.; (USA) Alltech Associates Inc., Analtech Inc., Camag Scientific Inc., Supelco Inc.; (Switzerland) Camag.

2.1 Adsorbents

2.1.1 Silica gel

Silica gel (Kieselgel) is the adsorbent most frequently used for the analysis of lipids by TLC. It is a synthetic material manufactured to closely defined specifications. The particle size of the silica gel used for standard TLC is in the range 10 to 15 μm. The particles are manufactured to contain pores of a

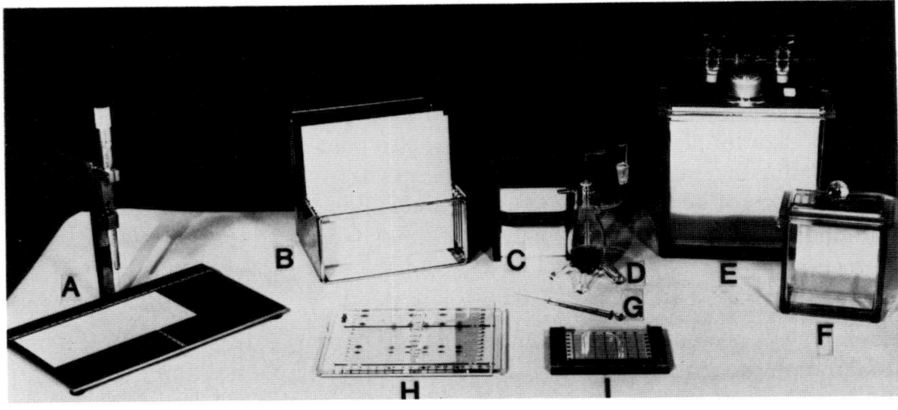

Figure 1. Typical apparatus used in TLC.
A: applicator with micrometer-controlled microsyringe containing 10 × 20 cm HPTLC plate. **B**: drying/storage rack containing standard 20 × 20 cm TLC plates. **C**: 10 × 10 HPTLC plates in drying/storage rack. **D**: spraying vessel. **E**: 20 × 20 cm plates being developed in standard chamber. **F**: 10 × 10 cm HPTLC plates being developed in HPTLC chamber. **G**: glass microsyringe. **H**: sample application template/R_f calculator overlayed on developed standard plate. **I**: HPTLC plate contained in sample application template.

definite diameter, 40, 60, 80, 100, or 150 Å. Silica gel 60 (the 60 denotes the pore size) is the most commonly used adsorbent in the TLC of lipids.

At a given humidity the amount of water absorbed by the silica gel increases as pore size decreases. The water content of silica gel determines the polarity of the adsorbent and hence its activity and chromatographic properties. For good separations by silica gel the water content of the gel must be carefully controlled. The silica gel on TLC plates is normally activated by heating the plates before use at temperatures above 100 °C to remove water.

Silica gel can be modified with various compounds to attain the separation of lipid components not resolved well by chromatography on standard silica gel. The modifications are described in later relevant sections.

2.1.2 Kieselguhr

Kieselguhr is diatomaceous earth based on silicaceous material. Since it is of fossil origin and non-synthetic its properties are more variable than silica gel. For lipid analysis, Kieselguhr is most frequently employed in reversed-phase TLC (Section 5.1).

2.1.3 Alumina

Aluminium oxide has been used as an adsorbent for TLC of lipids but it has considerably less sample capacity than silica gel. Furthermore, aluminium oxide of basic pH can cause hydrolysis of esters whilst acidic aluminium oxide may also cause undesirable reactions.

2.1.4 Others

Magnesium hydroxide mixed with an equal proportion of Kieselguhr can be used as an adsorbent specifically for the TLC separation of wax esters and steryl esters (2).

2.2 Supports

Glass is by far the most commonly used support for the adsorbent in the TLC of lipids. It is resistant to the acids and alkalis used in developing solvents or detection sprays and its rigidity makes it suitable for densitometry. Glass plates of 20 × 20 cm or 20 × 10 cm are usually employed in standard TLC and 10 × 10 cm or 10 × 20 cm in HPTLC (see later).

Aluminium and plastic sheets can also be used to support the adsorbent and have the advantage that they can be cut easily to make small plates if required. These non-glass supports are available only as pre-coated plates (Section 2.4).

2.3 Preparation of TLC plates

For the self-preparation of chromatographic layers, adsorbents are purchased in the form of powders. Silica gel G contains gypsum (calcium sulfate hemihydrate) which has been added as a binder to aid the adherence of the silica gel to the support. Silica gel H contains no binder, although a small amount of colloidal silica is added by some manufacturers to help the adsorbent stick to the glass support. A fluorescent indicator, usually with an excitation wavelength of 254 nm, can be included with the silica gel adsorbent. The inclusion of this indicator means that chromatograms are self-indicating when viewed under UV, as lipid components show up as dark spots against a green fluorescent background.

The chosen adsorbent powder is mixed with water or a solution of salts or buffering compounds to form a thick slurry and spread on to glass plates using a spreading device (*Protocol 1*). Several models of TLC plate coaters, both manually and automatically operated, can be purchased from most of the suppliers of TLC equipment. The thickness of the adsorbent layer is determined by the gap size between the spreader reservoir and the plates it is pushed over.

Protocol 1. Basic procedure for preparation of standard TLC plates

The quantities used are sufficient for five 20 × 20 cm plates.

1. Clamp clean glass plates (20 × 20 × 0.4 cm) into the spreading rack according to the manufacturer's instructions, and wipe with acetone.
2. Place the adsorbent applicator on top of the plates at one end of the rack, with the gap corresponding to the thickness of the final adsorbent layer next to the glass.
3. Shake 50 g of silica gel vigorously with 100 ml water in a stoppered flask to form a slurry, and pour quickly into the reservoir of the applicator.
4. Draw the applicator smoothly across the top of the plates to the other edge of the rack.
5. Allow plates to sit in rack for 2 min until gel has started to form.
6. Carefully remove the coated plates from the spreading rack and transfer to aluminium drying trays.
7. Allow plates to dry at room temperature for 1 h, then place in oven at 110–120 °C for 2 h to activate.
8. Remove plates and store in desiccator or air-tight box until required.
9. If stored for more than a few hours, reactivate the plates by heating at 110 °C for 30 min before using.

2.4 Pre-coated TLC plates

The use of pre-coated TLC plates rather than self-prepared plates can be worthwhile for several reasons. If the analyses are to be infrequent or exploratory, the purchase of spreading equipment is probably not justified. Furthermore, the batch to batch variation in pre-coated TLC plates is considered to be less than with self-prepared plates, due to carefully controlled manufacturing processes. Consequently, analyses performed with pre-coated plates are frequently more reproducible than with self-prepared plates. The pre-coated layers are also of superior optical quality in comparison with self-prepared layers. This is important when quantitative densitometry is to be applied to the developed chromatograms. Finally, in comparison with self-prepared plates, the layers on pre-coated plates are more resistant to abrasion.

2.4.1 TLC

An extensive range of pre-coated TLC plates manufactured by companies such Analtech, E. Merck, Macherey-Nagel & Co., and Whatman, is now offered by most commercial suppliers. Silica gel, Kieselguhr and aluminium oxide are all available commercially as adsorbents pre-coated on glass plates. Silica gel can also be obtained pre-coated on both aluminium and plastic sheets. For the analysis of lipids, silica gel 60, with or without fluorescent indicator, can be readily purchased as an adsorbent pre-coated in layers of various thickness from 0.25 to 2.0 mm on 20 × 20 cm glass plates. Some manufacturers incorporate polyacrylic acid as an inert organic binder, rather than gypsum in their pre-coated plates. Some companies use the letter K (for Kieselgel) rather than silica gel to denote supports coated with silica gel.

Pre-coated TLC plates can be purchased in which a pre-adsorbent or concentration zone some 25 mm wide is coated across the full width of the support along the bottom edge. The adsorbent in this zone is a silica of very high pore diameter and very small surface area. Consequently, this silica is very inactive and does not absorb material. The rest of the plate is covered in a normal adsorbent such as silica gel 60. A sample applied to the preadsorbent zone is concentrated by the developing solvent front into a narrow band and carried up on to the adsorbent silica gel. Separation of the components then proceeds as with conventional plates. Pre-channelled plates, in which narrow lines of adsorbent have been removed to divide the plates into 10 or 20 lanes, can also be purchased. The use of these plates prevents the spread of one sample into another. It is not necessary to specifically purchase pre-channelled plates since they can be readily prepared from conventional plates by drawing the edge of a spatula across the plate to remove a thin line of adsorbent.

Pre-coated 20 × 20 cm plates are available with score marks at regular intervals which allow the plates to be snapped apart to produce narrower plates and hence economize on materials. A similar result can be achieved with unscored plates by careful use of a glass cutter on the uncoated side of the plate.

2.4.2 High-performance TLC

The silica gel adsorbent used for high-performance TLC (HPTLC) is only available on pre-coated plates and consists of silica gel 60 with a particle size of only 4.5 or 5.0 μm. The particle characteristics of this material are such that the separation efficiency closely resembles that of HPLC. The adsorbent layer in HPTLC is slightly thinner (0.15–0.20 mm) than in standard analytical TLC. Glass of size 10 × 10 or 10 × 20 cm is the most common support used in HPTLC, although pre-coated aluminium sheets are also available. HPTLC plates with a pre-adsorbent zone at the bottom of the plate can also be obtained commercially.

HPTLC has several advantages over standard TLC in the analysis of lipids. Less sample is necessary, a smaller volume of developing solvent is used, the analysis time is shorter, and the resolution of components is superior. These features all make HPTLC very useful for the analytical separation of lipid mixtures.

2.5 Development chambers

For the development of 20 × 20 plates, rectangular glass chambers of approximate dimensions 10 × 21 × 21 cm are commonly used. In addition to being inert to the developing solvents, glass is transparent and allows the progess of the ascending solvents to be monitored. Stainless steel chambers with glass fronts are also available. Classical chambers have flat bottoms and sides and allow a maximum of two plates to be developed simultaneously. Chambers having vertical grooves in the end walls allow up to five plates to be developed simultaneously.

Although HPTLC plates can be developed in tanks designed for 20 × 20 cm plates, this is wasteful of solvent and smaller tanks specifically intended for HPTLC are better employed. Glass chambers of internal dimensions 10.8 × 20.4 × 21 cm which have grooves in both end walls and side walls can also be purchased (Camlab). These allow the simultaneous development of up to five 20 × 20 or ten 10 × 10 cm plates.

Regardless of type, all chambers have a heavy glass or stainless steel lid which acts to seal the chamber and hence maintain an atmosphere saturated with solvent vapour within the chamber. Lids with facilities for gassing the chamber with an inert gas such as nitrogen can also be obtained. These allow developments to be carried out under an inert atmosphere.

3. Application of sample and development of plates

A general scheme for TLC and HPTLC is outlined in *Protocol 2*.

Protocol 2. General scheme for TLC of lipids

Values are for standard 20 × 20 cm glass plates coated with silica gel 60 in a layer of 0.25-mm thickness. Values in parenthesis are for 10 × 10 cm HPTLC plates coated with silica gel 60.

1. Pre-wash TLC plate in chloroform or the developing solvent system to be used.
2. Dry plates at room temperature, and heat at 110 °C for 30 min. Allow to cool.
3. Dissolve lipid mixture in a suitable solvent at a concentration of 20 (10) mg/ml.
4. Apply 10–80 (2–10) μg sample as a spot or small streak 1.5 (1.0) cm from bottom edge.
5. Place plate in development chamber containing 100 ml (35 ml for HPTLC chamber) of fresh solvent. Lean plate against the side at a steep angle.
6. When solvent front has reached approx. 1.5 (1.0) cm from top of plate, remove plate from chamber and quickly mark position of solvent front with pencil or scalpel blade at edge of plate.
7. Allow solvent to evaporate in fume cupboard or vacuum desiccator.
8. Spray developed chromatogram with visualizing reagent or view directly under UV light if self-indicating.

3.1 Pre-washing

For best analytical results, the TLC plate should be pre-washed before use to remove impurities which could darken the background in the visualization process or cause a dark line along the solvent front. This is done by developing the plate to the top in either the solvent system to be used, or chloroform, or chloroform/methanol (1:1, v/v). Impurities are moved to the top of the plate, and can be completely removed by scraping off a 1-cm band of adsorbent. After removal of this band the plate is turned through 90 degrees before sample application to retain maximum development distance. When TLC is being used simply for the preparative separation of lipid classes pre-washing of the plate is not necessary, except where very small quantities of lipid are to be isolated.

3.2 Sample capacity

The maximum amount of lipid that can be applied as a spot or a full-width streak to a TLC plate, depends on the nature of the components as well as the thickness of adsorbent layer. Analytical 20 × 20 cm plates with layers 0.25 mm thick can resolve 20 to 80 μg of total lipid applied as a spot and up to 15 to

20 mg as a full-width streak. Plates of the same dimensions but adsorbent thickness 0.5 mm can separate up to 40 mg overall, whilst preparative plates of thickness 1 mm and 2 mm can accommodate even larger sample amounts. However, as a general rule, loss of resolution occurs as the thickness of the layer increases. Much less polar lipid than simple neutral lipid can be separated by the thick adsorbent layers. For 10 × 10 HPTLC, the maximum amount of a lipid mixture than can be applied as a spot of 1 mm diameter without loss of resolution is around 10 μg.

3.3 Sample application

The quality of lipid analysis by TLC depends greatly on the careful application of the sample mixture to the adsorbent. Care must be taken not to damage the adsorbent layer during this procedure, and TLC plates should only be handled by the edges, to avoid contamination from fingers.

For analysis by standard 20 × 20 cm TLC the sample should ideally be applied at a distance of approximately 1.5 cm from the lower edge of the plate. The smaller the area of application, the sharper the resolution. A spot of diameter not more than 2 mm, or a narrow streak of 5 to 10 mm length, both give good resolution. For the preparative separation of larger amounts of lipids, the sample can be applied as a narrow streak along almost the full width of the plate. In HPTLC, the sample should ideally be applied as a spot of 1 mm diameter or less, but in practice excellent resolution is also obtained when samples are applied as 4-mm streaks. Samples can be identified by lightly writing with a pencil at the top of plate above where the solvent front will be.

The sample is applied in as non-polar a solvent as possible. Unless sample solubility requires the use of chloroform/methanol (2:1, v/v), methanol should not be used for sample application as it tends to produce large spots and wide streaks.

3.3.1 Manual techniques

The sites of application at the origin can be marked by lightly touching the adsorbent with a sharp pencil or needle. Glass microsyringes with blunt needles held just above the adsorbent can be used to deliver drops of sample solution to the adsorbent at the origin. Disposable capillaries calibrated in the volume range 0.5 to 16 μl are also available commercially, as are Nanopipette capillaries which dispense 0.1 to 0.2 μl and are specifically designed for use with HPTLC. A stream of nitrogen supplied via a Pasteur pipette connected to the end of tubing can be used intermittently during sample application to evaporate the solvent as soon as it is applied to the adsorbent. This helps to maintain the applied sample as a narrow band or small spot at the origin. The use of nitrogen is particularly useful in preparative TLC when larger volumes are applied as a narrow streak to the plate.

Microsyringes should be rinsed out with a suitable organic solvent between different samples and sample application should be performed in a fume cupboard to minimize the inhalation of evaporating solvents.

Spotting guides or templates are available commercially for both 20 × 20 and 10 × 10 cm plates which ensure that samples are applied at defined regular distances. This is useful if a densitometer which can automatically change lanes is to be used for subsequent quantification. In their simplest form, spotting guides are plates of transparent Perspex with holes at set distances (frequently 1 cm for 20 × 20 cm plates and 2 mm for HPTLC) arranged near the lower edge through which samples are applied. Many models of spotting templates are also marked with scales which can be used for the calculation of R_f values on developed chromatograms.

With both standard and HPTLC plates equipped with a pre-adsorbent zone the sample can be applied to the zone about 1 cm from the bottom edge of the plate in a larger volume of solvent than is used for plates without such a zone.

3.3.2 Mechanical techniques

Although manual techniques such as those outlined above are most commonly used for applying lipid samples to TLC plates, various pieces of apparatus exist for the mechanical application of samples. Among these, the Nano- and Micro-Applicators (Camag) utilize micrometer-controlled micro-syringes to apply samples of volume in the range 50 nl to 2.3 μl without damaging the adsorbent layer. Semi-automatic TLC spotters such as the PSO1 by Camlab (Desaga) can dispense the sample from a microsyringe and automatically move on to the next spotting position, but still require the syringe to be filled manually. Apparatus such as the Camag Linomat can also be used to apply larger volumes in the region of 250 μl as streaks in preparative TLC.

More sophisticated automated applicators which apply many samples simultaneously are also available for use in the routine analysis of large numbers of samples but have generally found little use in lipid analysis.

3.4 Standards

Although lipid classes can be identified by reference to published R_f values, the application of authentic lipid standards, either as a mixture or individually, alongside the lipid being analysed, greatly aids in the identification of the components present in the lipid sample. Within any laboratory, the R_f values of lipid classes in a given solvent system are not always constant due to day-to-day variations in temperature, humidity and perhaps even the batch of plates used. By routinely analysing lipid standards alongside samples such variations can be taken into account.

The standards employed depend on the nature of the sample being analysed. The neutral lipid classes which are most commonly encountered in natural organisms, such as triacylglycerols, cholesteryl esters, cholesterol, and partial acylglycerols, are all readily available commercially either individually or as components of standard mixtures (Sigma Chemical Co. Ltd., UK; Sigma Chemical Company, Supelco, and Nu Chek Prep. Inc., USA; and Bast of Copenhagen, in Europe). The most naturally abundant phospholipid classes can also be purchased from the same suppliers, as can some galactolipids and glycerol-ether lipids. Standards of acyl esters are usually synthetic and contain only one fatty acid component, whereas the polar lipids sold as standards are frequently extracted from some abundant material such as egg yolk or bovine brain. Differences in fatty acid compositions between standards and the corresponding class in the sample being analysed can mean that their R_f values are not exactly identical.

The use of standards in the quantitation of separated components is dealt with in Section 7.

3.5 Development of plates

After the application of samples and standards, the plate is transferred to a development chamber containing sufficient solvent to cover the adsorbent up to a level of about 5 mm from the bottom edge. For chambers designed to accommodate 20 × 20 cm plates, 100 ml of solvent is adequate. Chambers for 10 × 10 cm HPTLC plates require approximately 35 ml of solvent. The solvent system to be used should be added to the chamber about 15 min before the plate is introduced. The component solvents should be thoroughly mixed.

When solvent systems containing large proportions of polar solvents such as methanol are employed, the chambers can be lined with filter paper to help saturate the atmosphere. This will prevent the so-called edge effects occasionally observed in TLC, which arise from non-uniform saturation of the chamber atmosphere. However, with non-polar solvents such as hexane and diethyl ether, the lining of chambers is not necessary.

The time taken for a TLC plate to develop depends on the ambient temperature and the solvent system employed. For instance, with a hexane/diethyl ether system a standard 20 × 20 cm plate will be developed fully in approximately 1 h at room temperature, whereas twice as long is typical for systems containing polar solvents. A batch of solvent system should not be used more than once for development.

4. TLC on standard silica

In its normal form, silica gel is a polar adsorbent. Consequently, polar lipids are more tightly adsorbed than non-polar lipids due to polar interactions. This is frequently termed normal-phase TLC. In the TLC separation of lipids using standard silica gel the most non-polar lipids therefore migrate at the fastest rates (high R_f values) and the polar lipids at the slowest rates (low R_f values). By increasing the polarity of the developing solvent system the R_f values of components can be increased. The choice of a suitable solvent system is critical in the separation of lipid classes.

4.1 Neutral lipids

The solvent systems used to separate simple neutral lipid classes most commonly contain hexane, diethyl ether, and acetic or formic acid in various proportions although other non-polar solvents have also been used.

The separation of neutral lipid components can often be enhanced by developing the plate a set distance in one solvent system and then redeveloping it to full length in a second system of different solvent composition. The plate must be thoroughly dried at room temperature, preferably under vacuum, between developments.

Table 1 shows the expected R_f values of different neutral lipid classes in

Table 1. R_f values of neutral lipid classes separated on silica gel 60 in various solvent systems

	1	2	3	4	5
Hydrocarbons	0.95	–	–	–	–
Trialkylglyceryl ethers	0.90	–	–	–	–
Steryl esters	0.90	0.94	0.94	0.95	0.59
Wax esters	0.90	0.92	0.94	0.91	–
Dialkyl monoacylglycerols	0.70	–	–	–	–
Alkenyl diacylglycerols	0.65	–	–	–	–
Fatty acid methyl esters	0.65	0.81	0.94	0.75	–
Alkyl diacylglycerols	0.55	–	–	–	–
Fatty aldehydes	0.55	–	–	–	–
Triacylglycerols	0.35	0.73	0.86	0.61	0.61
Fatty acids	0.18	0.33	0.39	0.35	0.35
Fatty alcohols	0.15	0.28	0.29	0.21	–
Sterols	0.10	0.24	0.24	0.16	0.32
1,2-O-dialkylglyceryl ethers	0.09	–	–	–	–
1,3-diacylglycerols	0.08	0.24	0.26	0.19	0.45
1,2-diacylglycerols	0.08	0.21	0.24	0.09	0.54
Monoacylglycerols	0.0	0.03	0.03	0.01	0.05
Chlorophyll/carotenoids	–	0–0.2	0–0.23	0–0.06	0.07–0.19 0.49–0.69*
Complex polar lipids	0.0	0.0	0.0	0.0	0.0

Solvent systems:

1. Petroleum ether (b.p. 60–70 °C)/diethyl ether/glacial acetic acid (90:10:1, by vol.) (3).
2. Hexane/diethyl ether/glacial acetic acid (80:20:2, by vol.).
3. Hexane/diethyl ether/glacial acetic acid (70:30:3, by vol.).
4. Hexane/heptane/diethyl ether/glacial acetic acid (63:18.5:18.5:1, by vol.) to 2 cm from top then full development in carbon tetrachloride (4).
5. Benzene/propan-2-ol/water (100:10:0.25, by vol.) (5)

– = not determined; * pigments run in two separate regions in this solvent.

various solvent systems using silica gel 60 as adsorbent. The actual R_f values obtained for a given solvent system will vary from laboratory to laboratory due to differences in apparatus and environmental conditions, but should still resemble those presented in the table. With all these systems, complex polar lipids remain at the origin.

Hexane may be substituted for petroleum ether without any great change in R_f values of the most non-polar lipids although those of the more polar lipids may change slightly. Similarly, the use of benzene in solvent systems can be avoided by using toluene instead although this may change R_f values. For a given solvent system and silica gel 60 the R_f values of separated lipid classes are only slightly different for standard 20 × 20 cm TLC plates and 10 × 10 cm HPTLC plates.

No solvent system will completely separate all naturally occurring neutral lipid classes in a single development. As is evident from *Table 1*, increasing the proportion of diethyl ether whilst decreasing that of hexane or petroleum ether causes the more polar neutral lipid classes to migrate further from the origin. However, this also causes loss of resolution of the least polar classes towards the top of the plate. The initial use of a solvent system such as hexane/diethyl ether/glacial acetic acid (80:20:2, by vol.) will establish the range of components present in an unknown mixture (*Figure 2*). Subsequent analyses can then be performed with other solvent systems to improve the separation of components in areas shown to contain several components.

Hydrocarbons always run very close to the solvent front in hexane/diethyl ether systems but can be better resolved by development with heptane/benzene (9:1, v/v) when they have an R_f value of around 0.8 (6). Steryl esters and wax esters are very difficult to separate on silica gel 60. However, when magnesium hydroxide/Kieselguhr (1:1, w/w) is used as adsorbent with 1% ethyl acetate in hexane as developing solvent, wax esters have an R_f value of 0.87, in comparison with 0.54 for cholesteryl esters (2).

The resolution of sterols from diacylglycerols is particularly difficult using hexane/diethyl ether mixtures but can be achieved by solvent system 5 in *Table 1*. Lipid-soluble pigments such as chlorophyll and carotenoids render the total lipid extracted from photosynthetic and some invertebrate sources, coloured. Although these pigments are visible on plates immediately after development, they are usually rendered colourless or become much fainter, during the staining and charring process.

Although lipid classes are given a single R_f value in *Table 1*, some classes such as triacylglycerols and steryl esters are better represented by a range of values since in natural samples they frequently exhibit more than one band. This is due to their partial resolution into molecular species on the basis of fatty acid composition. The degree of spreading depends on the solvent system used. The separation of individual lipid classes on the basis of fatty acid composition can be achieved by TLC using modified silica gel (Section 5).

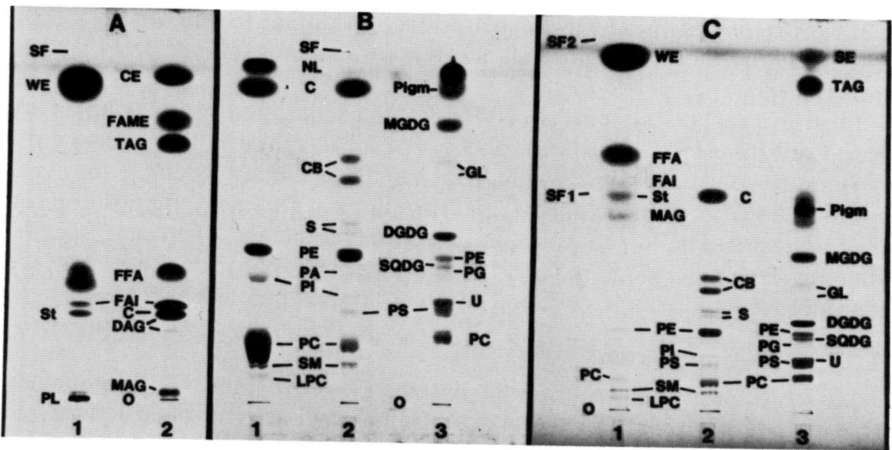

Figure 2. HPTLC chromatograms of lipid mixtures developed in various systems. **A**: Single development in hexane/diethyl ether/glacial acetic acid (80:20:2, by vol.). **1**, zooplankton lipid; **2**, standard authentic mixture, all based on oleic acid. **B**: single development in methyl acetate/propan-2-ol/chloroform/methanol/0.25% aqueous KCl (25:25:25:10:9, by vol.). **1**, cod-roe total lipid; **2**, rat-brain total lipid; **3**, algal (*Chroomonas salina*) total lipid. **C**: Development to half final distance in solvent system of B followed by full development in solvent system of **A**. **1**, zooplankton total lipid; **2**, rat brain total lipid; **3**, algal (*C. salina*) total lipid. All plates were stained with 3% copper acetate in 8% orthophosphoric acid.
Abbreviations: as for *Table 2* except C, cholesterol; CE, cholesteryl ester; DAG, diacylglycerol; FAI, fatty alcohol; FAME, fatty acid methyl ester; FFA, free fatty acid; GL, unknown glycolipid; MAG, monoacylglycerol; O, origin; Pigm, pigment; PL, polar lipid; SF, solvent front; St. sterol; TAG, triacylglycerol; U, unknown; WE, wax ester.

4.2 Polar lipids

The separation of phospholipid classes was traditionally achieved using plates coated with silica gel H rather than silica gel G, since the calcium sulfate present in the latter adsorbent prevented good separation of acidic phospholipids. However, plates coated with silica gel H require very careful handling, due to the lack of binder. More recently, pre-coated silica gel plates containing an inert organic binder have become more widely used in lipid analysis, and have been employed for the separation of polar lipids.

The types of polar lipid present in a lipid extract depends very much on its source and very obvious differences exist between plant and animal sources. On account of the large differences which exist in polarity among the various classes of polar lipids, no single TLC system exists which can completely separate all known polar lipid classes by development in one dimension. The spots or bands which are observed when polar lipid mixtures are separated on TLC plates developed in one direction may contain more than one component. To separate completely all components, two-dimensional TLC can be performed in which the plates are developed to full length in one direction, dried, turned through 90 degrees, and developed in a second direction.

Most of the TLC systems described for the separation of polar lipids are based on the use of chloroform and methanol as major components of the developing solvent system. Commercial standards are not available for every complex lipid occurring in nature, and variations in fatty acid composition can lead to a spread of R_f values for an individual lipid class. The use of specific staining reagents (Section 6.2) is therefore of particular use for the identification and characterization of polar lipids separated by TLC.

4.2.1 One-dimensional

None of the most commonly used systems (*Table 2*; refs. 1–4) completely separates all polar lipid classes. A system that is suitable for polar lipids from one biological source may not give satisfactory resolution with those from a different biological source. Small variations in the proportions of the solvents can be tried to improve the separation after initial development in one of the systems listed.

A useful solvent system for the analysis of polar lipids by TLC and HPTLC is that devised initially by Vitiello and Zanetta (10) (*Table 2*, system 5) for the complete separation in one dimension of the major galactolipids and phospholipids of brain tissue. This system works well with pre-coated plates

Table 2. Relative migration of polar lipid classes in one-dimensional TLC on silica gel in different solvent systems

Relative migration:

1. MGDG > CB > PA > CL > LPA > PE > = DGDG > PDE > PG > SQ > SU > PC > PI > SM > = PS
2. CL,PA > PE,PG > PDE > PME > PS > PI > PC > SM > LPC
3. CB > PE > CL > = PG > PC > SM > LPE > SU > = PI > LPC > = PS,PA,LPA
4. CL > = CB > PG > SU > PA > = PE > PS > PI > PC, LPE > SM > LPC
5. NL > CB > SU > PE > PA,CL > PI > PS > PC > SM > LPC > GS

Solvent systems used:

1. Chloroform/methanol/water (65:25:4, by vol.) (7).
2. Chloroform/methanol/acetic acid (25:15:4.2, by vol.) (8).
3. Chloroform/methanol/28% ammonia (65:25:5, by vol.) (9).
4. Chloroform/acetone/methanol/acetic acid/water (6:8:2:2:1, by vol.) (9).
5. Methyl acetate/propan-1-ol/chloroform/methanol/0.25% aqueous KCl (25:25:25:10:9, by vol.) (10).

Abbreviations: CB, cerebrosides; CL, cardiolipin; DGDG, digalactosyldiacylglycerol; GS, gangliosides; LPA, lysophosphatidic acid; LPC, lysophosphatidylcholine; LPE, lysophosphatidylethanolamine; MGDG, monogalactosyldiacylglycerol; NL, neutral lipids; PA, phosphatidic acid; PC, phosphatidylcholine; PDE, phosphatidyl-*N*,*N*-dimethylethanolamine; PE, phosphatidylethanolamine; PG, phosphatidylglycerol; PI, phosphatidylinositol; PME, phosphatidyl-*N*,*N*-methylethanolamine; PS, phosphatidylserine; Su, sulfatides; SM, sphingomyelin; SQ, sulfoquinovosyldiacylglycerol.

and can be applied to polar lipids from various sources including plants (*Figure 2*). The galactolipids characteristic of plants can also be separated from phospholipids on plates coated with silica gel 60 using diisobutyl ketone/acetic acid/water (40:25:3.7, by vol.) (11) or acetone/glacial acetic acid/water (100:2:1, by vol.) (12) as developing solvents.

Mixtures of neutral glycosphingolipids can be separated by TLC or HPTLC on silica gel 60 using chloroform/methanol/water (60:30:5, by vol.) (13). For the separation of gangliosides by standard TLC or HPTLC, the use of solvent systems containing acid must be avoided. For this reason basic solvent systems such as chloroform/methanol/2.5 M aqueous ammonia (50:40:10, by vol.) or a neutral system such as chloroform/methanol/water/1% $CaCl_2$ (55:45:8:2, by vol.) must be used (14).

Plasmalogens comigrate with their phospholipid analogue in one-dimensional TLC, but can be detected by treatment of the sample with acid reagents directly on the TLC plate (15). Phosphonolipids are not separated from their corresponding phospholipids using the solvent systems in *Table 2*. However, phosphonolipids can be completely separated by one-dimensional TLC on silica gel G, using methanol/water (2:1, v/v) as developing solvent (16). The phosphonolipids migrate close to the solvent front, whereas phospholipids and neutral lipids remain close to the origin.

4.2.2 Two-dimensional

Two-dimensional TLC and HPTLC is particularly useful for establishing which polar lipid classes are present in a sample. Although the separated components can be quantitated after recovery from the adsorbent, the technique is not suitable for *in situ* quantitation by scanning densitometry. It is generally found that two-dimensional TLC is less reproducible than one-dimensional systems.

The solvent systems used in the application of two-dimensional TLC to the analysis of phospholipids and glycolipids are reviewed elsewhere (17) and will not be detailed here. In general, the plate is developed with a water/organic solvent system containing acetic acid in one direction and a base such as ammonia in the other.

The amount of sample applied in two-dimensional TLC tends to be larger than that used for separations in a single direction. A basic procedure for the separation of a lipid mixture by two-dimensional HPTLC is outlined in *Protocol 3*. Typical chromatograms are shown in *Figure 3*. The same procedure can be used for standard TLC on layers 0.25 mm thick by increasing the amount of sample applied to 500 μg and the volume of the developing solvent system to 100 ml. The solvent front is allowed to migrate to within 2 cm of the top of standard TLC plates.

Protocol 3. Separation of polar lipids by two-dimensional HPTLC on silica gel 60

1. Pre-wash plate. Dry in air and heat at 110 °C for 30 min.
2. Make up 35 ml of methyl acetate/propan-2-ol/chloroform/methanol/ 0.25% aqueous KCl (25:25:25:10:9, by vol.) in an HPTLC development chamber, and line two sides with filter paper.
3. Apply 50 μg sample as spot at the lower right-hand corner of the plate, 1.0 cm from each edge.
4. Place plate in development chamber and allow solvent front to migrate to 1.0 cm from top of plate (approx. 40 min).
5. Remove plate and dry at room temperature in vacuum desiccator for 30 min.
6. Turn plate through 90 degrees clockwise from first development direction and place in lined chamber containing 35 ml of chloroform/methanol/7 M aqueous ammonia (65:35:5, by vol.).
7. Remove plate when solvent front is 1.0 cm from top edge (approx. 30 min).
8. Dry in vacuum desiccator for 30 min.
9. Visualize components, using general or specific detection reagents.

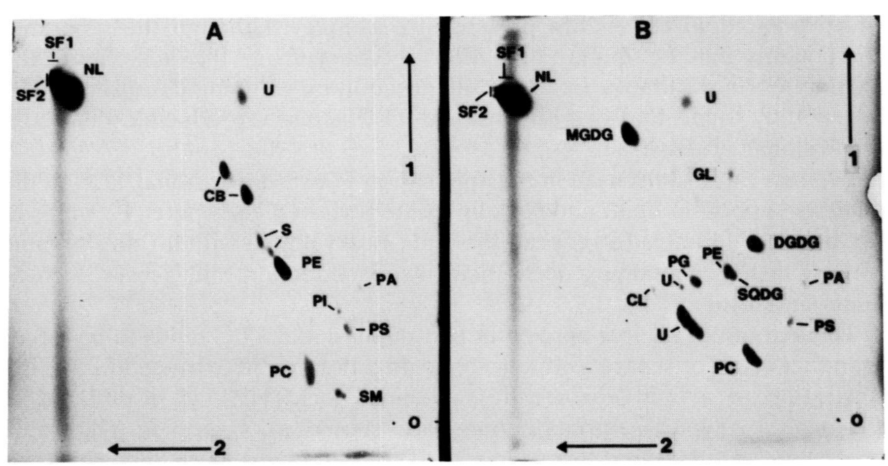

Figure 3. Two-dimensional HPTLC of total lipid from **A**, rat brain, and **B**, the alga *C. salina*. Plates were first developed in direction 1 with methyl acetate/propal-2-ol/ chloroform/methanol/25% aqueous KCl (25:25:25:10:9, by vol.) and then in direction 2 with chloroform/methanol/7N NH_4OH (65:35:5, by vol.). Developed chromatograms were stained with 3% copper acetate in 8% orthophosphoric.
Abbreviations: as for *Figure 2* and *Table 2*.

It is important that all traces of the solvents used for development in the first dimension are evaporated from the plate before it is developed in the solvent system of the second dimension. Although this can be achieved by use of an unheated stream of air from a hair dryer, vacuum desiccation is less likely to damage the lipids on the plate.

With all solvent systems used for two-dimensional separation of polar lipids, neutral lipids are observed as a single zone near the top corner of the developed plate (see *Figure 3*) when total lipid is analysed.

4.3 Combined neutral and polar lipid systems

Two-dimensional methods have been described in which the neutral lipid classes can be separated by developing the plate in the reverse direction of the second development after removal of a band of silica from the edge beneath the polar lipid classes to prevent their migration (18). Although this three-directional TLC does allow the resolution of both polar and neutral lipid classes on the one plate it does not lend itself to quantitation by densitometry and only one sample can be analysed per plate. More useful separations of neutral and polar lipids on the same plate can be achieved by using multiple developments in one dimension. In this technique the plate is first partially developed in a solvent system known to separate polar lipids. The neutral lipids which are deposited approximately half-way up the plate by this first development are then resolved by developing the plate fully in a second solvent system designed to separate neutral lipids.

Multi-development can be used with both standard TLC and HPTLC. The sample is applied to the adsorbent in the same manner as for single development. Between developments the plate must be thoroughly dried, preferably in a vacuum desiccator. A procedure which separates most of the common lipids found in plant and animals is outlined in *Protocol 4* (see also *Figure 2*). Lipids separated by this method can be quantitated by scanning densitometry bearing in mind the limitations of densitometry with respect to available standards. Systems using three or more developments can also be used. The R_f values obtained for some lipids using such systems are presented in *Table 3*.

Protocol 4. HPTLC separation of neutral and polar lipids by double development in one dimension (19)

1. Pre-wash plate with chloroform. Dry in air, and heat at 110 °C for 30 min.
2. Apply sample as spot or streak 1.0 cm from bottom edge.
3. Mark adsorbent with pencil at edge of plate 4 cm and 8 cm from origin.
4. Add 35 ml of methyl acetate/propan-2-ol/chloroform/methanol/0.25%

Protocol 4. *Continued*

aqueous KCl (25:25:25:10:9, by vol.) to development chamber and line with filter paper.

5. Place plate in development chamber and allow solvent front to migrate to the 4-cm mark (approx. 15 min).
6. Remove plate and dry in desiccator for 30 min.
7. Discard used solvent system and ensure chamber is dry. Add 35 ml of hexane/diethyl ether/glacial acetic acid (80:20:2, by vol.). Do not line the chamber.
8. Place plate in development chamber and allow solvent front to migrate to 8-cm mark (approx. 25 min).
9. Remove plate and dry in air.
10. Visualize separated components with general or specific detection systems.

Table 3. R_f values of lipid classes separated by TLC using silica gel 60 and multi-development

Component	HPTLC[a]	Standard[b] TLC
Alkane	0.97	–
Squalene	0.92	–
Cholesteryl esters	0.86	–
Fatty acid methyl esters	0.81	–
Triacylglycerols	0.77	0.89
1,3-diacylglycerols	0.71	0.82
1,2-diacylglycerols	0.68	–
Free fatty acids	0.65	0.72
Cholesterol	0.60	–
Monoacylglycerols	0.47	0.78
5-HETE	–	0.64
5,12-diHETE/PGE_2	–	0.63
Prostaglandin E_2	–	0.62
Thromboxane B_2	–	0.61
Prostaglandin F_{2a}	–	0.55
Prostaglandin F_{1a}	–	0.48
Monogalactosyldiacylglycerols	0.39	–
Galactocerebroside	0.35	–
Hydroxygalactocerebroside	0.33	–
Sulfatide	0.28	–
Hydroxysulfatide	0.26	–
Phosphatidylethanolamine	0.24	0.32
Phosphatidylglycerol	–	0.34
Cardiolipin	0.22	–
Trihexosylceramide	0.20	–
Phosphatidylinositol	0.18	0.18
Phosphatidylserine	0.17	0.17
Globoside	0.14	–

Component	HPTLC[a]	Standard[b] TLC
Phosphatidylcholine	0.13	0.23
Sphingomyelin	0.10	0.15
Phosphatidic acid	–	0.14
Lysophosphatidylcholine	0.06	–
CDP-diacylglycerol	–	0.08

Systems used:
[a] Methyl acetate/propan-1-ol/chloroform/methanol/0.25% aqueous KCl (25:25:25:10:9, by vol.) to 5 cm above origin, benzene/diethyl ether/ethanol/acetic acid (60:40:1:0.23, by vol.) to 8 cm, full length (9 cm) in hexane/diethyl ether (94:6, by vol.) and full length in hexane alone (20).
[b] Diethyl ether/hexane/ammonia (50:50:0.25, by vol.) to 18 cm above origin, chloroform/methanol/water/ammonia (70:30:3:2, by vol.) to 10 cm and ethyl acetate/acetone/water/glacial acetic acid (40:40:2:1, by vol.) to 16 cm (21).

– = not determined.

5. TLC on modified silica gel

5.1 Reversed-phase

For reversed-phase TLC (RPTLC) the silica gel (usually silica gel 60) is rendered apolar by impregnating the silica layer with long-chain hydrocarbons or by silanization with agents such as dimethyldichlorosilane (DMCS) or more aliphatic alkylsilanes. Chromatographic separation is consequently based on apolar interactions and, in contrast to normal-phase TLC, polar lipids exhibit higher R_f values than non-polar lipids, which are more tightly bound. Solvents of high polarity are used for development and to increase the R_f values of components the solvent system is made less polar.

Reversed-phase TLC and HPTLC plates can be purchased ready made or prepared as described in *Protocol 5*. The pre-coated reversed-phase plates frequently used are C_8 and C_{18}, where the number refers to the chain-length of the aliphatic hydrocarbon bound to the silica. As before, pre-coated reversed-phase plates can be purchased with or without fluorescent indicator and with or without pre-adsorbent zones. Sample application is the same as for normal-phase TLC.

Protocol 5. Preparation of reversed-phase plates

1. Dissolve 10 g of silicone oil DC 200 (Fluka Chemika-BioChemika) or *n*-undecane (Sigma Chemical Co.) in 100 ml diethyl ether, and place solution in development chamber.
2. Stand TLC or HPTLC plate coated with silica gel in chamber and allow the solution to rise to top of plate.

Protocol 5. *Continued*

3. Remove plate from chamber and allow solvent to evaporate at room temperature in fume cupboard.
4. Apply sample to plate and develop in the same direction as for the impregnation solution.

NB: All eluting solvents must be saturated with respect to the reverse phase, i.e. silicone or undecane.

RPTLC is of particular value in the separation of lipophilic substances, which do not separate well in TLC with standard silica. It has been used for the analysis of fatty acid methyl and phenacyl esters (22, 23), triacylglycerols (24), and cholesteryl esters (25).

5.2 Silver nitrate

Silver ions form complexes of varying stability with the double bonds of the fatty acid components of lipids. This feature is employed in silver nitrate (argentation) chromatography to separate lipids according to the number of double bonds present in the component fatty acids. The separation also depends to a lesser extent on whether the bonds are *cis* or *trans*, their location in the aliphatic chain and the chain-length itself. Lipid classes to which it has been applied include fatty acid methyl esters (26), phospholipids (27), triacylglycerols (28), cholesteryl/wax esters (29) and galactolipids (30).

Although TLC plates can be purchased pre-coated with silver nitrate impregnated silica gel (Alltech), it is more common to prepare such plates oneself. To prepare them silver nitrate is normally incorporated at a level of 3 to 10% on a weight basis, with the silica gel adsorbent used for spreading plates as described in *Protocol 1*. Exposure of the plates to high light intensities should be avoided during their preparation, and after activation the plates should be stored desiccated in the dark. Alternatively, commercial plates pre-coated with silica gel can be impregnated with silver ions by spraying with a solution of silver nitrate in acetonitrile as described in *Protocol 6*. Although aqueous solutions can also be used, these tend to cause the adsorbent layer to crack and lift off the glass support during activation. The method can also be scaled down for use with HPTLC plates.

Protocol 6. Preparation of impregnated silica gel plates

A. *Silver nitrate*

1. Dissolve 2 g silver nitrate in 10–20 ml acetonitrile.
2. In subdued lighting spray all of the solution evenly on to a commercial 20 × 20 cm glass-based TLC plate pre-coated with silica gel 60 in a layer 0.25 mm thick.

3. Place plates on paper towelling to soak up excess liquid, and allow plates to dry in air in dark container.
4. Activate plates by heating at 110 °C for 30 min. Ideally, use for analysis immediately or if necessary store in dark in desiccator.

B. *Urea*

1. Mix 50 g silica gel H and 25 g $CaSO_4$.
2. Dissolve 25 g urea in 50 ml water and add to the above mixture with sufficient additional water to make a pourable slurry.
3. For self-charring chromatograms add 5 g ammonium sulfate in 10 ml water to the slurry.
4. Spread slurry on to glass plates as described in *Protocol 1*.
5. Allow to dry at room temperature for 15 min, then dry in oven at 70 °C for 30 min, followed by activation at 105 °C for 1 h.

Since TLC plates coated with silver nitrate-impregnated adsorbent rapidly darken when exposed to light, sample application is best carried out in dim light and by keeping the plate other than the origin region covered. Similarly, the actual development of the chromatograms should be carried out in chambers placed in a darkened room or cupboard. Separated components on developed chromatograms are best visualized by spraying with 2′,7′-dichlorofluorescein as described in Section 6.1.2. When separated lipids are being recovered from the adsorbent, silver ions and the 2′,7′-dichlorofluorescein must be removed as described in *Protocol 7*.

Protocol 7. Recovery of lipids from silica impregnated with silver nitrate and urea

A. *Silver nitrate*

1. Scrape band of adsorbent containing separated lipid class from plate on to sheet of paper, using scalpel or razor blade. Transfer adsorbent to a test-tube.
2. Add 3.5 ml hexane/diethyl ether (1:1, v/v) to tube and mix contents by shaking or vortex mixer.
3. Add 1 ml of 20% (w/v) aqueous NaCl. Stopper tube and mix contents thoroughly.
4. Centrifuge tube at 1000 *g* for 5 min.
5. Transfer the upper organic layer to a second test-tube by Pasteur pipette.
6. Repeat steps 3 to 5 twice and combine the organic layers in the second tube.

Protocol 7. *Continued*

7. Add 2.5 ml 5% (w/v) $KHCO_3$ and shake tube. Centrifuge tube as in step 4.
8. Transfer organic layer to another test-tube and evaporate solvent to yield pure lipid component.

(Step 7 is required if 2′,7′-dichlorofluorescein is used to visualize separated components.)

B. *Urea*

1. Transfer band of adsorbent to test-tube as above.
2. Add 3 ml ethanol to tube. Stopper tube and mix contents thoroughly by agitation.
3. Add 5 ml hexane and mix contents again.
4. Centrifuge tube at 500 *g* for 5 min to precipitate silica.
5. Add water slowly in small volumes down the side of the tube using a Pasteur pipette until a two-phase system forms.
6. Transfer the upper organic layer containing the lipid to another tube and evaporate solvent.

5.3 Boric acid

Boric acid or borates form reversible non-polar complexes with adjacent free hydroxyl groups and thereby increase the mobility of the molecules on the adsorbent. The plates are prepared by combining 5 g of boric acid with 45 g silica gel in the slurry used to coat the glass plates as described in *Protocol 1*. Plates pre-coated with silica gel can be dipped into a solution of 1.2% boric acid in absolute ethanol/water (1:1, v/v) followed by air-drying and activation at 100 °C for 15 min.

TLC with boric acid impregnated silica is useful for the separation of partial acylglycerols. In particular, 1,2- and 1,3-diacylglycerols and 1- and 2-monoacylglycerols are resolved using chloroform/acetone (96:4, v/v) as the solvent system. Borate impregnation can also be applied to phospholipids with chloroform/methanol/water/ammonium hydroxide (120:75:6:2, by vol.) as developing solvent to allow the clear resolution of phosphatidylserine and phosphatidylinositol (31), two classes which are very difficult to separate by TLC on standard silica gel 60.

5.4 Urea

Molecules containing unbranched hydrocarbon chains form adducts with the urea in the adsorbent layer and their migration is consequently retarded. Plates coated with urea impregnated silica gel cannot be purchased pre-coated and are made according to the procedure outlined in *Protocol 6*. Silica gel G containing 13% gypsum cannot be used since the adsorbent flakes off the plate when treated with urea.

Using butyl acetate as developing solvent free fatty alcohols (R_f 0.15–0.35) can be separated from cholesterol (R_f 0.70–0.75), and wax esters (R_f 0.1–0.3) from steryl esters (R_f 0.80–0.90) (32). Triacylglycerols remain close to the origin. Separated components are recovered from the adsorbent by the method outlined in *Protocol 7*.

5.5 EDTA and oxalic acid

To improve the separation of anionic phospholipids such as phosphatidylinositol, phosphatidylserine, and phosphatidic acid, the silica gel adsorbent can be impregnated with EDTA or oxalic acid. These sequester any calcium present in the silica which might associate with the phospholipids and hinder their migration.

EDTA impregnated plates are prepared by using a slurry of 30 g of silica gel H in 70 ml of 1 mM EDTA in the method shown in *Protocol 1*. Alternatively, HPTLC plates coated with silica gel 60 can be sprayed with 3 ml of 1 mM EDTA followed by heating at 110 °C for 1 h. The EDTA solution is best applied gradually otherwise the adsorbent flakes off. Development with chloroform/methanol/acetic acid/water (75:45:3:1, by vol.) resolves phosphatidylinositol and phosphatidylserine, but phosphatidylglycerol and cardiolipin comigrate with phosphatidylethanolamine (33).

Silica gel 60 impregnated with oxalic acid can be applied to the separation of polyphosphoinositides (34). Pre-coated 20 × 20 cm plates are soaked by their complete immersion for 90 sec in a 0.054 M oxalic acid solution adjusted to pH 7.2 with KOH, and then dried at 110 °C for 15 min. After application of sample, the plates are developed in chloroform/methanol/acetic acid/water (55:43:3:4, by vol.) to within 1 cm of the top. After drying in a desiccator for 30 min the plates are developed to 25% of the distance of the first development in chloroform/methanol/ammonium hydroxide/water (40:70:10:20, by vol.). Relative to the first solvent front, the approximate R_f values of separated phospholipids are phosphatidylethanolamine 0.85, phosphatidic acid 0.79, phosphatidylinositol 0.48, phosphatidylserine 0.39, phosphatidylcholine 0.30, phosphatidylinositol-4-monophosphate 0.15, and phosphatidylinositol-4,5-bisphosphate 0.09.

6. Detection systems

After development, the plates must be treated or stained by some method to reveal the position of lipids. In general, these stains can be divided into two main categories; general stains that will enable virtually all lipids to be visualized non-specifically, and specific stains that will only stain certain types or classes of lipids. Some of the stains can be used quantitatively and this is covered in Section 7.

Developed plates can be simply immersed in a tank of stain or the reagent can be applied as a spray. Specific borosilicate glass bottle/jar atomizers that utilize compressed air, an air pump, or hand bellows can be purchased for this purpose (see *Figure 1*; Camlab Ltd.). It is also possible to purchase spray guns containing a refillable glass reservoir and driven by compressed gas in replaceable canisters. Many of the stains contain corrosive materials, such as strong acids, or solvents and so dipping and, especially, spraying must be performed in a suitable fume cupboard. Ideally, there should be a separate spray hood in the fume cupboard containing a crude filter (glass wool) to minimize the damage by corrosive mists.

There are a considerable number of different stains and different formulations of the same basic stain. Listed below is a variety of the most common stains and formulations. A further list of stains, formulations, and procedures can be found elsewhere (35).

6.1 General stains

The methods for visualizing lipids non-specifically can be grouped into three categories as follows:

(a) spraying the plate with a strong oxidizing agent followed by charring,

(b) spraying the plate with a fluorescent reagent and examining under UV light, and

(c) other miscellaneous methods including reaction with iodine vapour.

Charring is very sensitive and amounts as low as 1 μg of lipid can be detected, but the method is destructive, i.e. the samples are destroyed by the visualization process. Although sometimes not as sensitive, the fluorescence methods have the advantage of often being non-destructive, and so the sample can be subsequently used after removal of the indicator. A comparison of various reagents for detection (and quantification) of lipids on silica gel and C_{18} reversed-phase plates can be found elsewhere (36).

6.1.1 Oxidizing agents and charring

Plates can be sprayed with simple solutions of strong oxidizing acids, then heated in an oven at 180 °C for times ranging from 30–60 min and the lipids appear as black deposits of carbon. Sulfuric acid (5% in ethanol or 50% in water) or chromic acid (0.25% or 0.6% potassium dichromate in 15% sulfuric acid) are the most often used. However, phosphoric acid solutions containing copper salts are now more widely used using the same procedure except that charring gives brown/black spots in about 20–30 min. Copper acetate (3% in 8% orthophosphoric acid) (37) and copper sulfate (10% in 8% phosphoric acid) are the common formulations. The vapours given off using these reagents are corrosive and so ammonium bisulfate (20% in water) has also been used to char lipids, producing less corrosive vapours in the process. These methods can detect lipids down to 1–2 μg. Bitman and Wood (38) have compared the effectiveness of various charring formulas for detection and quantification.

6.1.2 Fluorescent stains

Probably the most commonly used fluorescent stain is 2′,7′-dichlorofluorescein (DCF) (0.1–0.2% in ethanol or 95% methanol). After spraying with this reagent lipids appear as bright yellow spots on a yellow/green background under UV light. It can be advantageous to place the plate in a vacuum desiccator for 15 min prior to viewing to reduce the background staining. Approximately 5 μg of lipid can be detected easily by this method, which is the preferred fluorescent stain when acidic solvents are used. When alkaline solvents are used it is preferable to use rhodamine 6G (0.01% in water or acetone). After spraying, lipids appear as red-pink spots against a pale pink background under UV light.

There are several other fluorescent stains that have been used, although not as widely as those above. These include aqueous ANS (0.1% 1-anilo-8-naphthalene sulfonate as the ammonium salt) or TNS (1 mM 6-*p*-toluidino-2-naphthalenesulfonic acid in 50 mM Tris-HCl, pH 7.4) both of which are more sensitive than DCF or rhodamine 6G, detecting lipids in the nanogram range. The reagent 8-hydroxy-1,3,6-pyrenetrisulfonic acid trisodium salt (10 mg/100 ml methanol), developed for the staining of prostaglandins, detects other lipid classes as purple fluorescent spots under UV light and is equally as sensitive (39). Plates dipped in Nile red (9-diethylamino-5H-benzo[a]pheno-xazine-5-one) solution (8 μg/ml of methanol/water 80:20, v/v) followed by a dilute aqueous solution of bleach to reduce background staining will reveal nanogram amounts of lipids as red fluorescent spots under UV (40). The plate must be dried at 100 °C after each dip.

With rhodamine B (0.05–0.25% in ethanol) lipids appear as red-violet spots on a pale pink background. Primuline reagent is prepared by diluting a stock solution (0.1% primuline in water) 1 in 100 with acetone/water (4:1, v/v). After spraying, plates must be viewed immediately under UV and lipids (1 μg limit) appear as purple spots on a green background. *N*-phenyl-1-naphthylamine (NPN), diphenylhexatriene (DPH) (0.03% in chloroform) and BBOT (2 mg 2,5-bis-[5-tert.-butylbenzoxazolyl (2′)]thiophene in 100 ml methanol) all produce bright fluorescent spots against a dark or weakly fluorescent background.

The above compounds are basically fluorescent dyes that bind to the lipid molecules and they generally fade with time. However, a more recent development are the fluorescence-inducing reagents. These reagents react directly with the lipid on the TLC or HPTLC plate converting them into fluorescent derivatives. These derivatization methods are very sensitive and nanogram (10 ng) quantities of lipids can be detected. A representative example (4, 20) of these methods is as follows. After developing and air-drying, the plate is dipped in the manganese chloride-sulfuric acid derivatizing reagent (3.2 g $MnCl_2$, 32 ml concentrated H_2SO_4, 480 ml water, 480 ml methanol) for 20 sec. The solvents are evaporated and the plate heated at 110 °C for 40 min before viewing under UV light.

6.1.3 Miscellaneous non-specific stains

Iodine has been used for many years to visualize lipids on TLC plates. The plate is sprayed with an iodine solution (1–3% in chloroform) or placed in a tank of iodine vapour. Lipids appear as brown spots on a yellow background. With brief exposure it is possible to remove the iodine effectively under vacuum but care must be taken as the iodine may irreversibly bind to polyunsaturated fatty acids. It should also be noted that glycolipids are poorly stained by iodine.

The phosphomolybdic acid (5–10% in ethanol) reagent, although sometimes regarded as a specific phospholipid stain, can be used as a general lipid stain. The dry chromatogram is sprayed with, or dipped in, the reagent and then heated to 120 °C for 10 min. Some versions of the method suggest that the plate should then be placed in a chamber containing concentrated ammonia. Most lipids, including sterols and antioxidants, appear as blue spots on a yellow-white background.

The common protein stain, Coomassie brilliant blue, has also been used to stain lipids on TLC plates (41). Plates are dipped in 0.03% brilliant blue R in an aqueous solution of 20% methanol for 10-30 min before destaining in 20% methanol for up to 5 min and air-drying. Submicrogram quantities of lipids can be detected this way. However, care should be taken to avoid dislodging the silica gel layer from the plates. This reagent has the advantage of being non-corrosive, and in this respect it should be noted that water alone can show up large amounts of lipids which appear on wet TLC plates as white spots against a translucent background.

6.2 Specific stains

These reagents, and the methods for using them, are usually more complicated than the non-specific stains. They generally involve a chemical or chemicals in the reagent reacting with specific groups in the lipids that results in the lipid being stained or made visible in some way. There are reagents for most of the reactive groups found in lipids, but it should be noted that there is no specific spray for simple glyceride derivatives. Some of the most commonly used reagents and methods are summarized below.

6.2.1 Free fatty acids

The developed plate is sprayed to dampness in turn with 2′,7′-DCF, 1%

aluminium chloride in ethanol, and 1% aqueous ferric chloride, warming the plate to 100 °C after each spray. Free fatty acids give a rose-violet colour.

6.2.2 Esterified fatty acids

Esterified fatty acids can be visualized using the method described in *Protocol 8*.

Protocol 8. Visualization of esterified fatty acids

1. Solution 1: Dissolve 10 g hydroxylamine hydrochloride in 125 ml of water/ethanol (4:1, v/v).
2. Solution 2: Add 26 ml saturated aqueous sodium hydroxide to 200 ml ethanol.
3. Reagent A: Combine solutions 1 and 2 and filter.
4. Reagent B: Grind 10 g ferric chloride ($FeCl_3.6H_2O$) in 20 ml concentrated hydrochloric acid in a mortar. Transfer to flask containing 300 ml diethyl ether and shake.
5. Spray plate with reagent A. Dry plate in stream of hot air from hair dryer and then respray with reagent B.
6. Fatty acid-containing lipids appear as purple spots on a yellow background.

6.2.3 Phospholipids

The phosphorus or 'Zinzadze' reagent reacts with lipid phosphorus and has had many different modifications, of which one of the most commonly used is described in *Protocol 9* (42). As little as 1 μg of phospholipid can be detected by this reagent.

Protocol 9. 'Zinzadze' reagent for visualization of lipids containing phosphorus

1. Solution A: Dissolve 8 g molybdic oxide in 200 ml of 70% sulfuric acid by gentle boiling, with constant stirring, and allow solution to cool.
2. Solution B: Add 0.4 g powdered molybdenum to 100 ml of solution A and reflux for 1 h. Allow solution to cool.
3. Solution C: Mix solutions A and B with 200 ml water and filter.
4. Combine solution C with water and glacial acetic acid in the ratio 1:2:0.75 to produce the spray reagent.
5. Set aside spray reagent for 3–4 days before use. The reagent will remain stable for about 6 months.

Protocol 9. *Continued*

6. Spray TLC plate with reagent. After 10 min, phospholipids appear as blue spots on a white background.

Other common stains for phospholipids include molybdenum blue (1.3% molybdenum oxide in 4.2 M sulfuric acid) and bromothymol blue (0.04 g in 100 ml of 0.01 M sodium hydroxide). With molybdenum blue, plates are sprayed, dried, and heated, and phospholipids (1–2 μg) appear as blue-green spots. With bromothymol blue, most lipids appear as yellow zones on a blue background.

6.2.4 Cholesterol and cholesteryl esters

Acidic ferric chloride can be used to detect cholesterol and its esters. Ferric chloride (50 mg $FeCl_3.6H_2O$) is dissolved in 90 ml water with 5 ml each of acetic and sulfuric acids. After spraying the plates are heated at 100 °C for 2–3 min. Cholesterol and cholesteryl esters appearing as red-violet spots.

6.2.5 Glycolipids

There are several reagents that can react with the sugar groups in glycolipids (for example, cerebrosides, sulfatides, and gangliosides).

One of the simplest reagents is α-naphthol (0.5 g α-naphthol dissolved in a 1:1 solution of methanol and water). After spraying, the plate is air-dried, and then sprayed with 95% sulfuric acid. Heating at 120 °C reveals glycolipids as purple-blue spots and other complex lipids as yellow spots. Diphenylamine is also commonly used as described in *Protocol 10*.

Protocol 10. Diphenylamine reagent for the visualization of glycolipids

1. Dissolve 2 g diphenylamine in 20 ml ethanol.
2. To prepare the spray reagent, add 100 ml concentrated hydrochloric acid and 80 ml glacial acetic acid to the above solution.
3. Spray plate with reagent and heat to 105 °C for 15 to 30 min.
4. Glycolipids appear as blue-grey zones on a light grey background.

With orcinol–sulfuric acid reagent (200 mg orcinol in 100 ml 75% sulfuric acid), glycolipids appear as blue spots on a white background after spraying and heating the plate at 100 °C for 10–15 min. The reagent is stable for 1 week if kept in the dark and refrigerated.

For the specific detection of gangliosides, resorcinol–hydrochloric acid can be used as described in *Protocol 11*.

Protocol 11. Resorcinol-hydrochloric acid reagent for the visualization of gangliosides

1. Solution 1: Dissolve 2 g resorcinol in 100 ml distilled water. Store this stock solution at 4 °C.
2. Solution 2: Add 0.25 ml of 0.1 M copper sulfate solution to 80 ml concentrated hydrochloric acid.
3. Spray reagent: Add 10 ml of solution 1 to solution 2, and make up to 100 ml with water.
4. Spray the plate with reagent, and heat at 110 °C for a few minutes.
5. Gangliosides appear violet-blue, and other glycolipids appear as a yellow colour.

A novel approach for detecting gangliosides using immunostaining techniques has recently been described (43). After separation on HPTLC plates, the gangliosides are treated with neuraminidase *in situ*, followed by incubation with asialoganglioside-specific antibodies. Alkaline phosphatase-conjugated second antibodies are used to visualize bound first antibodies by generating a blue dye from 5-bromo-4-chloro-3-indolylphosphate. This method is very specific and sensitive but may be limited to lipids that are antigenic.

6.2.6 Choline-containing lipids

Phosphatidylcholine, lysophosphosphatidylcholine, and sphingomyelin can be detected using the Dragendorf reagent (44). Immediately before use, 5 ml of a 40% aqueous solution of potassium iodide is mixed with 20 ml of 1.7% bismuth subnitrate in 20% acetic acid, and 75 ml water. The choline-containing lipids appear as orange-red spots a few minutes after spraying especially if the plate is warmed gently.

6.2.7 Amino groups in lipids

The most commonly used stain for amino groups is ninhydrin (0.2% in ethanol or water-saturated butanol). After spraying and heating at 100 °C for 5–10 min in a humid atmosphere, phosphatidylethanolamine, phosphatidylserine, and their lyso-derivatives appear as red-violet spots. The detection limit is about 1–2 μg.

The specific fluorescent stain fluorescamine (4-phenylspiro-[furan-2(3H),1,phthalan]-3,3′-dione), trade name 'fluram' (0.05% in acetone), reveals amino group-containing lipids as fluorescent white spots under UV light.

6.2.8 Vicinal diol groups

Periodate-Schiff's reagent detects lipids that contain vicinal diol groups which includes phosphatidylinositol, phosphatidylglycerol, 1-monoacylglycerol, and glycolipids. Schiff's reagent itself is produced by dissolving 0.2 g pararosanaline–fuchsin in 85 ml water, and then adding 5 ml of a 10% solution of sodium bisulfite. The mixture is left overnight before decolorizing with charcoal and filtering. To visualize the lipids, the plate is sprayed with 0.2% aqueous sodium periodate solution, and after 15 min is treated with sulfur dioxide to destroy excess reagent. The plate is then sprayed with Schiff's reagent and the sulfur dioxide treatment repeated. The vicinal diol-containing lipids appear as blue-purple spots after a short time, with other complex lipids appearing as yellow spots. Plasmalogens are also detected but their presence must be confirmed separately. Care must be taken to prevent the formation of autoxidized lipids as they are also detected.

6.2.9 Alkenyl lipids

Neutral plasmalogens can be visualized by spraying plates with a reagent made by dissolving 0.4 g 2,4-dinitrophenylhydrazine in 100 ml 2 M hydrochloric acid. Upon gentle heating the released aldehydes appear as yellow-orange spots. Exposing the TLC plate to the fumes of concentrated hydrochloric acid with also split the vinyl ether bond. The plate can then be sprayed with a 2% aqueous solution of 4-amino-5-hydrazino-1,2,4-triazole-3-thiol in 1 M NaOH. The released aldehydes appear as purple spots. Biebrich scarlet (0.2% in 0.05 M sulfurous acid) is another reagent that can be used to identify specifically alkenyl lipids on TLC plates.

6.2.10 Saturated lipids

A solution of α-cyclodextrin (1% in ethanol/water, 3:7 v/v) is sprayed on the plate and air-dried. After 10–20 min in a humidifying chamber, the plate is placed in a tank of iodine where saturated straight-chain lipids appear white on a purple background.

6.3 Radiolabelled lipids

A quite different method for the detection of lipids on TLC plates is the use of radiolabelled lipids. In general, ^{14}C and tritium are the radiolabels in common use for lipids, although phosphorus-32 can also be used to label phospholipids. The detection of ^{32}P and ^{14}C on TLC plates presents little problem and so are ideal radiolabels, but the detection of ^{3}H (tritium) after TLC requires more time and effort. Irrespective of the isotope used, it should be emphasized that with radioactive lipids, all TLC and subsequent manipulations of the plate, should be performed in an appropriate fume cupboard to prevent the spread of radioactive silica dust.

6.3.1 Scanner

Radioactivity scanners suitable for TLC plates have been available for many years, but are still not in widespread use. Despite the improved performance of modern instruments, the method remains limited by low sensitivity. For optimum sensitivity scan times can also be very long and be measured in hours rather than minutes. These factors combined with its high cost in comparison with the alternative, autoradiography, have prevented the scanner becoming a routine instrument for lipid analysis.

6.3.2 Autoradiography

Autoradiography for the detection of radioactive lipids after TLC is a very simple procedure. After development, the plate should be placed in a vacuum desiccator for 15–30 min to remove the solvent. The plate should not be sprayed with any other detection spray before autoradiography as this may lead to complications with quenching or chemiluminescence. It is often necessary to mark the plate, (a) for identification if more than one plate is to be autoradiographed, and (b) to enable correct alignment of the plate and autoradiogram after development. An ink-pad spiked with an aqueous ^{14}C solution (1 μCi ^{14}C per millilitre) and a small stamp produces simple identifying marks. If specific information is required on the plate/autoradiograph then a blunt stylus dipped in radioactive ink (as above), or a commercial non-radioactive (phosphor) autoradiograph marker can be used to write on the TLC plate, although this requires some care.

Once dry the plate is covered with the X-ray film, many types of which are suitable. Most X-ray films have emulsion on both sides, but with ^{14}C very few of the β-particles can reach the emulsion on the other side of the support, and with ^{3}H, none of them will. This is not a problem but the signal to background ratio may be improved with the use of single-sided film. However, Kodak X-Omat AR and Konica A2 have both been extensively and successfully used in the authors' laboratory. To ensure a good contact between the plate and the film, the use of an autoradiograph cassette is recommended. Cassettes also have the advantage of being light-tight provided no more than one layer of TLC plates is used. If cassettes are not available the film can be clamped between the TLC plate and a plain glass plate using foldback clips, and the sandwich wrapped in a black plastic bag and placed in a light-tight box. After exposure, the film should be developed in accordance with the manufacturer's instructions.

The exposure time must be determined empirically, as this will depend upon the isotope used, the amount of radioactivity in each lipid band, and the specific requirements of the experimenter. In general, exposure time is a compromise between detecting minor bands and overexposing major bands. As a guide to exposure length, bands easily visible above background are obtained with 4000 d.p.m./cm^2 of ^{14}C after 2 days' exposure. The energy of the tritium β-particle is very low (E_{max} = 18.6 keV) and has a short path-length. Self-absorption in the sample on silica, the air gap between the plate and the film and the protective anti-abrasive layer on the film all contribute to greatly reduce the number of particles that reach the emulsion. Therefore, large amounts of radioactivity (approximately 500 times more than ^{14}C) are required to cause blackening of the film.

If relatively large amounts of tritium are present on a plate, then it may be sufficient to simply use a specific X-ray film designed for tritium autoradiography (LKB Ultrofilm; Pharmacia LKB Biotechnology). This is a single-coated rapid film without an anti-abrasive layer that can be used for direct tritium autoradiography. However, extra care must be taken with this film to prevent damage to the emulsion. Fluorography, where the energy of the β-particle is converted to a photon by the use of a scintillator, is the usual method for detecting tritium on TLC plates. The scintillator, which is usually 2,5-diphenyloxazole (PPO) (5–10% PPO in chloroform), can be applied to the developed plate by dipping or spraying, although with the latter an even coating may be difficult to obtain giving variations in efficiency which would affect quantitation, if this is required. After evaporation of the solvent the plate can be autoradiographed as above except that exposure should be at −70 °C. Low temperature increases the efficiency of this technique as it increases the light output of the scintillator and improves the performance of the X-ray film. Overall the technique of low-temperature fluorography can increase the sensitivity for tritium by almost 100-fold. For quantitative fluorography it is necessary to pre-expose the X-ray film by using a single, short (less than 1 msec) flash from an electronic flash unit at a distance of 60–70 cm. Three filters must be used to reduce and diffuse the light output; an IR-absorbing filter, an orange filter (Kodak Wratten No. 22 or 21) and Whatman No. 1 filter paper. A combination of orange filters and the distance between the flash unit and the film are chosen to give a pre-exposure of 0.1–0.2 A_{540} units.

7. Quantitation of separated lipid classes

7.1 Scanning densitometry

Scanning photodensitometry of charred lipids directly on TLC or HPTLC plates is the method of choice when large numbers of samples have to be quantitated on a routine basis. Pre-coated TLC and HPTLC plates such as the Whatman Diamond series are commercially available and are specifically recommended by the manufacturers for densitometry. Microprocessor-controlled scanning densitometers are produced by various firms, including Shimadzu Corporation, Japan (UK agents, Dyson Instruments Ltd.), Pharmacia LKB Biotechnology AB, and Camag (*Figure 4*). Most of these can

Figure 4. Scanning densitometer used for quantitation of lipid classes separated by TLC and HPTLC. An HPTLC plate is visible, fixed to the scanning stage within the machine.

be linked to sophisticated computing integrators or personal computers thereby allowing the storage and manipulation of data.

During the operation of the densitometer, a light beam in the form of a slit variable in width and length is moved over the charred sample zones to be quantitated. The beam can either move in a straight line or zigzag to sample the background adsorbent at either side of the component zone. In general, charring with copper salts seems to be the most useful for densitometry. Copper sulfate is slightly more sensitive than copper acetate but the latter yields less background colouring. Reflectance measurements give lower background readings than measurements done in the transmission mode. Maximal absorbance values for lipids stained with copper acetate in phosphoric acid are achieved, using a single wavelength scan at 350 nm. If the width of the light beam covers the component spot then a linear scanning mode can be used but for smaller beam widths zigzag mode is preferable. Zigzag mode provides more reproducible integrated peak areas for spots with irregular shape and sizes. A beam width of 0.2 mm and scanning speed of 30 mm/min are commonly employed.

A major problem is the relationship between lipid concentration and absorption. For equal amounts, cholesterol always gives a larger response than other lipid classes when charred with copper acetate in phosphoric acid, the order being cholesterol > free fatty acids > cholesteryl ester > diacyl- and triacylglycerols > monoacylglycerols > phospholipids. Within the phospholipids, phosphatidylcholine > sphingomyelin > phosphatidylethanolamine > phosphatidylinositol > phosphatidylserine, in terms of response. For a given solvent system and charring technique, the absolute peak area obtained for a particular lipid varies from analysis to analysis, but the relative distribution of classes is constant. For amounts of lipid in the range 0.125–10 μg on standard plates and 0.02–1 μg on HPTLC plates, the relationship between integrated response values and amount is usually linear but at higher values becomes curvilinear. For accurate quantitation, the response of a given lipid class should be compared with a standard curve of integrated response plotted against mass of the lipid class in question. However, this approach is limited by the fact that not all lipid classes can be obtained commercially or at least inexpensively. Furthermore, differences in fatty acid composition between purchased lipid classes and those in the sample being analysed may cause differences in their response to staining and hence their quantitation. Standards and unknowns should be run on the same plate to eliminate interplate differences.

The plasmalogen content of phospholipid samples can be estimated, using scanning densitometry to compare on the same plate duplicate samples, one of which has been treated with a trichloroacetic acid–hydrochloric acid reagent to cleave the vinyl ether linkage (15).

7.2 Scintillation quenching

More commonplace in laboratories than the scanning densitometers described above are liquid scintillation counters used for the measurement of radioactivity. These can also be employed for the quantitative determination of lipids separated by TLC (45).

Basically, the lipid spots are visualized with a charring spray and the darkness of the spots determined by measuring the quenching produced when the adsorbent containing the charred lipid is suspended in a scintillation gel in the presence of a radioactive sample or, more simply, when the external standard mode is used in the scintillation counter (*Protocol 12*).

Protocol 12. Quantitation of separated neutral lipid classes by liquid scintillation counting

(Values refer to 20 × 20 cm plates coated with silica gel in layer of 0.25 mm thickness)

1. Apply up to 1 mg of lipid mixture as a 1.5 cm streak.
2. Develop the chromatogram in hexane/diethyl ether/glacial acetic acid (80:20:2, by vol.).

3. Spray the developed chromatogram with 3% (w/v) cupric acetate in 8% (w/v) orthophosphoric acid until translucent.
4. Heat in oven at 180 °C for 15 min.
5. Scrape charred bands into scintillation vials. Add 4 ml water and sufficient volume of an emulsifying scintillation fluid to form a gel.
6. Place vials in liquid scintillation counter and count each in the external standard channel ratios (ESCRs) mode for 1 min.
7. Refer to standard curves of ESCRs v. mass prepared previously, using the same method and known masses of authentic lipid class standards.

As with densitometry, standard curves must be prepared for the lipid classes being studied. The same limitations regarding the availability of standards applies as with densitometry. With liquid scintillation counters manufactured recently, the transformed spectral index rather than ESCR may be used. The technique has been applied to simple neutral lipid classes (45), cerebrosides (46), and sphingomyelin (47), but does not appear to have found much use in the quantitation of phospholipids in natural samples.

7.3 Gravimetry

For TLC on standard 20 × 20 cm plates, the crudest approach to quantitation is by weighing the separated fractions after their elution from the adsorbent. This is only really practicable with large samples applied as streaks across the whole plate.

To this end, developed chromatograms are exposed to iodine vapour and the areas of separated classes circled in pencil. After sublimation of the iodine, the bands of adsorbent containing classes are scraped from the glass support, using a scalpel or razor blade, on to a sheet of paper and transferred to a stoppered test-tube. With neutral lipid classes chloroform is added to the tube, or chloroform/methanol/water (5:5:1, by vol.) in the case of polar lipids. After vigorous shaking, the tube is centrifuged to compact the adsorbent and the solvent decanted into a previously weighed test-tube. The extraction with solvent is repeated and the extracts combined. The solvent is evaporated and the tube desiccated under vacuum to yield the pure lipid class, and the tube reweighed. Inaccuracies arise from differences in the efficiency of extraction between lipid classes and the presence of traces of adsorbent.

7.4 Absorbance measurements

Phospholipids and glycolipids, identified by iodine staining and eluted from the adsorbent in tubes as described above, can be estimated for phosphate and sugar contents respectively. When measured by the method of Bartlett (48), a minimum of 1 μg P per phospholipid component is required. Sugar determination by the method of Dubois *et al.* (49) requires at least 10 μg sugar for each glycolipid. For both these methods the lipid components must be eluted from the silica as above prior to their estimations to avoid high blank values.

Neutral lipid classes can be estimated after elution from the silica gel by digestion with potassium dichromate or concentrated sulfuric acid and measurement of absorbance at 350 or 375 nm respectively (50, 51).

Calibration curves obtained with standards must be employed in all of these procedures.

7.5 Gas–liquid chromatography

The amounts of lipid classes separated by TLC can be estimated from the amount of fatty acids present in them. This is done by adding a known amount of a non-abundant fatty acid, such as heptadecanoate or non-adecanoate, as an internal standard to each of the separated classes in test-tubes and transesterifying them to form fatty acid methyl esters. After analysis by gas liquid chromatography, the amount of fatty acids present in each lipid class can be calculated by reference to that of the internal standard. Various correction factors which take account of the amount of non-fatty acid material present in different lipid classes are then employed to convert the amounts of fatty acids to actual weight of lipid class. Full details of this somewhat time-consuming quantitation method can be found elsewhere (52).

8. TLC on coated quartz micro-columns

The separation of the components in complex mixtures of lipid classes has been the most successful aspect of conventional TLC. However, whereas detection of the separated components has been relatively simple, the accurate quantitation of the same has been a difficult problem. In contrast, GLC was characterized by having the highly successful flame ionization detector (FID). Around the mid-1970s these two factors were combined in the development of a new TLC system whereby material could be separated on re-usable silica-coated quartz micro-columns (rods) which could then be analysed by passing through an FID. The Iatroscan TLC-FID system thus enabled the separated components to be quantified on the universal basis of ionizable carbon.

Over the last 15 years the apparatus and techniques have been greatly improved and refined, including the addition of a nitrogen (and halogen) detector, the flame thermionic ionization detector (FTID). Despite this, TLC-FID has not become the predominant method for lipid analysis although it has been extensively used. Reproducibility problems and the fact that chromatographic separations are often not as good as those achieved with conventional TLC plates have contributed to this along with, perhaps, the high initial cost of the instrument. However, the Iatroscan TLC-FID system can have major advantages, especially speed of analysis, and so can be an extremely useful technique, particularly in routine situations where a large number of qualitatively similar samples have to be quantified (53). This section describes the instrument, the associated apparatus, and the procedures involved with the Iatroscan, indicating the potential advantages and problems associated with them.

8.1 Micro-columns

The quartz micro-columns or rods, termed chromarods, 150 mm × 1 mm o.d., are uniformly coated over the central 146 mm with a 75-μm-thick layer of a soft-glass frit based on silica (chromarod-S) or alumina (chromarod-A) of particle size 10 μm. Chromarods-SII, which have a 68-μm-thick coating of a finer silica-based frit of 5 μm particle size, offered improvements in resolution and superior reproducibility and superseded the S-type. More recently, chromarods SIII have replaced both S and SII rods, which are now no longer available. The SIII and current A rods are made by an automated coating process that gives the final rod a coating of more uniform density, thickness, and smoothness, with a resulting increase in reproducibility. Lipid analyses are performed almost exclusively with the silica-based rods.

The chromarods are mounted in a metal frame that serves as a combined storage, developing and scanning unit (*Figure 5*). The Iatroscan system is designed for the sequential analysis of up to ten rods, i.e. the frames hold 10 rods, which are also purchased in packs of ten. For optimum reproducibility it is necessary to regard the whole frame as one unit and subject the rods contained to the same treatment, i.e. number of analyses, etc. In this way inter-rod variation due to operational parameters can be minimized (see later).

Once in the frame, rods should never be handled and care should be taken to avoid damaging the rods; for instance, when applying the sample. For short-term storage, it is recommended that S and SII rods be placed in a humidified chamber, which may be a TLC tank containing some distilled water with a pair of glass rods or blocks on the bottom to keep the frames above the surface of the water. However, storage in a desiccator has been recommended by the manufacturer for the new chromarods-SIII and A. Before first-time use the rods must be decontaminated by passing them through the FID twice. Subsequently, the rods should be passed once through the FID to 'activate' and clean off atmospheric particles or absorbed organic material before every use. An analytical scan normally serves this purpose, too, when the rods are in constant use.

The total number of analyses that a set of rods can be used for is variable and highly dependent upon the composition of the samples and the resolution required. At minimum, chromarods-S can be used at least 25 times, but good

Figure 5. Iatroscan TH-10 Mk. IV TLC-FID Analyser with a rack of chromarods *in situ* for scanning and another rack in the process of development.

separations can be obtained with rods that have been used for over 150 developments. Our own experience is that about 100 developments is the upper limit to the life-time of chromarods-SII, although this can be as low as 50 with some applications. Rods can be cleaned by immersing in chromic or nitric acid overnight in a glass frame, followed by gentle rinsing with tap water, then distilled water, and finally reactivation by passing through the FID.

8.2 Instruments

The analytical instrument, the Iatroscan TH-10, is manufactured by Iatron Laboratories, Inc., Japan, and is distributed world-wide by Newman-Howells Associates Ltd., UK (*Figure 5*). The FID requires an external hydrogen supply, but the instrument is supplied with an air pump unit that is normally satisfactory for the operation of the FID. However, if the laboratory atmosphere is not clean a pre-filter may be required in the air supply line, or else bottled air can be used—although this is an expensive option. In an earlier review Ackman (54) described the basic mechanics of the scanning frame and the operation of the FID in the Mark II Iatroscan, and so this will not be repeated here as the basic principles also apply to the current microprocessor controlled Mark IV version. We will briefly describe the basic scanning operation of the instrument indicating the significant improvements with the Mark IV instrument.

When ready for analysis the frame of rods is slotted into the moving frame in the instrument. When the scan is started the moving frame carries out a fixed set of movements automatically which results in each rod in turn being passed through the fixed flame from top to bottom with the frame returning to the top of each new rod with the flame in a central position between the rod just scanned and the rod to be scanned. This normal detection mode can be modified in many ways. In the microswitch-controlled Mark II instrument these manipulations had to be performed semi-manually, but the current instrument can easily be programmed to carry out several variations automatically. Six different scan speeds are available, from 25 to 60 sec/scan. Specific zones on the rods can be scanned selectively, leaving other zones unscanned. The rods can be blank-scanned or the sample origin zone can be blank-scanned repeatedly if badly contaminated. Parameters and procedures can be programmed to change automatically from one rod to the next in an individual frame. All scans can also be controlled manually, allowing great flexibility in analysis.

The Iatroscan can simply be connected up to any recorder with variable ranges and chart speeds. For simultaneous recording and integration there is now a dedicated instrument available: the Iatrocorder TC-11. This has the dual advantages of being a modern programmable microprocessor controlled recording integrator and of being directly interfaced with the Iatroscan. As a result, the functions of the integrator are specifically synchronized with the various operating modes of the Iatroscan. However, it is possible to use non-dedicated integrators such as Hewlett-Packard, Shimadzu, and Spectra-Physics Instruments. For all non-dedicated integrators an additional auto-start controller is required, but it should be noted that, although this will synchronize the integrators start/stop functions with the Iatroscan, they will still not be compatible with the various operating modes of the Iatroscan.

Mark III and IV Iatroscans can be interfaced with an IBM-PC or compatible computer which allows the acquisition of data in digital form and transfer of the data into disk storage. The system requires an analogue–digital (A–D) card for the computer, an external interfacing box, and the TLC-FID data acquisition software. In addition, it is possible to add on to the system a chromatography data manipulation program which enables full processing of TLC-FID-FTID data.

8.3 Detection

The FID consists of a H_2 burner with a collecting electrode above, with the rods passing through the gap between the two during scanning. The current Mark IV instrument is fitted with a modified collector electrode which allows a much higher voltage to be applied resulting in improved reproducibility and linearity. Early Mark IV models can be updated with a conversion kit, but this is not possible in the earlier models. The FID can be used effectively for the whole spectrum of lipids.

More recently, it has been possible to fit an alternative detector, which enables simultaneous FID and thermionic (FTID) detection. In this detector, the combustion products first pass through a lower electrode which collects the FID signal and then continue through an upper electrode, surrounding a central thermionic source, which collects the TID signal. A ceramic ring separates the electrodes insulating the two independent signals. The FTID only responds to nitrogen and halogen-containing compounds and so may only be useful in the analysis of N-containing phospholipids. Therefore, it remains to be seen whether FTID will be anything other than of strictly limited use for lipid analysis.

8.4 Essential ancillary equipment

A certain amount of essential ancillary equipment is required for TLC-FID, including developing tanks, a storage tank, an oven, and a means of sample application. All of these items, custom-designed and built for use with the Iatroscan, can be purchased from the distributor, although it is possible to use the equivalent items normally found in a laboratory where lipid analysis is routinely performed. However, certain constraints do apply when using general laboratory equipment. The developing tanks must be of the correct size. General tanks for HPTLC plates are usually too small and we have found that normal TLC tanks are too large and give poor reproducibility. In this instance, the custom-built tanks have proved necessary. Any small bench-top oven can be used, but we have found it advantageous to suspend the frames in the oven rather than have the frames touching the base or sides. Storage chambers (see above) can easily be improvised.

For optimum performance the volume of sample spotted should be as small as possible, ideally about 0.5–1.0 μl. Microcap disposable micro-pipettes (Drummond Scientific Co., USA) or microsyringes (from, for example: the Hamilton Company, USA; Hamilton Bonaduz AG, Switzerland; and V. A. Howe & Co. Ltd., UK) are both suitable. The micro-pipettes have the advantage of the sample being visible but the disadvantage of requiring slightly more manual dexterity in their use. In both cases, care must be taken to avoid touching the surface of the rod and also to avoid the sample running up the outside of the pipette or syringe needle by capillary action. Accurate and reproducible sample application may be better achieved by the use of a semi-automatic autospotter that has been specifically designed for use with the Iatroscan (Trivector Scientific Ltd., UK).

8.5 Procedures

The basic practical procedure for TLC-FID is given below followed by a brief description of various modifications and refinements that have proved useful.

If rods have been stored in a humidified atmosphere they may require a brief period in the oven at 110 °C to dry before activation by passing through the Iatroscan FID. Samples, ideally in the range of 5–25 μg, are applied in a tight spot approximately 5–10 mm from the bottom of the frit. This can be achieved by addition of the sample in 0.2 μl aliquots and no rotation of the rod is necessary. The sample application can be performed in a covered spotting tray or under a conventional spotting guide, if the space permits, to prevent particulate material (dust, etc.) contaminating the rods. The rods are then transferred directly to the developing tank, which can be lined or unlined. Complete development of the rods takes approximately 20–30 min depending upon the solvent system employed and the temperature of the room and solvents. Once a method with a particular solvent system is established, subsequent developments can be timed (provided temperature is constant) as the solvent front can be difficult to see on chromarods in certain solvent systems. A cold light source behind the tank can assist viewing of the solvent front.

When developed, the frame is transferred to the oven, where the developing solvent is evaporated at 110 °C for 2 min. The frame is then transferred directly into the Iatroscan and analysed. The spotting area, developing tanks, oven, and Iatroscan should ideally be located beside each other in sequence to minimize the possibility of the rods being contaminated by external solvent vapours or particulate material present in the laboratory. General operating conditions for the Iatroscan are an air-flow of 2 litres.min^{-1} and a H_2 flow-rate of 160–180 ml.min^{-1} to the FID and a scan speed of 30 cm.min^{-1}. After analysis the rods are normally ready for spotting immediately, without further cleaning or activation.

Modifications to this basic method include pre-saturation, solvent focusing, multiple developments, and partial combustion, followed by redevelopment as described below.

In some instances it may be necessary to pre-saturate the rods, after sample application and prior to development, for optimum resolution (55). This is most easily done in a special developing tank (Newman-Howells Associates Ltd.) which contains a second solvent reservoir for adding vapour and a device for suspending the frame in the tank to pre-saturate the rods before lowering the frame for development. Solvent focusing can be used to narrow the point of application when large volumes of sample must be applied. After the sample is applied the rods are pre-run briefly in an appropriate solvent system to just above the point of application thereby focusing the sample (56).

The sequential use of different solvent systems can separate a range of components that a single solvent cannot, and may also improve resolution (54). Between developments in each solvent system the mobile phase is evaporated in the oven at 110 °C. Therefore, in essence this procedure is similar to that for conventional TLC, but in practice the much shorter development time of chromarods increases the benefit. There are many different examples of this method applied to lipids in the literature (54, 55). However, this method is limited by the number of different components that can be accommodated along the length of the chromarod. An alternative method takes advantage of the fact that the Iatroscan, if operated as described, analyses the rods from the top (solvent front end) down. This allows the rods to be developed in one solvent and partially combusted to some predetermined position on the rod and then redeveloped in another solvent before complete combustion or a further partial combustion step (54). Care must be taken with this procedure to avoid unwanted pyrolysis of components close to the point where partial combustion ends. However, this method in effect lengthens the chromarod and has proved useful for a variety of applications in lipid analysis (55).

8.6 Solvent systems

A great variety of biological samples containing virtually all classes of lipids have been separated by TLC-FID. The solvent systems used in TLC-FID are often very similar to, or at most slight modifications of, solvent systems used to separate the same samples by conventional TLC. Therefore, for separating neutral lipid mixtures (with phospholipids as an unresolved band), including plasma and serum lipids, mixtures of hexane or petroleum ether and diethyl ether (85:15, 90:10, etc.) have been used. Small amounts of organic acid such as formic or acetic are often added to this basic mixture. For polar lipids, especially phospholipid classes, chloroform/methanol/water mixtures (65:25:3 or 4, 65:35:4, 70:35:3.5, 80:20:2, 80:25:3, etc.) are the most common systems. Ammonia solutions have been used to replace the water component in this system (see *Figure 6*). Combinations of these two basic formulations can be used to good effect in multiple development or partial combustion and redevelopment methods to give a full polar and neutral analysis of a single sample.

These basic systems represent only a starting point for any Iatroscan user, and virtually all organic solvents are suitable because they will effectively be removed by evaporation. For a more detailed list of solvent systems used for lipid analysis by TLC-FID see Ackman (54). Since that review there have been a considerable number of applications of TLC-FID for the analysis of

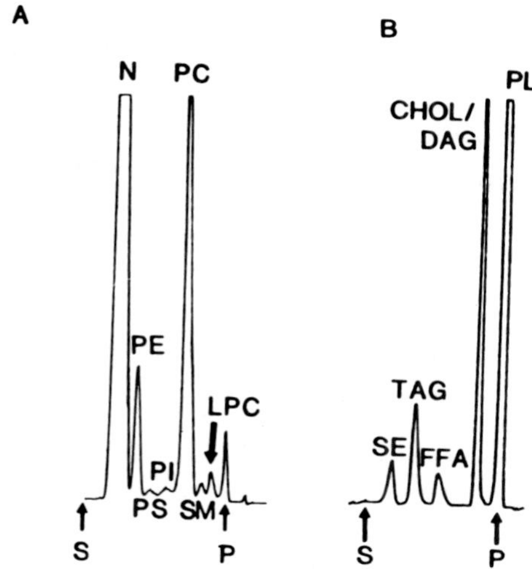

Figure 6. TLC-FID analysis of polar and neutral lipids from herring larvae total lipids on chromarods SII: **A**, polar lipid developing solvent, chloroform/methanol/water (70:35:3.5, by vol.); **B**, neutral lipid developing solvent, hexane/diethyl ether/formic acid (85:15:0.04, by vol.). *Abbreviations* are as for *Figure 2*, except for CHOL, cholesterol; N, total neutral lipids; P, point of application; PL, total polar lipids, S, solvent front; SE, steryl ester.

biological lipid samples, using numerous solvent systems and methods. Indeed, the list is so great that it is outside the scope of a general review of TLC methods in lipid analysis.

8.7 Chemical impregnation of rods

Silver nitrate impregnation of rods has been used to separate triacylglycerols (54, 57), phospholipids (57), and fatty acid methyl esters (58) by degree of unsaturation. Chromarods (SII or SIII) are activated by passing through the FID, immersed in 2.5% (w/v) silver nitrate in acetonitrile and then dried at 120 °C for 3 h. After use, the rods can be regenerated by washing the rods in concentrated nitric acid, rinsing with distilled water, drying, and reimpregnating with silver nitrate. This procedure limits the use of the chromarods, as once impregnated it is impossible to wash off all the silver salts.

Boric acid impregnation has proved useful in the separation of partial acylglycerols, in particular the isomers of monoacylglycerol (59). The rods (SII or III) are passed through the FID once or twice, dipped into a 3% aqueous boric acid solution for 5 min, and then dried in an oven at 120 °C for 5 min. Immediately before use the rods are again passed through the FID. The impregnated rods cannot be used repeatedly without deterioration of resolution, but can be washed (nitric or chromic acid overnight) and regenerated. However, regenerated rods do show some variation in separation characteristics.

Oxalic acid impregnation of chromarods has been used to improve the separation of phospholipids, in particular the separation of phosphatidic acid, phosphatidylethanolamine, and phosphatidylinositol/phosphatidylserine (60). Rods (SII or SIII) are washed by immersion in 30% nitric acid overnight, rinsed with distilled water and activated by passing through the FID. The rods are dipped in 0.25 M oxalic acid in acetonitrile for 15 min and then dried at 110 °C for 60 min. The rods are then ready for sample application. As with silver nitrate and boric acid impregnation, the rods must be reimpregnated before subsequent use.

In contrast to the more specialized applications of the treatments above, impregnation of rods with copper(II) sulfate may offer a more generalized improvement in performance of TLC-FID (61). It is claimed that impregnation gives a more uniform response for various lipid classes, the rods more uniform chromatographic behaviour, less tailing on polar lipid peaks, and a more visible solvent front. Chromarods-SIII are washed in concentrated sulfuric acid, rinsed thoroughly with distilled water, and dried at 120 °C for 30 min. The rods are immersed in 5% (w/v) aqueous copper(II) sulfate solution, dried at 120 °C for 30 min, passed through the FID, and are then ready for sample application. Apparently, impregnation may not be required before each run and so the life-time of the rods may not be significantly shortened.

It should be noted that partial combustion techniques may not be suitable with impregnated rods as pyrolysis may remove any benefit during redevelopment.

8.8 Reproducibility

Variation in the response in TLC-FID can be attributed to three main sources; the chromarods, the analyser, and the lipids themselves. The first two variables can be controlled largely by strict adherence to correct operational procedures which can optimize responses and minimize variations. The third variable can be controlled by accurate calibration and the use of internal standards, but there is no single calibration method that can be used universally, and so calibration must be related to the samples being analysed and the procedures employed.

Chromarods display inter-rod variation, primarily in absolute response and reproducibility of response. The rods may also age differentially and, as well as affecting the response as above, the separation and resolution of components may also be affected. Ideally, rods should be tested with the samples to be analysed routinely and sorted into matched sets which should then be treated identically. This does require some additional investment in several sets of rods and time initially, but is worthwhile. These procedural controls and the use of the new SIII chromarods and copper(II) sulfate impregnation will minimize the possible inter-rod variations. Furthermore, the speed of the technique encourages the use of sufficient rods to give the required confidence limits. However, the use of appropriate internal standards (see later) is always necessary for greatest accuracy.

With the Iatroscan instrument itself, variation arises from the FID gas settings, the scan speed and the geometry of the rod to the FID. The recommended gas flows were given above, but it is possible to vary both to optimize the FID response for particular analytical situations (62). In general, the higher the H_2 flow rate (up to about 200 ml/min), the greater, and more reproducible, the response—but this is offset by an increased risk of damaging the rods through overheating. Response and reproducibility also increase with increasing scan speed, but this may lead to problems due to evaporation losses or incomplete pyrolysis. The response decreases with increasing distance from the burner nozzle to the rod, although from 0.8 (the minimum) to 2.0 mm the decrease is relatively insignificant (62). Clearly, therefore, it is important that the settings for the Iatroscan and the FID are kept constant throughout an investigation. However, with optimization of all these variables, reproducible detection of 0.05 μg samples of lipids is achievable (62).

As with all methods for quantitation of lipids after TLC, different lipid classes give quantitatively different responses in TLC-FID. This is a result of a number of factors, including different degrees of volatility and differences in the amount of ionizable carbon produced during pyrolysis of different lipid classes (53, 54). Therefore, accurate quantitation requires the calculation of relative response factors for each lipid class, preferably in relation to an additional internal standard. However, the response factors of some lipids vary with the amount of sample (63) and so factors calculated at one loading may not apply at other loadings. The responses of lipid classes can also be affected by running the sample in solvent and by the presence of other components in a mixture. We have found that accurate quantitation can be achieved by calibration, using different loadings of a composite standard mixture containing the components of the experimental samples in the same relative proportions as in the experimental samples (53). This is relatively simple, but the preparation of the composite standard is time-consuming and only calibrates the instrument for one specific purpose, and calibration has to be repeated for samples of different composition. Therefore, this method is most suited to situations where large numbers of similar samples are to be analysed, such as routine screening. However, when high degrees of accuracy are required, this application-specific calibration is generally necessary irrespective of the calibration procedure adopted.

References

1. Kirchner, J. G. (1978). *Techniques of chemistry*, Vol. XIV, *Thin-layer chromatography*. John Wiley, New York.

2. Stewart, M. E. and Downing, D. T. (1981). *Lipids*, **16**, 355.
3. Mangold, H. K. and Malins, D. C. (1960). *J. Am. Oil Chem. Soc.*, **37**, 383.
4. Conte, M. H. and Bishop, J. K. B. (1988). *Lipids*, **23**, 493.
5. Janero, D. R. and Barrnett, R. (1981). *J. Chromat.*, **216**, 417.
6. Haahti, E. and Nikkari, T. (1960). *Acta Chem. Scand.*, **17**, 536.
7. Lepage, M. (1964). *J. Lipid Res.*, **5**, 587.
8. Skipski, V. P., Peterson, R. F., and Barclay, M. (1964). *Biochem. J.*, **90**, 374.
9. Riuser, G., Fleischer, S., and Yamamoto, A. (1970). *Lipids*, **5**, 494.
10. Vitiello, F. and Zanetta, J.-P. (1978). *J. Chromat.*, **166**, 637.
11. Nichols, B. W. (1963). *Biochim. Biophys. Acta*, **70**, 417.
12. Gardner, H. W. (1968). *J. Lipid Res.*, **9**, 139.
13. Kundu, S. K. (1981). In *Methods in enzymology*, Vol. 72 (ed. J. M. Lowenstein), p. 185. Academic Press, London.
14. Mullin, B. R., Poore, C. M. B., and Rupp, B. H. (1983). *J. Chromat.*, **278**, 160.
15. Touchstone, J. C., Snyder, K. A., and Levin, S. S. (1984). *J. Liquid Chromat.*, **7**, 2725.
16. Demopoulos, C. A., Moschidis, M. C., and Kritikou, L. G. (1983). *J. Chromat.*, **256**, 378.
17. Zakaria, M., Gonnord, M.-F., and Guiochon, G. (1983). *J. Chromat.*, **271**, 127.
18. Kramer, J. K. G., Fouchard, R. C., and Farnworth, E. R. (1983). *Lipids*, **18**, 896.
19. Olsen, R. E. and Henderson, R. J. (1989). *J. Exp. Mar. Biol. Ecol.*, **129**, 189.
20. Yao, J. K. and Rastetter, G. M. (1985). *Anal. Biochem.*, **150**, 111.
21. Irvine, R. F., Letcher, A. J., Meade, C. J., and Dawson, R. M. (1984). *J. Pharmacol. Methods*, **12**, 171.
22. Heckers, H. and Melcher, F. W. (1983). *J. Chromat.*, **256**, 185.
23. Gattavecchia, E., Tonelli, D., and Bertocchi, G. (1983). *J. Chromat.*, **260**, 517.
24. Nikolova-Damyanova, B. and Amidzhin, B. (1988). *J. Chromat.*, **446**, 283.
25. Kaufmann, H. P. and Makus, Z. (1961). *Fette Seifen Anstrich*, **63**, 235.
26. Inomata, M., Takaku, F., Nagai, Y., and Saito, M. (1982). *Anal. Biochem.*, **125**, 197.
27. Kennerly, D. A. (1986). *J. Chromat.*, **363**, 462.
28. Chobanov, D., Tarandjiska, R. and Chobanova, R. (1976). *J. Am. Oil Chem. Soc.*, **53**, 48.
29. Aasen, A. J., Hofstetter, H. H., Iyengar, B. T. R., and Holman, R. T. (1971). *Lipids*, **6**, 502.
30. Nichols, B. W. and Moorehouse, R. (1969). *Lipids*, **4**, 311.
31. Fine, J. and Sprecher, H. (1982). *J. Lipid Res.*, **23**, 660.
32. Tiffany, J. M. (1982). *J. Chromat.*, **243**, 329.
33. Allan, D. and Cockcroft, S. (1982). *J. Lipid Res.*, **23**, 1373.
34. Medh, J. and Weigel, P.H. (1989). *J. Lipid Res.*, **30**, 761.
35. Vioque, E. (1984). In *Handbook of chromatography*, *Lipids* Vol. II (ed. H. K. Mangold), p. 309. CRC Press, Boca Raton, Florida.
36. Sherma, J. and Bennett, S. (1983). *J. Liquid Chromat.*, **6**, 1193.
37. Fewster, M. E., Burns, B. J., and Mead, J. F. (1969). *J. Chromat.*, **43**, 120.
38. Bitman, J. and Wood, D. L. (1982). In *Advances in thin layer chromatography* (ed. J. Touchstone), p. 187. John Wiley, New York.
39. Goswami, S. K. and Kinsella, J. E. (1981). *Lipids*, **16**, 759.

40. Fowler, S. D., Brown, W. J., Warfel, J. and Greenspan, P. (1987). *J. Lipid Res.*, **28**, 1225.
41. Nakamura, K. and Handa, S. (1984). *Anal. Biochem.*, **142**, 406.
42. Dittmer, J. C. and Lester, R. L. (1964). *J. Lipid Res.*, **5**, 126.
43. Muthing, J. and Muhlradt, P. F. (1988). *Anal. Biochem.*, **173**, 10.
44. Wagner, H., Horhammer, L., and Wolff, P. (1961). *Biochem. Z.*, **334**, 175.
45. Shand, J. H. and Noble, R. C. (1980). *Anal. Biochem.*, **101**, 427.
46. Selvam, R. and Radin, N. S. (1981). *Anal. Biochem.*, **112**, 338.
47. Radin, S., Deshmukh, G. D., Selvam, R., and Hospattankar, A. V. (1982). *Biochim. Biophys. Acta*, **713**, 474.
48. Bartlett, G. R. (1959). *J. Biol. Chem.*, **234**, 466.
49. Dubois, M., Gilles, K. A., Hamilton, J. K., Rebers, P. A., and Smith, F. (1956). *Anal. Chem.*, **28**, 350.
50. Skipski, V. P. and Barclay, M. (1969). In *Methods in enzymology*, Vol. 14 (ed. J. M. Lowenstein), p. 530. Academic Press, London.
51. Marsh, J. B. and Weinstein, D. B. (1966). *J. Lipid Res.*, **7**, 574.
52. Christie, W. W. (1982). *Lipid analysis*. Pergamon Press, Oxford.
53. Fraser, A. J., Tocher, D. R., and Sargent, J. R. (1985). *J. Exp. Mar. Biol. Ecol.*, **88**, 91.
54. Ackman, R. G. (1982). In *Methods in enzymology*, Vol. 72 (ed. J. M. Lowenstein), p. 205. Academic Press, London.
55. Parrish, C. C. and Ackman, R. G. (1985). *Lipids*, **20**, 521.
56. Harvey, H. R. and Patton, J. S. (1981). *Anal. Biochem.*, **116**, 312.
57. Tanaka, M., Itoh, T., and Kaneko, H. (1979). *Yukagaku*, **28**, 96.
58. Sebedio, J.-L. and Ackman, R. G. (1981). *J. Chromat. Sci.*, **19**, 552.
59. Tanaka, M., Itoh, T., and Kaneko, H. (1980). *Lipids*, **15**, 872.
60. Banerjee, A. K., Ratnayake, W. M. N., and Ackman, R. G. (1985). *Lipids*, **20**, 121.
61. Kaimal, T. N. B. and Shantha, N. C. (1984). *J. Chromat.*, **288**, 177.
62. Terabayashi, T., Ogawa, T., Kawanishi, Y., Tanaka, M., Takase, K., and Ishii, J. (1986). *J. Chromat.*, **367**, 280.
63. Kaitaranta, J. K. and Nicolaides, N. (1981). *J. Chromat.*, **205**, 339.

4

Gas chromatography of lipids

RICHARD P. EVERSHED

1. Introduction

Gas chromatography is used extensively in the analysis of all the major lipid classes. In many instances GC analyses are performed on intact molecular species. However, for involatile or thermally unstable substances, such as phospholipids, chemical or enzymatic degradations must be used to release their simpler lipid moieties, as these are more amenable to GC analysis. (See Chapter 2 for accounts of degradation and derivatization methods commonly employed in the analysis of lipids.)

This is a practical account of GC techniques used in lipid analysis. Wall-coated open tubular (WCOT) capillary columns offer substantial improvements in separation power compared to that attainable using packed GC columns. However, as packed-column GC is still used in many laboratories, reference is made to packed-column GC methods where appropriate. Theoretical treatment of GC has been more than adequately covered elsewhere (for example in ref. 1) and so will not be repeated here. Readers requiring further information on the gas chromatography of lipids are directed to a recent monograph (2) which contains an extensive bibliography covering the period up to 1989.

2. Applications of the GC analysis of lipids

With the exception of some high-molecular-weight isoprenoid compounds (for example, carotenoids and dolichols), GC has an important role to play in the analysis of all lipid classes found in biological materials. In due course, some applications may be superseded by high-performance liquid chromatography (HPLC), as universal detectors become more widely available (HPLC and lipids is discussed in detail in Chapter 5). Supercritical fluid chromatography (SFC) also has potential for the analysis of lipids. However, a limiting factor in its general use may be the relatively high cost of capillary SFC apparatus, when compared to GC and HPLC equipment.

GC methods will be presented here for analysis of the major acyl lipid classes, i.e. fatty acids, triacylglycerols, diacylglycerols, monoacylglycerols, wax esters, steryl esters, etc. Consideration will also be given to the GC analysis of other long-chain alkyl compounds which occur in total lipid extracts, i.e. ether lipids, hydrocarbons, alcohols, aldehydes, and ketones. Brief reference will also be made to methods for the GC determination of steroids, prostaglandins and related compounds which also occur commonly in lipid extracts.

3. Basic principles

Gas chromatography (GC) is used to separate mixtures of compounds which exhibit appreciable vapour pressures (volatilities). The GC analysis of lipids involves:

(a) Introduction of a small volume (typically 1 μl) of a dilute solution (microgram to nanogram quantities of lipid per microlitre, dissolved in a suitable organic solvent, such as cyclohexane) into the GC column via a gas-tight injection port.

(b) Volatilization of the diluting solvent and dissolved lipids into the inert carrier gas, flowing continuously through the GC column. (NB: At, or near, ambient GC oven-temperatures, lipids exhibit negligible vapour pressures, and so would condense on to the first region of the column they encounter. In order to achieve elution within a reasonable analysis time the lipid vapour pressures must be raised; and this is brought about by increasing the GC oven temperature. Where the vapour pressure of a substance has been effectively reduced through interactions (hydrogen bonding and/or dipole–dipole interactions with a polar stationary phase, then changing to an apolar stationary phase may increase the vapour pressure at a given oven temperature. Changing the stationary phase may lead to the loss of selectivity (resolution) in the GC separation. The importance of stationary phase selectivity in the GC analysis of lipids is discussed in greater detail in Section 5.3.1. Temperature programming is commonly used in the GC analysis of mixtures of lipids of widely varying volatility (vapour pressure). In temperature programming the analysis starts at a low GC oven temperature, which allows elution of the more volatile substances. The oven temperature is then increased gradually, usually at a constant rate (for example, 5–10 °C/min), in order to sequentially increase the volatility of the various components of the mixture; and so effect their separation and elution in a convenient analysis time. GC run times are typically 20 to 60 min, depending upon the complexity of the mixture under investigation and the nature of the separation(s) required. Clearly, there is a limit to how far the oven temperature can be raised to increase the vapour pressures of involatile components. At GC oven temperatures which are too high, thermal decomposition of either the analytes and/or the stationary phase coating the GC column will occur.

Further consideration will be given below (Section 5.3) to the thermal stability of the various GC stationary phases used in lipid analysis.

(c) Separation of mixtures of volatile and semi-volatile lipids according to their different vapour pressures. (NB: Those compounds exhibiting highest vapour pressures will spend relatively longer in the mobile phase and so progress through the column more rapidly, i.e. elute at shorter retention times (t_R). Compounds progressing through the column more slowly, i.e. eluting at longer t_Rs, do so either because they possess lower vapour pressures, owing to their higher boiling points, or because they are interacting with the stationary phase in such a way as to effectively lower their vapour pressures.)

(d) Detection of the compounds eluting from the GC column (the commonly used detectors used in the GC analysis of lipids are discussed below in Section 4.3.)

4. Basic instrumentation

The gas chromatograph consists of:

- a sample inlet system (the injector);
- the analytical column containing the stationary phase;
- a supply of inert, gaseous mobile phase (carrier gas);
- a thermostatted oven;
- a detector;
- a recording device (printer/plotter).

Most commercial GC systems available will meet the operating specifications required for lipid analyses. Variations between different instruments relate to their overall level of sophistication, i.e. range of injector and detector systems, mode of carrier gas-flow control, and oven-temperature operating range. All modern GC instruments are microprocesser-controlled. A range of data recording/collecion options also exists; ranging from simple chart recorders, through integrator/plotters to computerized work-stations, capable of controlling the operation, data acquisition, and processing for several instruments simultaneously. Column selection is distinct from the consideration of basic GC instrumentation and will be considered later in a specialist section. In the subsequent section, key elements of the chromatographic instrumentation are considered in terms of their relative suitabilities for lipid analysis.

4.1 Sample injection

4.1.1 Sample size

Capillary GC is a sensitive technique, from which optimum results are obtained when sample sizes per analytical run lie in the low nanogram range for each individual component of a mixture. Higher sample loadings can be accommodated by packed GC columns and wide-bore (0.53 mm i.d.) capillary columns, containing thick films of stationary phase (more than 1 μm). Hence, with concentrated sample solutions considerable dilution, with a suitable solvent, will be required prior to injection. When interest focuses on a minor constituent, sample solutions can be concentrated and analysed to reveal the component of interest, while accepting that major components will display undesirable chromatographic performance, due to column overload (*Figure 1*; see Section 8 for further discussion of the column overload

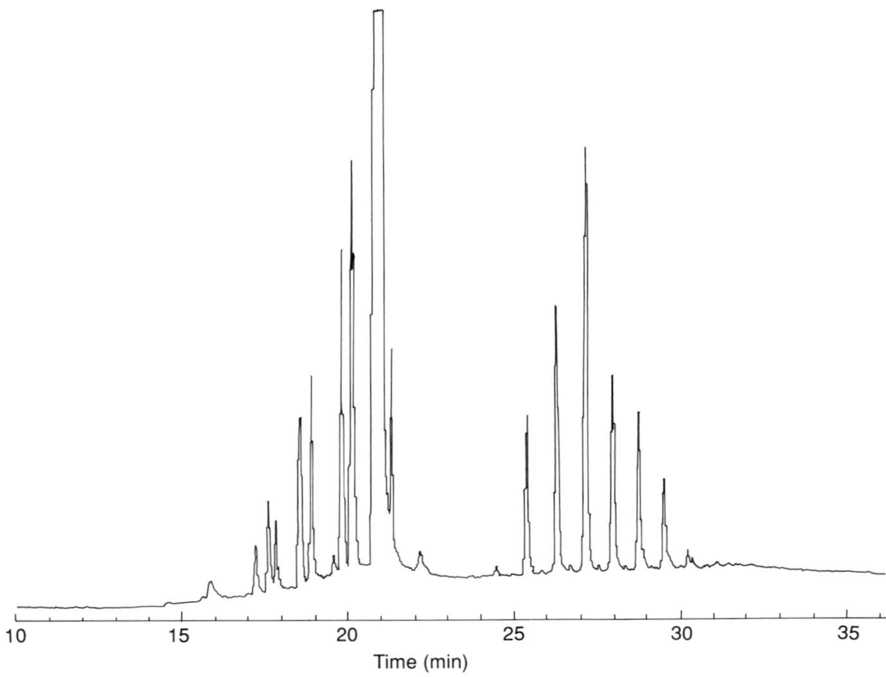

Figure 1. Capillary GC analysis of the hexane soluble constituents of the epicuticular leaf wax of *Allium porrum* (leek) showing sample overload to reveal minor components. Analysis performed on a 12 m × 0.22 mm i.d. BP-1 (OV-1 equivalent, cross-linked dimethyl polysiloxane) coated (0.1 μm film thickness) fused silica capillary column. After on-column injection the GC oven-temperature was held at 50 °C for 2 min, then programmed to 350 °C at 10 °C/min followed by a 10-min isothermal hold. Helium was the carrier gas at a column head pressure of 25 p.s.i.

phenomenon). All sample injection techniques use graduated high-precision syringes, typically of 5 or 10 μl capacity.

4.1.2 Packed-column injection

The sample injection technique for packed-column GC analyses is straightforward and similar for all the major lipid classes. Aliquots (1 to 5 μl) of sample solution are introduced directly into the heated injection port (200 to 300 °C) via the rubber septum inlet. Sample solutions can either be volatilized into the carrier gas stream in the space above the GC column packing, or injected directly into the packing material. The latter approach is to be preferred in situations where sample transformations can occur; for example, dehydration of sterols or double-bond isomerizations in fatty acids. Blockage of the syringe needle, by column packing, is a common problem of this latter approach; injection into a plug of deactivated (silanized) glass wool provides a satisfactory alternative.

4.1.3 On-column injection (3)

On-column injection offers the most versatile means of sample introduction currently available for the capillary GC analysis of lipids. Syringes incorporating narrow-gauge (for example, 0.17 mm o.d.) needles, constructed of fused silica, are used to introduce small volumes (less than μl) of dilute sample solutions directly into the interior of the capillary column. This eliminates problems of discrimination between high and low boiling sample components during injection. Although various configurations of on-column injector exist, all rely on depositing the sample solution directly on to the internal wall of the capillary column; while maintaining the GC oven at a low temperature (typically 50 °C) to avoid instantaneous 'flash' volatilization of the diluting solvent. Rapidly depressing the syringe plunger results in spray deposition of the sample solution on to the column. Quiescent evaporation of the solvent ensures that the sample components are deposited in a narrow continuous band at the start of the column. Following injection, the GC oven temperature is raised, by temperature programming, in order to elute the sample components which are of interest. Too large an injection volume can, in some circumstances, lead to undesirable chromatographic phenomena; for example, peak splitting of a pure component (see Section 8 for further discussion). Limiting the sample size to less than 1 μl generally minimizes such problems.

Larger injection volumes can be accommodated in on-column injection by use of retention gaps (4). A retention gap is an uncoated length of deactivated fused silica tubing, connected between the injector and the analytical column, by means of zero dead volume (5) or push-fit (6) connectors. The retention gap serves to re-concentrate sample bands, so eliminating adverse chromatographic effects. In addition, the retention gap functions as a protective guard-column, so minimizing the problem of contamination of the analytical column by involatile solutes. By frequent (weekly or daily) analysis of a mixture of authentic compounds possible deteriorations in chromatographic performance are readily monitored. Full chromatographic efficiency is usually restored by simply replacing the retention gap.

The use of wide internal diameter (0.53 mm i.d.) retention gaps allow on-column injection to be fully automated. These wide-bore retention gaps will accept the more robust, wider gauge (0.47 mm o.d.), stainless-steel syringe needles used in automated injection systems.

4.1.4 Split injection (7)

In the split injection mode the sample solution is introduced using a syringe equipped with a robust stainless steel needle, via a rubber septum inlet into a heated injection zone (200 to 300 °C). The high temperature ensures rapid volatilization of the sample solution. A large proportion of the carrier gas flow through the injector, and hence, a correspondingly large proportion of the sample and solvent, are vented directly to atmosphere. This factor, combined with the high flow rate (100 to 200 ml/min) of carrier gas through the injector area, ensures that the sample remains there only transiently, before delivery into the GC column as a narrow plug. Split injection is used where concentrated sample solutions are involved, to avoid problems of overloading capillary GC columns.

While the split injection technique is adequate for many lipid analyses (for example, simple fatty acid methyl esters and hydrocarbons), problems arise in the case of thermally-labile compounds, owing to the high temperature required to ensure instantaneous volatilization. The discrimination that can occur against relatively involatile, high molecular weight, components introduces imprecision into quantitative analyses. Hence, quantitative analyses of mixtures of lipids of widely differing volatility are best performed using the on-column injection technique (see above).

4.1.5 Splitless injection (7)

As with on-column injection, splitless injection is best-suited to the analysis of dilute solutions. In common with the split injection technique, samples are introduced using syringes equipped with robust stainless steel needles, via a rubber septum inlet, into a heated zone. The sample solution is volatilized and the vapour delivered, by the low carrier gas flow (*c*.2 ml/min), on to the GC column, which is held at a low temperature (typically 50 °C). This ensures that the solvent and sample components condense as a narrow plug at the head of the column. After a lapse of around 1 min, i.e. *c*.1.5 times that required to flush the injector volume, a flow of gas is introduced to purge any remaining sample from the injector area. In practice, the extent of the time lapse is established by trial and error, until optimum chromatographic performance (maximum solute delivery and minimum solvent peak tailing) is achieved. Splitless injection is the least reliable mode of injection for quantitative analysis of lipids, due to problems of discrimination between high- and low-molecular weight components.

4.1.6 Programmed temperature vaporizing (PTV) injection (8)

Recently introduced, this injection system combines split, splitless, and direct modes in a single module. The sample solution is injected at a temperature below the boiling point of the solvent, and condensed in the injector inlet. The temperature of the injector inlet is then increased rapidly, to sequentially volatilize solutes, which pass on to the column under the influence of the carrier gas flow. The major advantage of this mode of injection is the temperature programming of the injector inlet, which minimizes the thermal shock experienced by sample components. This injection system appears to be well-suited to the analysis of lipids of widely varying volatilities, including triacylglycerols.

4.2 GC oven requirements

All modern GC instruments are equipped with highly accurate temperature control mechanisms. The use of microprocessor-controlled feedback ensures excellent run-to-run reproducibility of oven temperature, and hence GC retention times (t_R), in both isothermal and temperature programming modes. Modern GC instruments are able to achieve the oven temperatures required for the most demanding of lipid analyses. For more than 95% of applications, operating temperatures in the range 50 to 320 °C will be adequate. Only for analyses of relatively involatile lipids, such as triacylglycerols, will higher GC oven temperatures (320 to 370 °C) be required. High-temperature stable GC stationary phases are required for such GC analyses (see later, Section 5.3.2).

4.3 Detection of eluting compounds

4.3.1 Flame ionization detector

The flame ionization detector (FID) is the most commonly used technique for the detection of lipids. Compounds eluting from GC columns are combusted in a hydrogen/air flame to yield ions, which produce an increase in the potential between the FID jet tip and a collector electrode. The resulting signal is amplified and recorded as the GC profile or chromatogram. The FID detector has a fast response, which is linear over a wide dynamic range, and very reliable in a long-term operation. As a result of minor configurational differences, FID detectors fitted to different models of gas chromatograph may operate under slightly different conditions (for example, mode of column installation, and hydrogen, air, carrier and make-up gas flow rates) and so should be set up and maintained according to the manufacturer's instructions.

4.3.2 Mass spectrometry

Mass spectrometry is very widely used as a detector in the GC analysis of lipids. The advantage of combined GC-MS is that mass spectra are recorded continuously during the GC analysis, with the result that, in many instances, complete structure identification can be made on eluting components, without recourse to other physico-chemical techniques. Mass spectometry will not be discussed further here, as it is the subject of Chapter 8 in this volume.

5. GC column selection

5.1 Capillary columns

Wall-coated open tubular (WCOT) capillary columns are the most common type of column now used in the analysis of lipids. Although glass capillary columns are still used, flexible-fused silica capillary columns coated externally, with either polyimide or aluminium, are more robust, and widely available commercially. The polymeric stationary phase is coated, as a thin film (0.1 μm to 10 μm film thickness), on the internal wall of the capillary. The common practice now is to 'bond' (or immobilize) the stationary phase by cross-linking the polymer chains (this will be discussed in more detail in Section 5.3.2). Typical column dimensions are 10 to 50 m × 0.1 to 0.32 mm i.d.

Advantages of capillary columns include:

- high column efficiencies, hence, high resolving power;
- enhanced sensitivity, due to improved signal/noise ratios as a result of narrow peak widths;
- compatability with mass spectometers, owing to low carrier gas flow rates (*c*. 2 ml min^{-1}).

Disadvantages of capillary columns include:

- long analysis times, often up to 1 h;
- low sample capacities, typically less than 100 ng/component;
- expensive and more fragile than packed columns.

Recently introduced open tubular columns, with an internal diameter of 0.53 mm, are now readily available. Most packed column instruments are economically and conveniently adapted for use with these columns. Conversion kits are available from several chromatographic suppliers. The thicker films (0.5 to 5.0 μm) of stationary phase that are used mean such columns offer enhanced sample capacity, compared to conventional capillary columns. While the chromatographic efficiencies are less than those of conventional capillary columns, they are superior to those of packed columns. Typical column lengths available are 15 and 25 m. The higher carrier gas-flow rates used with these columns preclude direct GC-MS interfacing.

5.2 Packed columns

Typical packed-column dimensions are 1 to 3 m × 4 mm i.d. While metal tubing was used in early work, this was superseded by deactivated glass. The column is filled with packing material, usually comprising fine particles of washed and deactivated diatomaceous earth (support), coated with polymeric stationary phase (typical loadings of stationary phase are in the range 1–10% w/w, depending on compounds under investigation).

Advantages of packed columns include:

- robust nature;
- cheap and readily re-packed when chromatographic performance deteriorates through overuse;
- sample capacity into the high microgram range depending on stationary phase loading;
- analysis times generally less than 30 min

Disadvantages of packed columns include:

- relatively low efficiencies, hence, limited separation power;
- difficult to interface to mass spectrometers (separators are required to remove excess carrier gas).

5.3 Stationary phase selection

5.3.1 Stationary phase selectivity and chromatographic efficiency

The nature of the stationary phase coated on the GC column is the single most important factor in dictating the selectivity achieved in GC analyses. Hence, selection of an appropriate stationary phase is essential if a desired separation is to be achieved. Also important is chromatographic efficiency, expressed in terms of numbers of theoretical plates per unit length of column (plates m^{-1}). The higher efficiencies of capillary columns means that they will generally be successful in resolving compounds that are unresolvable on packed columns. The lower efficiencies of packed columns may be overcome by taking advantage of the wider range of stationary phases, of varying polarity, and hence, varying selectivity for different solutes. In the case of WCOT capillary columns the range of stationary phases available is more limited, and demanding separations rely more on the very high separation efficiencies offered by these columns.

The definition of strict guide-lines for the selection of a suitable stationary phase for a particular analysis is problematical. Instead, the reader is referred to the examples presented below, where indication will be given of the factors affecting separations when employing particular stationary phases. Numerous other examples are available in the current literature and manufacturers' catalogues to enable a pragmatic, or empirical, approach to column selection to be adopted in the vast majority of cases. It should be considered that many gas chromatography suppliers offer free advice on column selection. In some instances, manufacturers will perform test analyses on mixtures of interest. This facility is well worth exploiting, as the purchase of a capillary column is a substantial investment for most laboratories. An important point to note in column selection is that different manufacturers market identical stationary phases under different trade names. *Table 1* lists the names and compositions of a selection of commercially available stationary phases.

The most important factor in the selection of a suitable stationary phase is the nature of the separation required. The most commonly used stationary phases are the polysiloxanes, particularly the apolar dimethyl polysiloxanes. However, as a result of separations relying exclusively on differences in the boiling points of solutes the usefulness of this latter phase is limited for resolving some lipid species. For example, saturated fatty acid methyl esters are not resolved from their monoenoic counterparts on packed columns containing a dimethyl polysiloxane stationary phase, such as OV-1. In order to achieve this separation a polar stationary phase must be employed, which affords separations on the basis of differences in dipole–dipole interactions. The polyethylene glycol stationary phase, Carbowax 20M, is widely used for this purpose (separations of fatty acid methyl esters on this stationary phase are discussed in detail in Section 7.6).

As can be seen from *Table 1*, a wide range of stationary phases of differing

Table 1. 'Equivalent' GC stationary phases

Type	J & W	HP	Quadrex	Chrompak	Supelco	SGE	RSL
1[a]	DB-1	Ultra-1	1	CP-Sil-1CB	SPB-1	BP 1	150
5[b]	DB-5	Ultra-2	2	CP-Sil-8CB	SPB-5	BP 5	200
17[c]	DB-17	HP-17	17		SP-2250		300
210[d]	DB-210						400
225[e]	DB-225		225	CP-Sil-43CB		BP 15	500
275[f]	DL-2330			CP-Sil-84	SP-2330		
1701[g]	DB-1701		1701	CP-Sil-19CB		BP 10	
PEG[h]	DB-Wax		CW	CP-Wax-51CB	Supelco wax-10	BP 20	

[a] 100% dimethyl polysiloxane (SE-30. OV-101).
[b] 5% phenyl. 1% vinyl methyl polysiloxane (SE-54).
[c] 50% phenyl methyl polysiloxane (OV-17. SP-2250).
[d] 50% trifluoropropyl methyl polysiloxane (OV-210. QF-1. SP-2401).
[e] 25% phenyl. 25% cyanopropyl methyl polysiloxane (OV-225. XE-60).
[f] 70–80% cyanopropyl methyl polysiloxane (OV-275. SP-2340. SP-2330).
[g] 6% phenyl. 6% cyanopropyl methyl polysiloxane (OV-1701).
[h] The various 'bonded' polyethylene glycol stationary phases are markedly different in (1) maximum and (2) minimum operating temperature limits and (3) their solubilities in (and compatibility with injections containing) water and low-molecular-weight alcohols.
(Reproduced from ref. 1 with permission.)

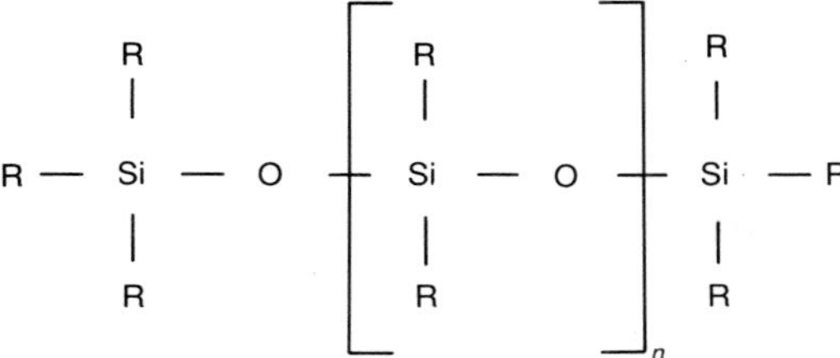

Figure 2. The oxygen–silicon lattice which forms backbone of polysiloxane stationary phases. Stationary phases of different polarity are produced by substituting different R-groups on the polysiloxane backbone.

chemical nature are available. While polyester and polyethylene glycol phases are still used extensively in packed column GC work, polysiloxane stationary phases are preferred, particularly for WCOT capillary GC analyses. These phases display enhanced thermal stability and resistance to the degradative effects of moisture and oxygen, consequently column life is extended. Polysiloxane stationary phases of widely varying polarity, and hence differing selectivity for different lipid species, are now available. As a result, polyethylene glycol and polyester phases are no longer necessarily the first choice for many applications.

Modifications of the polarity of polysiloxane stationary phases is achieved by substitution of different R groups to the polysiloxane back-bone (*Figure 2*) during the manufacturing process. The dimethyl polysiloxanes are the least polar stationary phase currently available. Polarity is increased by increasing the proportion of polar moieties such as phenyl-, cyanopropyl-, or trifloro-propyl- on the polysiloxane backbone. The relative suitabilities of the various stationary phases for particular lipid analyses will not be addressed here, but considered in more detail in the specific applications presented later in this chapter.

5.3.2 'Bonded' or immobilized stationary phases

The majority of commercially available WCOT capillary columns are coated with stationary phases of the 'bonded' variety. The term 'bonding' refers to the process of stationary phase immobilization in the capillary column (9). This is carried out during the manufacturing process and involves cross-linking the stationary phase polymer chains, and in some cases inducing covalent bonding between the stationary phase and the vitreous silica column wall. The development of immobilized stationary phases has had important implications for the GC analysis of many classes of lipid. For all applications, as alluded to above, the use of immobilized phases enhances durability and extends column life. Immobilized phases have the advantage of being amenable to washing with solvents, to remove contaminants arising from involatile sample impurities and deterioration of the stationary phase. Column washing will be discussed later in Section 8.

Enhanced thermal stability is the most significant advantage of 'bonded' stationary phases, compared to their 'non-bonded' counterparts. This enhanced thermal stability raises the possibility for the routine GC analysis of relatively involatile, high-molecular-weight lipids; for example, triacylglycerols, wax esters, etc. (see Section 7.2). While cross-linked, apolar dimethyl polysiloxanes were amongst the first 'bonded' phases to become commercially available, a more recent advance has been the development of high-temperature stable, polarizable, immobilized stationary phases. One such stationary phase is an immobilized OV-22 (65% phenyl methyl polysiloxane) polymer, which becomes more polar with increasing temperature, and is stable up to 360 °C. This stationary phase is capable of separating molecular species which exhibit only small differences in polarity; for example, triacylglycerols possessing different numbers of double bonds (the use of polarizable stationary phases will be discussed further in Section 7.5).

6. Compound identification

The extensive use of GC in the analysis of lipids makes gas chromatography/mass spectrometry (GC-MS) the method of choice for compound identification. Complete structure assignments can frequently be made by consideration of mass spectra and GC retention data alone. The availability of extensive computer library databases of mass spectra of many of the commonly occurring lipids provides very convenient means of mass spectral interpretation. However, the newcomer should be aware that the results of computerized library searches are by no means unambiguous, and they must be assessed critically, in conjunction with other analytical data; for example, retention times and elution orders. The mass spectrometry of lipids will not be discussed further here, as it is the subject of Chapter 8 in this volume.

Where GC-MS is not available, considerable progress can be made in assigning GC peaks by investigating retention-time relationships. Mixtures of authentic ('standard') compounds are commercially available for most of the commonly encountered lipids. Where the desired mixtures are not commercially available, customized standards are readily produced in the laboratory, from the pure components, or natural extracts. Frequent analysis of authentic mixtures serves not only as a test of t_R reproducibility, but also as a check of the chromatographic performance of the class of compounds under examination. Direct comparisons of GC t_Rs of unknowns with authentic compounds constitutes an important provisional means of compound identification. Retention time (t_R) comparisons may proceed as follows:

(a) By direct comparison of t_Rs of unknowns and authentic compounds, determined separately under identical GC operating conditions. Overlaying chromatograms using a light box is an effective means of t_R comparison, in the absence of a plotter/integrator print-out, or other data-handling facility.

(b) By co-chromatography of unknowns with authentic compounds. Only perfect co-elution, i.e. no GC peak deformity, is acceptable as a positive comparison. *Figure 3* presents the approach to the interpretation of co-injection data.

Retention time is not a unique property of a substance. Therefore, where reliance is placed on t_Rs for making compound identifications, the minimum requirement is for independent comparisons on two columns, coated with stationary phases of different polarity. The probability of making an incorrect assignment, as a result of two different compounds exhibiting coincident t_Rs, is greatly reduced on capillary columns, owing to their high efficiencies.

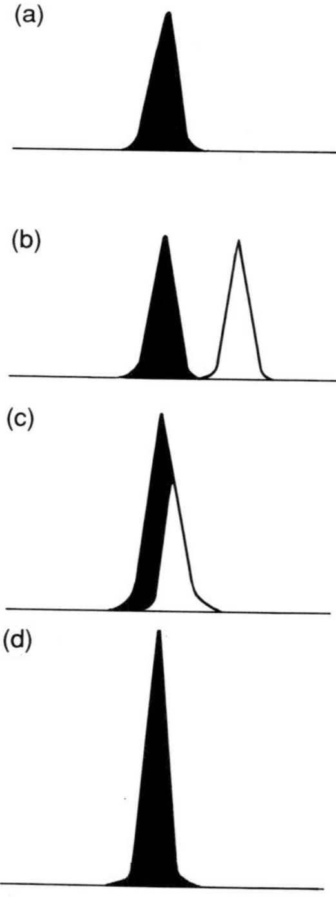

Figure 3. Schematic representation of the co-injection approach to GC peak identification. Chromatograms represent: GC peak of unknown compound under investigation (a); peaks well-separated (b), hence, the unknown compound is not the same as that co-injected; the co-injected substance elutes as a shoulder on the unknown (c), the unknown and the co-injected compound are not identical; perfect co-elution of the unknown and co-injected authentic, providing strong, though not definitive, evidence for the unknown and authentic compound being identical (d).

Other means of assigning compound identity from GC retention data will be discussed in the specific application sections below. Identification of novel compounds cannot be made by GC alone, supporting evidence must be obtained by other physico-chemical techniques, e.g. MS or NMR.

7. Applications

7.1 Methodological considerations

Lipids occur most commonly as complex mixtures of molecular species. The aim of GC is to resolve the individual molecular species in order to provide compositional information, such as compound identity and absolute abundance. The high molecular weight, and hence, low volatility, of the most commonly occurring lipids species (such as triacylglycerols, phospholipids, etc.) limits the usefulness of packed-column GC. Although triacylglycerols, diacylglycerols (released from phospholipids by enzyme treatment), steryl esters, wax esters, etc., can be analysed by packed-column GC (10), the low chromatographic efficiencies attainable by these columns limit resolution.

Thus, until recently, the most common approach to the analysis of high-molecular-weight acyl lipids has been to chemically or enzymatically hydrolyse the total lipid extract, or a specific lipid class isolated by adsorption chromatography; then analyse the liberated simpler lipid moieties by GC, with derivatization where necessary (see Chapter 2 for general approaches to the degradation and derivatization of lipids). The major disadvantage of this approach is the loss of important compositional information. For example, in the case of wax esters or steryl esters, hydrolysis may yield a mixture of fatty acids and a mixture of alcohols (long-chain alcohols for wax esters and sterols in the case of steryl esters) and there is no simple means of deducing which alcohol was bound to which fatty acid in the original mixture (*Figure 4*). As a result, biochemically important information relating to the selectivity of enzymes, or pools of lipid substrates, will be lost. Additionally, the similar mobilities of some commonly-occurring lipids on adsorbents can lead to inefficient class separations, which will result in erroneous conclusions concerning composition, if degradation was performed. Moreover, in clinical diagnosis, important quantitative or qualitative variations in the distributions of key intact acyl lipid species may be obscured when only fatty acid profiles are examined, following hydrolysis or trans-esterification.

Owing to these shortcomings, and the now wide availability of efficient capillary columns coated with high-temperature stable phases, GC analysis of many intact acyl lipid species can now be performed relatively routinely. It is therefore appropriate to present a description of high-temperature GC procedures before discussing methods for examining the various moieties released by hydrolysis and other degradative techniques. While this might be regarded as an unconventional approach, from the analytical standpoint,

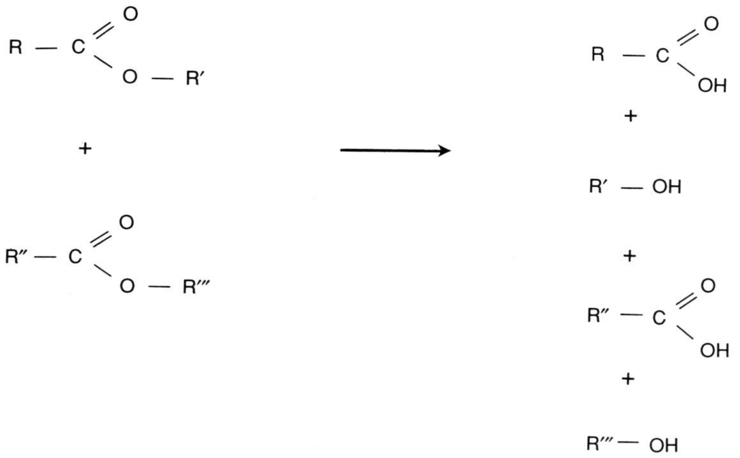

$$R{-}C(=O){-}O{-}R' \; + \; R''{-}C(=O){-}O{-}R''' \longrightarrow R{-}C(=O){-}OH \; + \; R'{-}OH \; + \; R''{-}C(=O){-}OH \; + \; R'''{-}OH$$

Figure 4. Schematic representation of the loss of compositional integrity resulting through the hydrolysis of acyl lipids. It cannot easily be deduced which alcohol was bound to which acid in the intact acyl lipid mixture.

every effort should be made to derive the maximum compositional information from the intact lipid species, employing procedures that require the minimum of sample manipulations. Degradative methodologies are then used to corroborate the high-temperature GC or GC-MS data, to provide information unobtainable by the analysis of intact molecular species alone.

7.2 GC of intact acyl lipids

The analysis of intact acyl lipids by GC is one of the most demanding applications of the technique. The major advantage of determining intact acyl lipids is the ability to study individual molecular species. Compound classes commonly studied in this way include triacylglycerols, diacylglycerols, ceramides, steryl esters, wax esters, etc.

Analyses of intact acyl lipids can be carried out on both packed (10) and capillary (11) GC columns. Packed columns suitable for such work are short (30 or 45 cm × 3 mm i.d.) and contain low stationary phase loadings (for example, 2.25% w/w SE-30). In this field, even more so than in other areas of the GC analysis of lipids, packed columns have been superseded by capillary columns.

Relatively short (10 to 15 m) fused silica capillaries, coated with thin (typically 0.1 μm) films of immobilized apolar stationary phase (for example, dimethyl polysiloxane) are widely used. *Figure 5* shows the GC analysis of butter-fat triacylglycerols on such a column. Separation occurs largely according to carbon number. Partial resolution of molecular species according to degree of unsaturation is seen in the later eluting components. Full

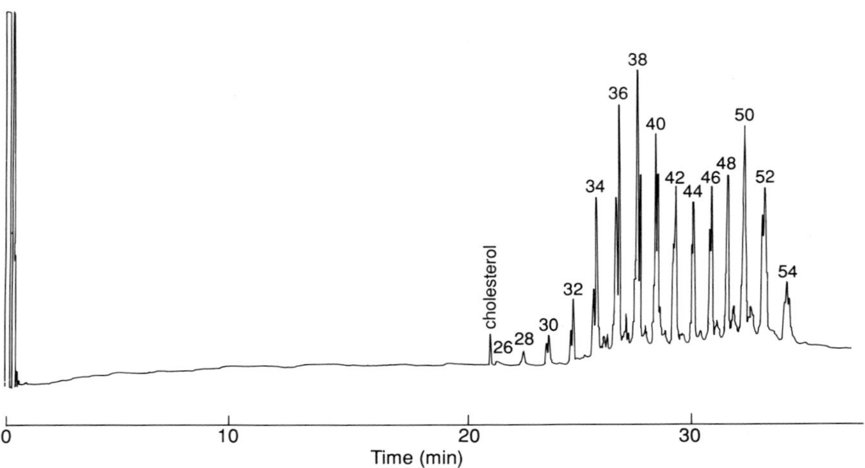

Figure 5. High-temperature GC analysis of butter-fat triacylglycerols. The experimental conditions were the same as those given in *Figure 1*. The numbers on the peaks correspond to the total number of acyl carbon atoms.

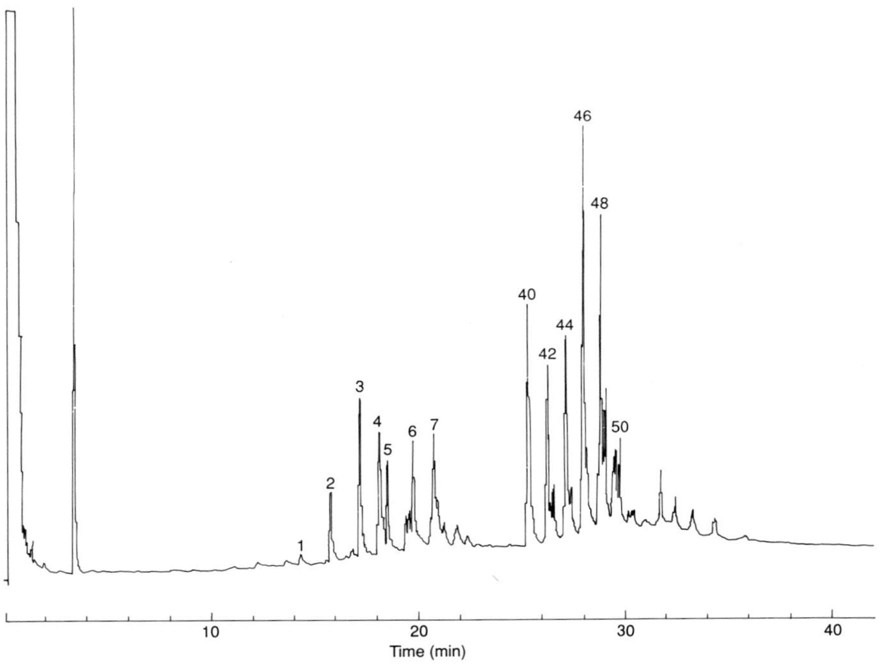

Figure 6. High-temperature GC analysis of beeswax treated with BSTFA. The experimental conditions were the same as those given in *Figure 1*. Peak identities are: 1 = *n*-tricosane; 2 = *n*-pentacosane; 3 = *n*-heptacosane; 4 = tetracosanoic acid (TMS ester derivative); 5 = *n*-nonacosane; 6 = *n*-hentriacontane; 7 = *n*-tritriacontane; peaks 42, 44, 46, 48, 50, and 52 are wax esters containing 42, 44, 46, 48, 50, and 52 carbon atoms respectively, palmitic acid is the predominant acyl moiety of these esters. The major fatty alcohols moieties range in chain length from 26–36 carbon atoms. The lower abundance compounds eluting with the wax esters are hydroxy acid wax esters.

experimental details are given in the figure caption. This same column can also be used to analyse high-molecular-weight steryl esters, wax esters, diacylglycerols, monoacylglycerols, glycerol ethers, ceramides, etc.

An example of the analysis of the wax esters in the hexane-soluble fraction of beeswax is shown in *Figure 6*. The later eluting components correspond to

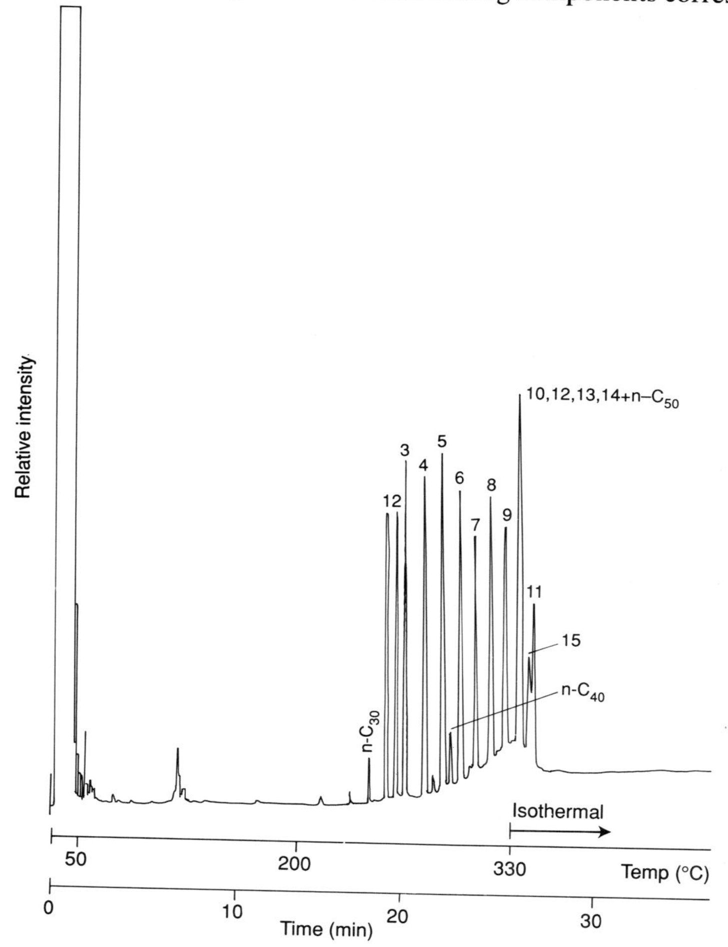

Figure 7. High-temperature GC of authentic cholesteryl fatty acyl esters bearing saturated and unsaturated acyl moieties together with co-chromatographed *n*-alkanes. The experimental conditions were the same as those given in *Figure 1* except that temperature programming was at 8 °C/min. Peak identities are: 1 = cholesteryl acetate; 2 = cholesteryl propionate; 3 = cholesteryl butyrate; 4 = cholesteryl caproate; 5 = cholesteryl caprylate; 6 = cholesteryl caprate; 7 = cholesteryl laurate; 8 = cholesteryl myristate; 9 = cholesteryl palmitate; 10 = cholesteryl stearate; 11 = cholesteryl arachidate; 12 = cholesteryl oleate; 13 = cholesteryl linoleate; 14 = cholesteryl linolenate; 15 = cholesteryl arachidonate. (Redrawn from ref. 34 with permission from *Biomedical Environmental Mass Spectrometry*.)

wax esters in the C_{38} to C_{52} carbon number range. The carbon number is readily assigned by retention time comparison or co-injection with a synthetic wax ester of known structure. Beeswax itself is well-characterized, and of such constant composition that it can be conveniently used as a standard mixture for carbon number assignment in the case of unknowns. In beeswax wax esters, palmitic acid is the predominant fatty acid moiety, hence the chain-length of the alcohol moieties range from C_{26} to C_{36}. The earlier eluting components in *Figure 6* are long-chain aliphatic hydrocarbons (the GC analysis of hydrocarbons is discussed in greater detail below).

Steryl esters can also be analysed by high-temperature GC; *Figure 7* shows the analysis of a mixture of synthetic cholesteryl esters. Separation is by carbon number, and to a lesser extent according to the degree of unsaturation; i.e. cholesteryl arachidonate elutes before cholesteryl arachidate. The effect of varying the nature of the sterol moiety follows the same general trends as for free sterols or their acetates, trimethylsilyl (TMS) ethers or other derivatives (see below in Section 7.11 'Sterols', for further discussions of elution orders).

7.3 Total lipid profiling by high-temperature GC

While the profiling of intact acyl lipid mixtures by high-temperature GC was first demonstrated in the clinical field (11), the potential exists for its application in any area in which complex mixtures of lipids, of widely varying molecular weight are encountered; for example, for chemistry, environmental chemistry, organic geochemistry, etc.

In clinical diagnosis the following procedure is adopted (11):

(a) total lipid extracts of human plasma are treated with phospholipase C in order to release diacylglycerols and ceramides from phospholipids.

(b) trimethylsilylation (see Chapter 2) of the total extract is carried to block protic sites; for example, hydroxyl and carboxylic acid moieties.

(c) GC analysis of the total lipid extract can be performed on short packed columns of the type referred to above. However, capillary GC analysis is to be recommended, as greatly improved separations are attainable.

The total lipid profile of human plasma is shown in *Figure 8* and was obtained using a capillary column coated with a thin film of immobilized, dimethyl polysiloxane-type, stationary phase. The profile shows the distribution of lipids, ranging from lower-molecular-weight free fatty acids (as TMS ester derivatives) to intact triacylglycerols containing up to 56 carbon atoms in the acyl moieties.

7.4 Thermally stable polarizable stationary phases

Although the high-temperature GC profiles obtained using capillary columns coated with apolar dimethyl polysiloxane-type stationary phases are very

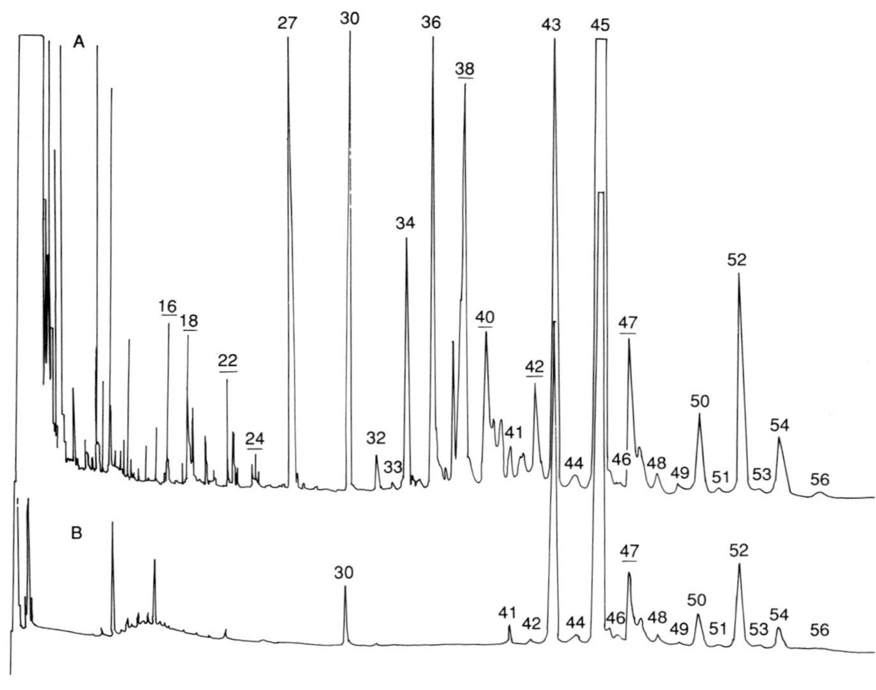

Figure 8. Plasma total lipid profile obtained by high-temperature GC using an 8 m × 0.3 mm i.d. fused silica capillary coated with SE-54 ('bonded' methyl phenyl polysiloxane), stationary phase. After on-column injection the GC oven temperature was programmed ballistically (30 °C/min) to at 150 °C, then to 230 °C at 20 °C/min, to 280 °C at 10 °C/min and to 340 °C at 5 °C/min. Hydrogen was the carrier at a column head pressure of 6 p.s.i. The detector temperature was maintained at 350 °C. Peak identities are: 16 and 18, TMS esters of fatty acids with 16 and 18 acyl carbons respectively; 27 is the TMS ether of cholesterol; 30 is the tridecanoylglycerol internal standard; 34 is the TMS ether of the palmitic acid amide of sphingosine; 36, 38, and 40, TMS ethers of diacylglycerols with 34, 36, and 38 acyl carbons, respectively; 41 and 42, TMS ethers of ceramides with 41 and 42 fatty carbons; 43, 45, and 47, cholesteryl esters with fatty acyl moieties containing 16, 18 and 20 carbons respectively; 44, 46, and 48–56, triacylglycerols with a total of 44, 46, and 48–56 acyl carbons respectively. The total lipid extact of plasma had previously been dephosphorylated by treatment with phospholipase C. Total run time was 30 min. (Redrawn from ref. 11 with permission of the authors and *Journal of Biochemical and Biophysical Methods*.)

useful, separations are largely by carbon number. A recent advance referred to above (Section 5.3 'Stationary phase selection') has been the development of thermally-stable polarizable stationary phases. Such phases become more polar as the GC oven temperature increases, with the result that capillary columns containing this type of stationary phase may possess sufficient

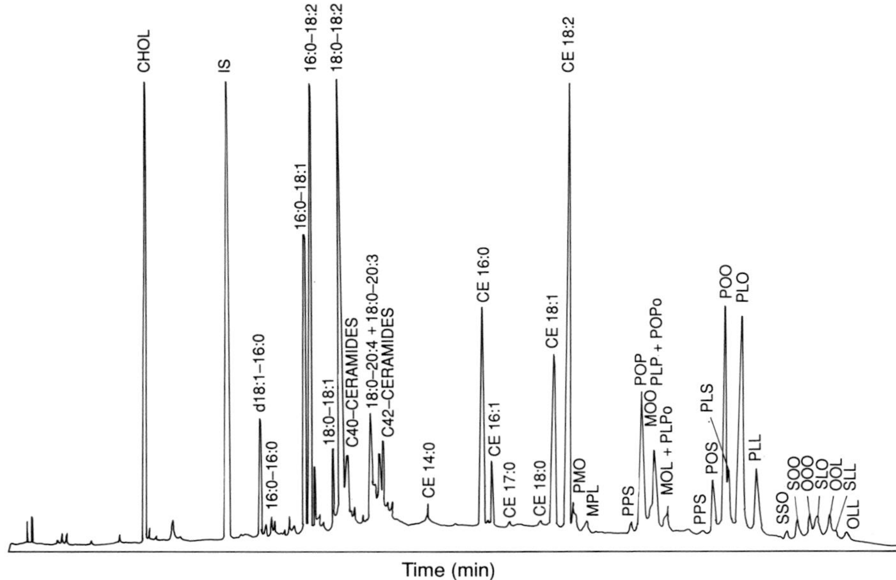

Figure 9. Capillary GC of plasma lipids from a hyperlipemic subject on a polarizable capillary column following dephosphorylation and trimethylsiliylation. IS = internal standard (tridecanoylglycerol); CHOL = cholesterol; 16:0–16:0 etc. = diacylglycerols as TMS ethers; CE 16:0 = cholesteryl palmitate, etc.; for the triacylglycerols M = myristic, P = palmitic, S = stearic, O = oleic and L = linoleic, hence, SSS = tristearin, etc. The analysis was performed on a 25 m × 0.25 mm i.d. fused silica capillary coated with 65% phenyl 35% methyl polysiloxane (0.1 µm film thickness, immobilized OV-22). Following on-column injection at 40 °C and a 1-min delay, the GC oven temperature was raised ballistically to 220 °C, then to 320 °C at 10 °C/min, then to 360 °C at 2 °C/min. Hydrogen was the carrier gas at a column head pressure of 1 bar. The retention time of tridecanoylglycerol with 24.6 min. (Redrawn from ref. 12 with the permission of the authors and *Journal of Chromatography*.)

resolving power for the separation of molecular species of intact acyl lipids, such as triacylglycerols. *Figure 9* shows the analysis of plasma lipids on a capillary column coated with a polarizable stationary phase (12). The greatly improved resolution for the ceramides and triacylglycerols is clearly evident (compare with *Figure 8*). In the latter part of the chromatogram individual triacylglycerols are resolved according to the number of double-bonds present in their fatty acyl moieties. Similar GC retention behaviour is observed for both cholesteryl esters and diaclylglycerols under these GC conditions.

7.5 Cautionary notes on the high-temperature GC of intact lipids

It is important to be aware that the high temperatures required for elution of intact lipids (generally above 300 °C) can cause the loss of some thermally labile compounds. Most thorough assessments of recoveries from high-temperature GC analyses have been carried out on triacylglycerols. Loss of material is dependent on the molecular weight and degree of unsaturation of a given compound (13). Hence, while good recoveries are obtained for analyses of many fats and oils, high-temperature GC is unsuitable for the analysis of fish oils, and the lipids of other marine organisms, owing to the high content of polyunsaturated fatty acids in their acyl lipids. Loss of steryl fatty acyl esters bearing polyunsaturated fatty acyl moieties (for example, cholesteryl arachidonate), also occur (13), probably as a result of thermal decomposition. The loss of saturated species that occur is attributed to irreversible saturation of the stationary phase. There is negligible loss of lower-molecular-weight lipids, such as wax esters, diacylglycerols, and short-chain triacylglycerols.

Losses are best assessed through the analysis of authentic compounds. Analysis of equimolar mixtures of pure analogues of the compounds to be analysed allows calibration, or correction, factors to be calculated to account for losses during the GC analysis. Where a given compound is not available then only approximate correction factors can be derived, from a closely related compound.

A useful test for the loss of polyunsaturated components during GC analyses of unknowns is to perform a microscale catalytic hydrogenation then repeat the GC analysis. Major losses of polyunsaturated components are revealed by the appearance of new peaks in the chromatogram for the repeat analysis. Alternatively, chemical or enzymatic degradation and subsequent analysis of the simpler lipid moieties released, can be used to check for components lost in high-temperature GC analyses. As pointed out in the introduction to this section ('Methodological considerations'), one must be confident of the purity of the lipid fraction under investigation when interpreting the results of analyses performed after degradation in this way.

As the loss of lipids containing polyunsaturated fatty acyl moieties relates to their inherently lower thermal stability, high-temperature GC is not the method of choice for their routine analysis. Analysis of highly unsaturated, intact acyl lipids should be carried out by HPLC (see Chapter 5) or by GC of their component fatty acid and alcohol moieties, following chemical or enzymatic cleavage. Methods for the GC analysis of these simpler moieties are given below.

7.6 Fatty acids

7.6.1 General approach

Fatty acids are the class of lipid most commonly analysed by GC. As many fatty acids occur widely in plant and animal kingdoms, the GC methods described in this section are applicable to samples from a wide range of natural materials. Although fatty acids possessing a wide carbon number range (C_4 to above C_{30}) are known, the most commonly occurring are those in the C_{14} to C_{22} chain-length range; analysis of fatty acids in this carbon number range is routinely carried out by GC.

GC analysis of fatty acids is most commonly performed following their conversion to apolar, methyl ester derivatives (see Chapter 2). Polar GC phases are employed in both packed and capillary column analyses, as these afford separation according to both carbon number and the degree of unsaturation in the alkyl chain. The polyethylene glycol stationary phase, Carbowax 20M, is that most often used for capillary GC analyses (14). *Figure 10* shows the GC separation of rape seed oil fatty acids (as methyl esters) on a Carbowax 20M coated capillary column, operated under the conditions given in the figure caption. This is a typical separation achieved without special steps being taken to optimize the chromatographic parameters.

The GC profile reveals a number of general properties that apply to both packed and capillary analyses. The pattern of peaks for the C_{16} and C_{18} components is typical. On the polar Carbowax 20M stationary phase, fatty acids with increasing numbers of double bonds exhibit incrementally increasing t_R's, owing to gradually increasing dipole–dipole interactions. Hence, methyl palmitate ($C_{16:0}$) elutes before methyl palmitoleate ($C_{16:1}$). The order of elution for the C_{18} components is:

$$\text{stearate}(C_{18:0}) < \text{oleate}(C_{18:1}) < \text{linoleate}(C_{18:2}) < \text{linolenate}(C_{18:3})$$

These elution orders are very reproducible, and will soon become familiar to analysts involved in the frequent GC analysis of fatty acids.

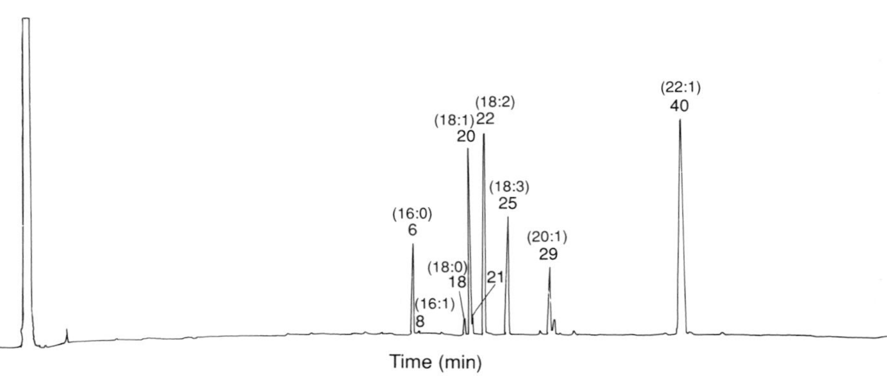

Figure 10. Capillary GC analysis of rape-seed oil fatty acid methyl esters on a 25 m ×0.32 mm i.d. HP-20M (Carbowax 20M) coated (0.3 μm film thickness) fused silica capillary. After on-column injection the GC oven temperature was held at 50 °C for 2 min, then programmed to 170 °C at 10 °C/min, then to 220 °C and 2 °C/min, followed by an isothermal hold for 10 min. Helium was the carrier gas at a column head pressure of 10 p.s.i. The numbers on the peaks refer to the numbers listed in *Table 2*. The total analysis time was 30 min.

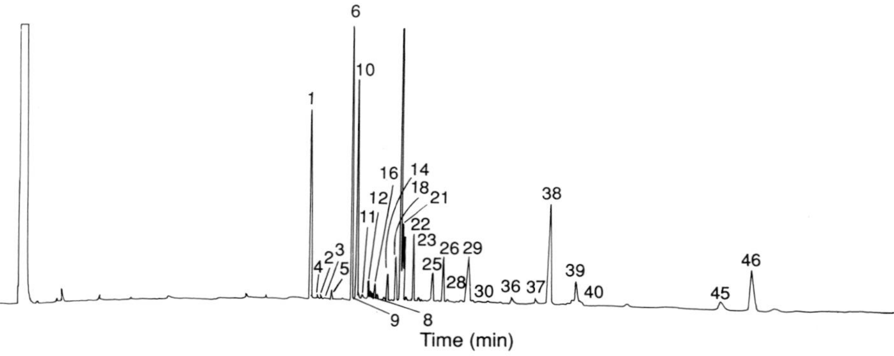

Figure 11. Capillary GC analysis of cod liver oil fatty acid methyl esters. The GC conditions were the same as those used for the analysis shown in *Figure 10*. The number on the peaks refer to the numbers listed in *Table 2*. The total analysis time was 30 min.

Figure 11 shows the analysis of the fatty acids (again as methyl ester derivatives) from cod liver oil, under the same conditions as those employed in *Figure 10*. This analysis demonstrates the separation of polyunsaturated fatty acid species commonly found in marine animals. The composition of cod liver oil is sufficiently well-established for it to be used as a reference material. The peak identities for the major components are taken from ref. 15.

7.6.2. Equivalent chain-lengths (*ECLs*; refs 16 and 17)

The equivalent chain-length (*ECL*) concept has been widely used to identify unknown fatty acid methyl esters, by expressing their elution positions relative to those of authentic straight-chain saturated fatty acid methyl esters (FAMEs). *ECL* is defined as:

$$ECL = n + \frac{\log t'_{R,x} - \log t'_{R,n}}{\log t'_{R,n+1} - \log t'_{R,n}}$$

where $t'_{R,x}$ is the corrected retention time of an unknown FAME x; $t'_{R,n}$ is the corrected retention time of the nearest saturated straight-chain FAME eluting ahead of x; $t'_{R,n+1}$ is the corrected retention time of the next highest homologue of n. NB: $t'_R = t_R - t_M$, where t_R is the uncorrected retention time and t_M is the column dead time (retention time of solvent peak). In practice, *ECL*s are measured in the following way:

(a) Determine the t'_Rs for a homologous series of straight-chain saturated FAMEs under isothermal GC oven conditions (this requires the split injection mode for capillary GC analyses).

(b) Plot a graph of log t'_R against carbon number (*ECL*) for the homologous series of authentic FAMEs.

(c) Determine log t_R for the unknown under identical GC operating conditions (i.e. same column, oven temperature, and carrier gas-flow rate). Exactly identical conditions are ensured by co-injecting the authentic mixture with the unknown, and making the t'_R measurements from the same chromatogam.

(d) The ECL for the unknown is determined directly from the *ECL* axis at its measured log t'_R value.

GC retention characteristics, and hence *ECL*s, are directly related to molecular structure. In the case of FAMEs the *ECL* increment changes that are observed will reflect the number, position, and geometry of double bonds, presence and position of alkyl chain branching and other functional groups, i.e. oxygenated moieties and ring systems.

Substantial ECL data exist for compounds of known structure which can be used for reference purposes when attempting to deduce the structures of unknown compounds (14, 15). Reference data are available for both packed and capillary columns and various stationary phase types. It should be realized that these published *ECL* values are not absolute and interlaboratory discrepancies arise for a number of reasons, including:

- differences in stationary phase loading;
- column condition;
- sample loading (more important in capillary than packed column GC measurements).

Although the measured *ECL* values vary somewhat, the elution orders generally remain constant for a given stationary phase. Consequently, *ECL* values are important in establishing a elution order ranking, which can be used to predict the identities of unknown peaks in chromatograms. The *ECL* values for some naturally occurring fatty acid methyl esters are presented in *Table 2*.

7.6.3 Analysis of fatty acids of unusual chain-length

i. Short-chain fatty acids

Care must be taken in the handling and analysis of short-chain fatty acids (C_4 to C_{10}) owing to them exhibiting appreciable vapour pressures even at relatively low temperatures. Milk fat is one of the most commonly analysed biological materials to contain appreciable quantities of these short-chain fatty acids, in addition to the more commonly observed long-chain species (C_{12} to C_{18} chain-length range). The use of on-column injection is essential in capillary GC in order to minimize losses of the more volatile short-chain constituents, where quantitative assessments are required. Temperature-programming from as near ambient temperature as possible (*c*.30 °C) is necessary to avoid the shorter chain, methyl butyrate, component co-eluting

Table 2. Equivalent chain-lengths of the methyl esters derivatives of some natural fatty acids

No.	Fatty acid	Stationary phase			
		Silicone	Carbowax	Silar 5CP	CP-Sil 84
1	14:0	14.00	14.00	14.00	14.00
2	14-isobr	14.64	14.52	14.52	14.51
3	14-ante-iso	14.71	14.68	14.68	14.70
4	14:1 (*n*-5)	13.88	14.37	14.49	14.72
5	15:0	15.00	15.00	15.00	15.00
6	16:0	16.00	16.00	16.00	16.00
7	16-isobr	16.65	16.51	16.51	16.50
8	16-anteiso	16.73	16.68	16.68	16.69
9	16:1 (*n*-9)	15.76	16.18	16.30	16.48
10	16:1 (*n*-7)	15.83	16.25	16.38	16.60
11	16:1 (*n*-5)	15.92	16.37	16.48	16.70
12	16:2 (*n*-4)	15.83	16.78	16.98	17.47
13	16:3 (*n*-3)	15.69	17.09	17.31	18.06
14	16:4 (*n*-3)	15.64	17.62	17.77	18.82
15	17:0	17.00	17.00	17.00	17.00
16	17:1 (*n*-9)	16.76	17.20	17.33	17.50
17	17.1 (*n*-8)	16.75	17.19	17.33	17.51
18	18:0	18.00	18.00	18.00	18.00
19	18:1 (*n*-11)	17.72	18.14	18.24	18.40
20	18:1 (*n*-9)	17.73	18.16	18.30	18.47
21	18:1 (*n*-7)	17.78	18.23	18.36	18.54
22	18:2 (*n*-6)	17.65	18.58	18.80	19.20
23	18:2 (*n*-4)	17.81	18.79	18.98	19.41
24	18:3 (*n*-6)	17.49	18.85	19.30	19.72
25	18:3 (*n*-3)	17.72	19.18	19.41	20.07
26	18:4 (*n*-3)	17.55	19.45	19.68	20.59
27	19:1 (*n*-8)	18.74	19.18	19.32	19.47
28	20:1 (*n*-11)	19.67	20.08	20.22	20.35
29	20:1 (*n*-9)	19.71	20.14	20.27	20.41
30	20:1 (*n*-7)	19.77	20.22	20.36	20.50
31	20:2 (*n*-9)	19.51	20.38	20.59	20.92
32	20:2 (*n*-6)	19.64	20.56	20.78	21.12
33	20:3 (*n*-9)	19.24	20.66	20.92	21.43
34	20:3 (*n*-6)	19.43	20.78	21.05	21.61
35	20:3 (*n*-3)	19.71	20.95	21.22	21.97
36	20:4 (*n*-6)	19.23	20.96	21.19	21.94
37	20:4 (*n*-3)	19.47	21.37	21.64	22.45
38	20:5 (*n*-3)	19.27	21.55	21.80	22.80
39	22:1 (*n*-11)	21.61	22.04	22.16	22.30
40	22:1 (*n*-9)	21.66	22.11	22.23	22.36
41	22:3 (*n*-9)	21.20	22.52	22.78	23.25
42	22:3 (*n*-6)	21.40	22.71	22.99	23.47
43	22:4 (*n*-6)	21.14	22.90	23.21	23.90
44	22:5 (*n*-6)	20.99	23.15	23.35	24.19
45	22:5 (*n*-3)	21.18	23.50	23.92	24.75
46	22:6 (*n*-3)	21.04	23.74	24.07	25.07

(Reproduced from ref. 15 with the permission of the author and *Journal of Chromatography*.)

with the solvent peak. Otherwise, comparable GC operating conditions to those given in the caption to *Figure 10* are suitable.

ii. Longer chain fatty acids

There is rarely the need to determine fatty acids with greater than 22 carbon atoms. However, several mammalian tissues, some marine organisms, bacteria, and many geological materials (peats, soils, sediments, etc.) have been found to contain fatty acids with greater than 30 carbon atoms. Methyl esters of fatty acids in this chain-length range are too involatile to elute from Carbowax 20M within the maximum operating temperature of this stationary phase (*c*.220 °C). The carbon number range of long-chain fatty acids is best determined using capillary or packed columns containing a polysiloxane stationary phase. A 12 m × 0.22 mm i.d. fused silica capillary with 0.1 μm film of BP-1 (immobilized OV-1 equivalent stationary phase) can be used to determine long-chain fatty acids with well in excess of 30 carbon atoms. On-column injection is preferred to avoid problems of discrimination. With temperature programming from 50 to 350 °C at 10 °C min and the helium carrier gas pressure set at 20 p.s.i., methyl triacontanoate (C_{30}) elutes after approximately 25 min. As this is well before the end of the temperature programme, there exists plenty of scope for the determination of still higher homologues.

7.6.4 Separation of isomers

As the requirement to separate isomeric fatty acids occurs relatively infrequently in analyses of natural mixtures, the subject is covered only briefly here. Readers requiring more specific details of separations of fatty acid isomers are referred to ref. 2.

i. Positional isomers

Variations in double-bond position are observed more commonly than variations in geometry in unsaturated fatty acids. Monoenoic fatty acids with unsaturation other than at the 9-position occur in many plant and animal fats. As can be seen from *Table 2*, best separations are achieved using capillary columns coated with polar stationary phases. Partial separation of 18:1(*n*-7) and 18:1(*n*-9) is evident in the cod liver oil fatty acid analysis shown in *Figure 11*, on the polar Carbowax 20M stationary phase. Separation of the 18:1 (*n*-11) from 18:1(*n*-9) is generally more difficult. Consequently, assessment of petroselenic acid [18:1(*n*-11)] content if required, is best performed by HPLC (see Chapter 5) or mass spectrometry (see Chapter 8).

Although linoleic acid is the most abundant fatty acid in nature other isomeric dienes do occur at low levels. Separations of positional isomers are represented by the *ECL* values listed in *Table 2*. A general trend derived from the analysis of a range of synthetic isomers showed that with both polar and apolar phases the *ECL* values increased with increasing distance of the double bonds from the carboxyl group (15). The presence of a conjugated double bond increases the fatty acid methyl esters t_R to a much larger degree, compared to a normal methylene interrupted ester.

ii. Geometric isomers

While *E*- and *Z*-isomers of fatty acid methyl esters are not easily separated on packed columns using polyester phases, such separations are readily attainable on cyanoalkyl polysiloxane phases, such as SP-2340 or Silar 10C. Baseline separations of geometric isomers can be achieved on capillary columns (18) using various stationary phases. The elution of the *E*-isomer before the *Z*-isomer on the polar column, compared to the elution of the *Z*-isomer before the *E*-isomer on the apolar column, holds true for a wide range of polar stationary phases. The latter trend in elution is also observed for dienoic fatty acid methyl esters. On polar cyanopropyl polysiloxane phases the elution order for methyl linoleate and its three geometric isomers is: 9-*E*, 12-*E* < 9-*Z*, 12-*E* < 9-*E*, 12-*Z*, < 9-*Z*, 12-*Z*. As the occurrence of fatty acids containing double-bonds of the *E*-configuration is rare, there is little requirement for such separations. However partially hardened fats used in commerce do contain *E* configuration double bonds.

Geometric isomers of polyenoic fatty acids bearing higher degrees of unsaturation can be separated on capillary columns coated with either polar or apolar stationary phases. Fused-silica capillary columns are to be preferred to packed columns, owing to their relative inactivities, ensuring elution of thermally labile components in an unchanged form.

7.6.5 Less commonly occurring fatty acids

i. Branched-chain fatty acids

It is a general property in GC that increasing alkyl-chain branching reduces t_R compared to a linear saturated counterpart. This is confirmed by the ECL values listed in *Table 2* for the monomethyl iso- and anteiso-branched chain fatty acid methyl esters. It is notable that the elution order is independent of the nature of the stationary phase. Further chain-branching greatly complicates compound identification, to the extent that GC-MS is the only reliable means of compound identification in the absence of suitable authentic standards for co-injection.

ii. Fatty acids containing carbocyclic rings

The methyl esters of cyclopropane and cyclopropene fatty acids elute at *ECL* values approximately one methylene unit higher than methyl esters of the corresponding monoenoic fatty acid from which they are bio-synthetically derived. It is important to be aware of the possibility of decomposition of cyclopropene fatty acids during GC analysis. For this reason, the inert fused silica capillary columns are preferred to packed columns for their analysis.

Where only packed columns are available simple chemical transformations, such as hydrogenation, can be used to produce the more stable cyclopropane structure, so enabling assessment of possible losses during GC analysis.

iii. Hydroxy fatty acids

The 2-hydroxy fatty acids observed in association with sphingoglycolipids have been extensively investigated. Best results are obtained in the GC analysis of hydroxy fatty acids, in terms of resolution and absolute recoveries, by employing fused silica capillary columns coated with apolar stationary phases. Earlier work on packed columns encountered substantial problems of loss of methyl esters of hydroxy fatty acids, through undesirable adsorption effects. Such problems are overcome in both capillary and packed-column analyses by blocking the hydroxyl group through derivatization; for example,

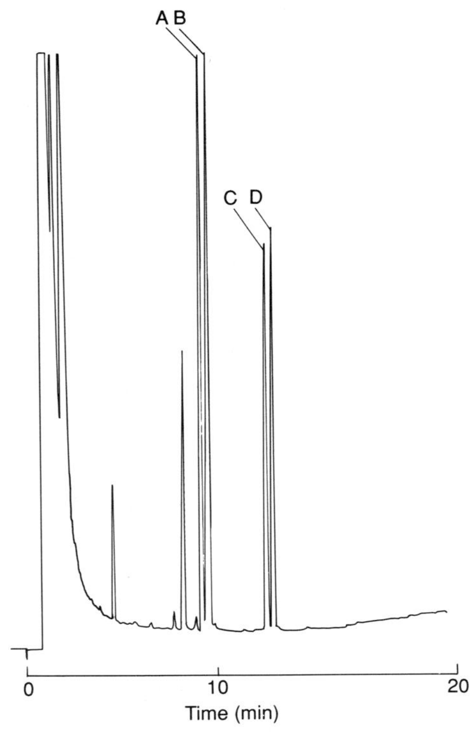

Figure 12. Capillary GC separation of diastereomeric methoxytrifluoromethylphenylacetate derivatives of methyl 2-hydroxypalmitate (**A,B**) and 2-hydroxystearate (**C,D**). The column was a 20 m × 0.32 mm i.d. fused silica capillary coated with OV-1 (0.2 μm film thickness). Splitless injection was employed, followed by temperature prgramming from 180 to 240 °C at 2 °C/min. Helium was the carrier gas . (Redrawn from the original paper with kind permission of *Journal of Chromatography*.)

acetylation or trialkylsilylation (see Chapter 2). Derivatization of the hydroxyl moiety also serves to enhance the structure information content of mass spectra (see Chapter 8).

The asymmetry that exists in hydroxy fatty acids raises a question of the stereochemistry at the chiral centre and enantiomer composition. The most convenient means of determining stereochemistry in these compounds is by the use of chiral resolving agents. Enantiomer determinations of the hydroxyl groups at various sites in the fatty acid alkyl chain can be achieved by formation of diastereomeric derivatives (19, 20) of the hydroxy acid methyl esters, such as: (−)-menthyloxycarbonyl, drimanoyl, chrysanthemoyl, (*R*)-1-phenylethyl urethanes and D-phenylpropionyl. Few descriptions exist of such analyses being performed on capillary columns. However, as the most important factor in adequately resolving diastereomers is column efficiency, analyses are best carried out on capillary columns. Reasonably high-temperature stable stationary phases (for example, dimethyl polysiloxanes), are generally preferred owing to the relatively high molecular weights of these derivatives. *Figure 12* shows near baseline resolution of diastereomeric methoxytrifluoromethylphenyl acetate derivatives of methyl 2-hydroxy-palmitate and 2-hydroxystearate, formed as follows (20):

(a) α-Methoxy-α-trifluoromethylphenylacetic acid chloride (200 μl, 1% solution in pyridine) added to 200 μg of the hydroxylated fatty acid.
(b) Heat reaction mixture for 12 h at 60 °C.
(c) Evaporate pyridine under a stream of dry nitrogen.
(d) Convert the diastereomeric derivatives to methyl esters by treatment with diazomethane according the method given in Chapter 2.
(e) Perform capillary GC analysis under the conditions given in *Figure 12*.

The diastereomers produced above are also resolvable by TLC, using phenylmethylvinylchlorosilane-treated silica gel-coated plates (20).

7.7 Prostaglandins and related compounds

Prostaglandins are 20 carbon unsaturated carboxylic acids, containing a cyclopentane ring, which are derived from arachidonic acid. Their occurrence at low concentration in a range of biological tissues and fluids, and ill-defined physiological roles, has prompted extensive investigation. A wide range of methods have been developed for their determination. Derivatization prior to GC is essential, due to the multiple functionality and lability of prostaglandins. Owing to their low abundances in biological materials, GC-MS with selected ion monitoring (SIM) is generally required to detect prostaglandins. Deuteriated internal standards are commonly used for the purposes of quantification. GC with FID of authentic compounds can obviously be employed in developing suitable GC strategies, or testing derivatization methods.

Numerous methods of derivatization have been employed in the determination of prostaglandins. The most commonly employed derivative is the methyl ester–TMS ether. Formation of the methyl ester is achieved by treatment with ethereal diazomethane (see Chapter 2). Conversion of free hydroxyl or carbonyl groups to TMS or enol–TMS ethers can be achieved by treatment with various reagents, one of the most effective of these is *bis*(trimethylsilyl)acetamide in pyridine. BSTFA and methyltrimethylsilyltrifluoroacetamide have also been successfully employed. A range of other derivatization procedures for prostaglandins and related compounds are to be found in ref. 21. The principal reason for exploring alternatives to the methyl ester–TMS ether derivatives is to enhance detection in GC-MS analysis.

GC (and GC-MS) analysis of prostaglandins is best carried out on capillary columns. The apolar derivatives of prostaglandins display exellent chromatographic behaviour on dimethyl polysiloxane coated columns. In trace analyses good quality columns are preferred, in order to minimize adsorption effects (22).

7.8 Alcohols

Although aliphatic alcohols do occur in the free form (for example, primary and secondary alcohols are common constituents of plant cuticular waxes) they are found commonly as constituents of wax esters, and ether lipids; for example, glyceryl ethers. The chain-length of the long-chain alcohols is often similar to that of the common fatty acids from which they are biosynthetically derived.

Similar GC conditions can be employed as for free fatty acids of comparable chain length. Both packed and capillary column GC can be used. GC behaviour is improved by preparing TMS, trifluoroacetate, acetate, or methoxy derivatives (see Chapter 2); the TMS and acetate derivatives are probably the most widely used. The factors affecting elution orders are largely analogous to those observed for fatty acids, and so dependent upon the nature of the stationary phase employed (see Section 7.6 above). Compound identifications are made by comparison (preferably by co-injection) of retention times with authentic compounds of known structure on at least two different stationary phases, or by mass spectrometry. The mass spectra of long-chain alcohols will be discussed later (Chapter 8).

7.9 Ketones and aldehydes

Long-chain ketones occur as constituents of higher plants, insects and bacteria, while free aldehydes are found in higher plants and insects. An additional source of aldehydes is the hydrolysate of the alk-1-enyl moieties in plasmalogens. Aldehydes are generally saturated or monoenoic compounds of the same carbon number range as the common fatty acids. Shorter chain aldehydes are also produced in structure elucidation studies, when ozonolysis is employed to cleave unsaturated alkyl compounds (23).

Aldehydes can be analysed without derivatization on similar columns to those used for fatty acid methyl esters (see above). As a result of the tendency of free aldehydes to polymerize on storage, it is recommended that more stable derivatives be prepared if analyses are to be delayed following lipid extraction. *N,N*-dimethylhydrazone (23) or dimethyl acetal (24) derivatives of aldehydes are conveniently prepared and display excellent GC properties.

7.10 Hydrocarbons

Hydrocarbons are more easily analysed by GC than other lipid classes, owing to their greater chemical and thermal stability. The only real concern is the possibility of degradation of unsaturated species during sample handling and analysis. Hydrocarbons are amenable to analysis on packed or capillary columns, coated with either apolar or polar stationary phases, depending upon the demands of the particular analytical problem. Capillary columns coated with relatively thin films of immobilized stationary phase offer greatest versatility. Such columns are well-suited to the analysis of low- and high-molecular-weight hydrocarbons alike.

Figure 6 shows the GC analysis of beeswax on a capillary column coated with a thin film of immobilized dimethyl polysiloxane stationary phase. The earlier eluting peaks correspond to predominantly saturated straight-chain alkanes in the C_{23} to C_{33} carbon number range. The chromatographic behaviour is excellent on this column, and clearly considerable scope exists for analysis of higher and lower homologues.

Trends in elution orders for unsaturated and branched compounds, compared to their saturated straight-chain counterparts, parallels that of the methyl esters of fatty acids, as discussed at length above. Briefly, on an apolar stationary phase, elution orders largely reflect differences in the vapour pressures, dictated by the molecular weight of the various components. While on polar phases (such as polyesters or polyethylene glycols), dipole–dipole interactions become important, serving to retain unsaturated components relative to their saturated counterparts. The effects of chain-branching, positional and geometric isomerism, follow the same trends described above for fatty acids.

7.11 Sterols

Sterols are commonly observed, minor constituents of lipid extracts of the majority of biological materials. GC is the most widely used technique for the separation and characterization of sterols. As with prostaglandins, the enormous diversity of sterols structures makes it impossible to cover the subject of their GC analysis comprehensively herein. Consequently, this section will concentrate largely on the analysis of the sterols occurring most commonly in lipid extracts.

WCOT capillary columns are now used almost exclusively in sterol analyses owing to their greater resolving power. This is of particular importance, as in the vast majority of analyses extracts may contain a number of sterols of similar structure. The special importance of GC in sterol analysis has resulted in thorough investigations of their GC properties. As in the case of fatty acids, attempts have been made to quantify the effects of changes in structural features on retention characteristics (retention times and relative elution orders). While the values that have been derived have some use in predicting the structures of unknowns in GC analyses, in practice retention time comparisons or co-injections of sterols of known structure are to be preferred. Most reliable peak identifications are made on the basis of mass spectral data derived from combined GC-MS analyses. The mass spectrometry of sterols will be described in more detail in Chapter 8.

Capillary columns suitable for sterol analyses are generally of 10 to 50 m × 0.22 or 0.32 mm i.d. The relatively low volatility of sterols require columns with a stationary phase film thickness in the range 0.1 to 0.25 μm, in order to achieve acceptable run times. Immobilized apolar dimethyl polysiloxane type stationary phases (see *Table 1*) are generally to be preferred. Where problems of resolution are encountered then switching to a more polar stationary phase, such as a methyl phenyl polysiloxane (for example, OV-17), can effect the required separations.

Figure 13 shows the analysis of a mixture of authentic sterols by capillary GC. These sterols were determined as their TMS ether derivatives (prepared using BSTFA according to the procedure given in Chapter 2). Sterols may be analysed under analogous chromatographic conditions without derivatization. However, we have found that analyses of sterol TMS ethers are less dependent on the condition and age of the column. Moreover, the functional life and the column is significantly increased by performing analyses of TMS ether derivatives. Sterols may also be determined by GC as acetate and trifluoroacetate derivatives; however, for convenience of handling and preparation the TMS ethers are generally to be preferred. The chromatogram shown in *Figure 13* serves to demonstrate several pertinent features of sterol analysis. 5α-Cholestane lacks a hydroxyl moiety and so elutes at shorter retention time than the 3β-hydroxy sterols; which is convenient as it is often used as an internal standard for quantitative analysis and relative retention time determinations.

Cholesterol, campesterol and sitosterol constitute an homologous series, and consequently elute at gradually increasing retention times; reflecting the decrease in vapour pressure brought about by increasing carbon number. Although stigmasterol differs in structure from sitosterol only by the presence of a $\Delta^{22,23}$ double-bond in the side-chain, this produces a marked shift to shorter retention time. The baseline separation of cholesterol from its

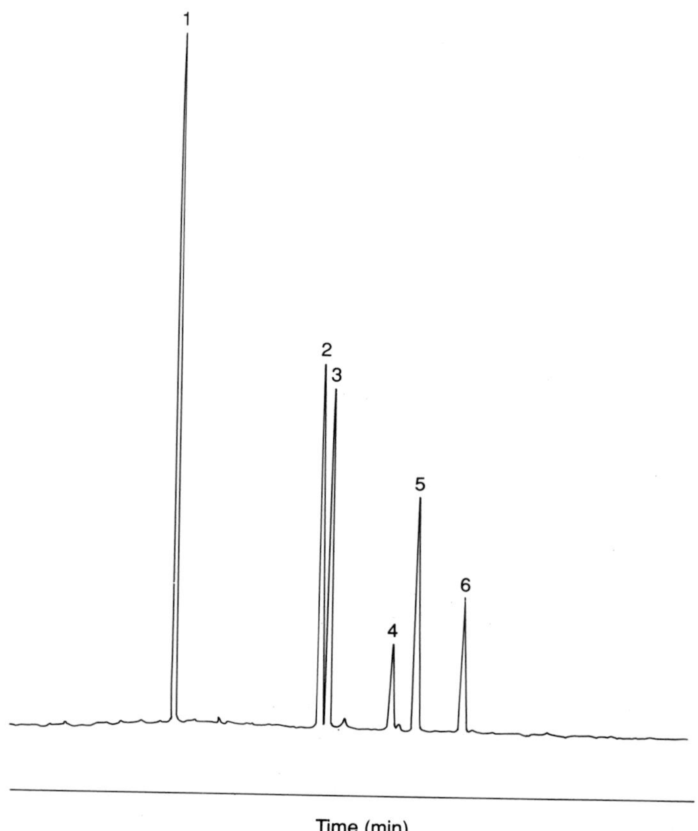

Figure 13. Capillary GC analysis of authentic sterols as their TMS ether derivatives, performed on a 20 m × 0.22 mm fused silica capillary coated with BP-5 (immobilized 5% phenyl 95% methyl polysiloxane, 0.25 μm film thickness). After on-column injection at 50 °C and a 2-min delay, the GC oven-temperature was programmed ballistically (50 °C/min) to 200 °C then maintained at that temperature for 0.5 min before programming to 290 °C at 5 °C/min. Helium was the carrier gas at a column head pressure of 15 p.s.i. Peak identities are: 1 = 5α-cholestane; 2 = cholesterol; 3 = 5α-cholestanol; 4 = campesterol; 5 = stigmasterol; 6 = sitosterol.

reduced counterpart, 5α-cholestanol, is also shown in *Figure 13*. Although not shown here, the separation of 5*ß*-cholestanol (coprostanol) from cholesterol is easier to achieve. Under the GC conditions shown in *Figure 13*, 5ß-cholestanol would elute approximately one minute earlier than cholesterol. The differing GC behaviour of the 5α and 5*ß* epimers results from the marked conformational differences arising from the differing stereochemistries at the A/B-ring junction.

Owing to the large number of known sterols, full discussion of the variations in GC behaviour with structure will not be presented in this chapter. More detailed discussions of this nature, and elution orders of a wide range of sterol structures, on several GC stationary phases, can be found in refs 25 to 29.

7.12 Steroid hormones

Steroid hormones are trace constituents of lipid extracts of animal tissues and fluids, readily analysed by GC under similar conditions to those described above for sterols. GC provides the most convenient means of steroid hormone profiling for use in the clinical setting, particularly for urinalysis, where concentrations are in the p.p.m. range. At this level, steroid hormones are readily detected by FID. At lower concentrations, GC-MS selected ion monitoring (see later in this volume, Chapter 8, Section 4.2.2) is required to detect trace components. Elaborate methods have been developed for the extraction and isolation of steroids for GC or GC-MS. However, full discussion of these is not possible in this volume. For useful reviews of the necessary methodologies the reader is directed to refs 29 and 30.

Where reasonable concentrations of steroid are involved GC can provide both qualitative and quantitative information in a single determination. Although the first analyses of steroid hormones were performed on packed columns, the high complexity of the mixture obtained from biological fluids necessitates the use of capillary columns, in routine application. The polyfunctional nature of steroids requires derivatization to ensure quantitative recoveries of the various molecular species. The most commonly employed approach is a two-step derivatization, involving methoxylamine and trimethylsilyl ether (MO-TMS) formation (31). The methoxylamine derivatization is required to block ketone groups, which occur commonly in steroids and are liable to yield a mixture of products as a result of incomplete enol–silyl ether formation. Procedures for derivatization are discussed in Chapter 2. While methoxylamine formation is straightforward, a variety of methods are available for producing TMS ethers. Trimethylsilyl imidazole (TMSI) is the most powerful TMS donor available and its use ensures effective trimethylsilylation of the sterically hindered hydroxy moieties commonly encountered in steroids. Unlike BSTFA, which can be injected directly into the GC, the residue remaining after derivatization with TMSI is harmful to GC columns. An effective method of purification involves passing the derivatization mixture through a Pasteur pipette containing Lipidex 5000 gel prior to GC (31).

The relatively involatile nature of derivatized steroid hormones requires the use of thermally-stable polysiloxane stationary phases. Columns 10 to 25 m × 0.22 to 0.32 mm i.d., coated with immobilized apolar dimethyl polysiloxane stationary phase (see *Table 1*), are well-suited to steroid hormone analysis. The use of other stationary phases of different selectivity is rarely required. The low levels of steroid generally present in natural extracts requires the use of splitless, or preferably on-column injection, to ensure quantitative transfer of all sample components on to the analytical column. Temperature-programming from 50 °C to 300 °C ensures elution of the majority of steroids likely to be encountered. The elution of the compounds of interest must be checked by the analysis of authentic compounds of known structure. Analysis of authentic compounds will reveal the efficiency of derivatization and work-up procedures. In addition, the analysis of authentic compounds enables peak identifications through retention times comparisons with unknowns. Mass spectrometry, particularly GC-MS, is used extensively in the identification of steroids as discussed later (Chapter 8).

It should be noted that high efficiency capillary columns will resolve the *syn*- and *anti*-isomers occurring at the MO moiety in MO-TMS derivatives of ketosteroids. While the production of this isomeric mixture will tend to reduce the sensitivity of the GC analysis, the appearance of pairs of closely-eluted peaks can be used diagnostically in confirming the presence of ketosteroids.

8. Trouble-shooting

Most GC operating manuals offer basic trouble-shooting hints. These hints will address problems as simple as the lack of a signal from the FID, as a result of the failure to ignite the flame, to defects in electrical circuitry, which required specialist remedial attention. Comprehensive coverage of all such problems and possible solutions would be very lengthy and so this section will concentrate on those which can be solved by the analyst.

Analysts often mistakenly attribute poor chromatographic performance to imperfections of the GC column. However, other factors are often responsible for poor GC peak shapes, rising baselines, etc. One of the most common causes of such effects arises from incorrect plumbing of the GC column into the injector or detector assembly. This is especially the case in capillary GC, where incorrectly positioned column ends can cause increased dead volume, and thus inefficient sample transfer effects. Different GC instruments and injector types often demand somewhat different modes of column installation. The advent of flexible fused silica capillary columns, and improved instrument designs, have greatly simplified column installation, but no system is foolproof, and the inexperienced operator may still encounter difficulties. Therefore, when poor chromatography is observed immediately following column installation, incorrect plumbing of the column into the injector and detector could be considered. The operating manual should give clear instructions on how to do this. Strict attention should be paid to ensuring that the correct lengths of column project into the injector and detector assemblies. Inadequate tightening of column end-fittings can mean that ferrules do not make gas-tight seals, resulting in lack of column head pressure, and thus longer than expected retention times.

Once confident that the GC column is correctly installed, column performance is best assessed by analysis of standard mixtures of compounds. In the case of new columns the performance of the chromatographic system can be judged by analysing the manufacturer's test mixture. As all manufacturers test their columns before shipment, poor GC performance will probably relate to the injector or detector.

The test mixture can also be used in the case of older columns. In these instances, poor chromatographic performance might well relate to deterioration of the stationary phase. More rigorous assessment of column performance can be based on the use of more sophisticated test mixtures of the type proposed by Grob (33). Use of such test mixtures will provide a precise indication of the nature active sites within the column. While to the purist this information is very interesting, from the purely practical standpoint the most pertinent question is whether or not a given column is capable of performing the required separation? This is best resolved by testing the column with an authentic mixture of the compounds under investigation. During routine analyses it is good practice to analyse periodically a standard mixture as a means of monitoring column performance.

Deterioration of chromatographic performance, as revealed by gradual loss of resolution or the development of peak tailing, during a series of analyses, can arise in several ways. One explanation is the presence of impurities in the carrier gas supply which degrade the stationary phase at the high oven-temperature required for the analysis of lipids. This problem can be minimized by employing the grades of gas recommended by suppliers. Resorting to cheaper, low grade gas is false economy, as relatively rapid deterioration of column performance will ensue. Even with the recommended grades of carrier gas require oxygen and water strippers to be installed in the supply line between the cylinder and GC.

The samples themselves are the commonest sources of impurities, which can diminish chromatographic performance. Involatile impurities, resulting from derivatizing reagents, can also have a detrimental effect on chromatographic behaviour, and so care should be taken to remove these. Biological extracts are a common source of involatile impurities which will accumulate on the glass liner in split or splitless injectors. Periodic cleaning of these liners will minimize loss of sample components through undesirable adsorption phenomena.

Owing to the mode of operation of on-column injection, involatile sample impurities will inevitably be deposited directly on to the GC column. The accumulation of these impurities over a period of time will lead to a loss of chromatographic performance. Simply removing a coil of the column will generally restore optimum performance. An alternative, already alluded to above, is to employ a pre-column of deactivated, uncoated, fused silica connected to the analytical column via an suitable zero dead-volume coupling (5, 6). These relatively inexpensive pre-columns are simply replaced once they become contaminated.

Changes in GC peak shape provide a good indication of the nature of problems arising in the chromatographic system. Loss of chromatographic efficiency is manifested by peak broadening, tailing, splitting, and fronting. Peak broadening and tailing most commonly occur due to the presence of absorbed impurities and/or active sites on the column, arising through degradation of the stationary phase. Peak splitting is most commonly observed in on-column injection, as a result of the injection of too large a volume of sample solution. This phenomenon can be overcome by employing long retention gaps (4) to re-focus sample components, or by reducing the volume of sample solution injected. 'Fronting', seen as the near vertical trailing edge of sample peaks, can result through the incompatibility of a solute and stationary phase. This incompatibility can be rectified through derivatization, or by changing to a stationary phase of different selectivity. However, a more common cause of 'fronting' is sample overload, which is rectified by dilution of the sample solution or by reducing the volume injected.

The phenomenon of excessive column bleed is revealed as excessive rising of the baseline in the latter stages of a temperature programmed GC run. Column bleed can arise from degradation of the stationary phase, or of involatile sample impurities, or contaminants present in detector or injector gas supplies. The possibility of the column bleed originating from the GC column can be confirmed one way or the other by replacing the column with a new, uncoated length of fused silica tubing. If column bleed is still evident, then the problem is one of contamination of the gas supply lines. Whether or not it is the detector or injector supply that is contaminated, can be deduced by removing the column, blanking off the detector inlet, and running the temperature programme. An absence of the column bleed at this stage suggests contamination of the injector. The occurrence of the column bleed suggests contamination of the detector supply lines. The contaminated assembly (either injector or detector) must be dismantled and cleaned according to the manufacturer's instructions.

If the column bleed is thought to arise from contaminants of the column itself, then an extended period (overnight) of reconditioning, with the column disconnected from the detector, will often eliminate the involatile impurities. If this approach is unsuccessful, then columns coated with immobilized stationary phases can be washed to remove impurities. As washing procedures are somewhat complicated, and to a certain extent dependent on the nature of the stationary phase advice on how to proceed should be sought from the manufacturers.

References

1. Jennings, W. (1987). *Analytical gas chromatography*. Academic Press, Orlando, Florida.
2. Christie, W. W. (1989). *Gas chromatography and lipids*. The Oily Press, Ayr, Scotland.
3. Grob, K. and Grob, K. Jr. (1978). *J. Chromat.*, **151**, 311.
4. Grob, K. (1984). *J. Chromat.*, **324**, 251.
5. Roeraade, J., Blomberg, S., and Flodberg, G. (1984). *J. Chromat.*, **301**, 189.
6. Rohwer, E R. and Pretorius, V. (1986). *J. High Res. Chromat. Chromatogr. Commun.*, **9**, 295.
7. Jennings, W. (1980). *Gas chromatography with glass capillary columns* (2nd edn). Academic Press, New York.
8. Poy, F., Visani, S., and Terrosi, F. (1981). *J. Chromat.*, **217**, 81.
9. Grob, K. (1986). *Making and manipulating glass capillary columns*. Huethig, Heidelberg, Germany.
10. Kuksis, A., Myher, J. J., Geher, K., Hoffman, A. G. D., Breckenridge, W. C., Jones, G. J. L., and Little, J. A. (1978). *J. Chromat.*, **146**, 393.
11. Myher, J. J. and Kuksis, A. (1984). *J. Biochem. Biophys. Methods*, **10**, 13.
12. Kuksis, A., Myher, J. J., and Sandra, P. (1990). *J. Chromat.*, **500**, 427.
13. Mares, P. (1988). *Prog. Lipid. Res.*, **27**, 107.
14. Ackman, R. G. (1986). In *Analysis of oils and fats* (ed. R. J. Hamilton and J. B. Rossell), p. 137. Elsevier Applied Science Publishers, London and New York.
15. Christie, W. W. (1988). *J. Chromat.*, **447**, 305.
16. Woodford, F. P. and van Gent, C. M. (1960). *J. Lipid Res.*, **1**, 188.
17. Miwa, T. K., Mikolajczak, K. L., Earle, F. R., and Wolf, I. A. (1960). *Anal. Chem.*, **32**, 1739.
18. Jager, H., Kloer, H. U., Ditschuneit, H., and Frank, H. (1981). In *Applications of glass capillary gas chromatography* (ed. W. G. Jennings), p. 289. Marcel Decker, New York and Basel.
19. Smith, C. R. (1976). *J. Chromat. Sci.*, **14**, 36.
20. Beneytout, J .L., Tixier, M., and Rigaud, M. (1986). *J. Chromat.*, **351**, 363.
21. Knapp, D. R. (1979). *Handbook of analytical derivatization reactions*. Wiley/Interscience Publication. John Wiley, New York and Chichester.
22. Gleispach, H., Moser, R., and Leis, H. J. (1985). *J. Chromat.*, **342**, 245.
23. Attygala, A. B., Zlatkis, A., and Middleditch, B. S. (1989). *J. Chromat.*, **472**, 284.
24. Beaummelle, B. D. and Vial, H. J. (1986). *J. Chromat.*, **356**, 187.
25. Patterson, G. W. (1971). *Anal. Chem.*, **43**, 1165.
26. Patterson, G. W. (1971). *Lipids*, **6**, 120.
27. Itoh, T., Tami, H., Fukushima, K., Tamura, T., and Matsumoto, T. (1982). *J. Chromat.*, **234**, 65.
28. Vanluchene, E. and Sandra, P. (1981). In *Applications of glass capillary gas chromatography* (ed. W. G. Jennings), p. 395. Marcel Decker, New York and Basel.
29. Goad, L. J. (1991). In *Methods in plant biochemistry*, Vol. 7 (ed. B. V. Charlwood and D. V. Barnthorpe) pp. 369–434. Academic Press, New York.

30. Shackleton, C. H. L., Merdinck, J., and Lawson, A. M. (1990). In *Mass spectrometry of biological materials* (ed. C. N. McEwen, and B. S. Larsen), p. 297. Marcel Decker, New York and Basel.
31. Thenot, J. P. T. and Horning, E. C. (1972). *Anal. Lett.*, **5**, 21.
32. Axelson, M. and Sjovall, J. (1974). *J. Steroid Biochem.*, **5**, 733.
33. Grob, K., Grob, G., and Grob, K. (1978). *J. Chromat.*, **156**, 1.
34. Evershed, R. P. and Goad, L. J. (1987). *Biomed. Environ. Mass Spectrom.*, **14** 131.

5

High-performance liquid chromatography

PETER A. SEWELL

1. Introduction

High-performance liquid chromatography (HPLC) is, in many areas, the main technique used for analysis. However, in the area of lipid analysis its use is still relatively uncommon, and thin-layer chromatography (TLC) and gas chromatography (GC) are the more commonly used techniques. Both of these techniques have advantages and disadvantages compared to HPLC: TLC is a very simple and cheap technique but it is a relatively inefficient process and is difficult to quantitate; GLC (gas–liquid chromatography) is easier to use and the equipment is cheaper than HPLC, but samples must be capable of being volatilized without decomposition. Thus, whilst it is unlikely that HPLC will ever displace GC as the preferred technique for the analysis of fatty acids the use of HPLC will increase in other areas of lipid analysis.

HPLC is the analogue of gas chromatography where the stationary phase, be it a liquid, a solid surface, an ion-exchange resin or a porous polymer, is held in a metal column and the liquid mobile phase is forced through under pressure. The analyte is injected into the mobile-phase stream and the separated components are detected as they elute from the column. The 'chromatogram' (a trace of the variation in component concentration against time) may be used to obtain both qualitative and quantitative information. For a discussion on the technique of HPLC the reader is referred to any of the several texts available on the subject (1–3). Only those topics which are essential to an understanding of the methodology presented later in the chapter will be discussed here.

2. Theory of HPLC

As a sample component passes through the chromatographic column it spends part of its time in the mobile phase (t_m) and part in the stationary phase (t_s). The total time (t_R) that a component takes to pass through the column is thus given by

$$t_R = t_m + t_s. \quad [1]$$

Since a substance which spends all its time in the mobile phase is a substance which is not retained at all by the stationary phase, t_m is known as the retention time of an unretained peak. It is assumed that components only move through the column when in the mobile phase. Since all components will spend the same time in the mobile phase, as determined by its linear flow-rate, components will not be separated unless there are differences in the time spent in the stationary phase. This is determined by the distribution of the component between the stationary phase and the mobile phase such that the distribution coefficient (K) is given by

$$K = C_s/C_m \quad [2]$$

where C_s and C_m are the component concentrations in the stationary phase and the mobile phases respectively.

The retention time and the distribution coefficient are related by

$$t_R = \frac{L}{v}(1 + KV_s/V_m) \quad [3]$$

where L is the column length, v is the linear velocity of the mobile phase and V_s and V_m are the volumes of the stationary and mobile phases respectively. The term KV_s/V_m is the amount of the component in the two phases and is known as the *capacity factor* (k'), so that

$$t_R = \frac{L}{V}(1 + k'). \quad [4]$$

From this it can be seen that the capacity factor is a retention parameter and retentions are often presented as k' values. It can be further shown that k' is related to t_R and t_m by

$$k' = \frac{(t_R - t_m)}{t_m}. \quad [5]$$

This gives a simple and practical way to describe retention in chromatography.

3. Mechanisms of separation

One of the advantages of liquid chromatography over gas chromatography is that several different mechanisms (or modes of chromatography) may be utilized to produce the separation. Differences in functional group composition within the molecules (for example, polar or polarizable groups, ionic centres, etc.) can thus be exploited by the proper choice of the chromatographic system. Such have been the developments in liquid chromatography in the last ten years that a simple classification of mechanisms oversimplifies the real situation, but nevertheless it does help in an understanding of the separation process. Thus, the main modes of chromatography of interest to the lipid analyst are adsorption, partition (both normal-phase and reversed-phase), ion-exchange and ion-pairing, and exclusion chromatography. Chiral-phase and complexation chromatography are additional modes which are being utilized to a greater extent.

3.1 Adsorption chromatography

Low-pressure column chromatography and TLC, using silica gel as adsorbent, have long been used by the lipid analyst, and now silica is widely used for the HPLC of lipids.

With a polar adsorbent such as silica or alumina the mobile phase used will be of low polarity (for example, hexane or chloroform), but with a non-polar adsorbent such as polymer beads or carbon the mobile phase will be polar (for example, water or methanol). Separations by adsorption chromatography are highly sensitive to the presence of different functional groups on the molecule but are less sensitive to differences in molecular weight. Complex samples can therefore be separated into classes of compounds having the same functional groups (such as free fatty acids, monoglycerides, diglycerides, and triglycerides), whereas molecules that differ only in alkyl chain length would be poorly separated.

The explanation of this selectivity lies in the mechanism of the adsorption process. Solvent molecules in the mobile phase are in competition with the solute (sample) molecules for adsorption sites on the adsorbent. With a polar adsorbent, non-polar molecules (for example, hydrocarbons) will have little affinity for the adsorbent and will not be able to displace an adsorbed solvent molecule and will not be retained. Molecules with polar functional groups will have a high affinity for the adsorbent because of dipole–dipole interactions and will be strongly retained, whereas molecules which are polarizable (for example, aromatic and high-molecular-weight compounds) will exhibit dipole-induced dipole interactions, and their retention will depend on the strength of these interactions. For non-polar adsorbents the dominant interactions are dispersion forces, and polar and polarizable molecules will be less strongly retained than non-polar molecules.

The activity of silica is due to the presence of silanol groups (Si–OH) on the surface and these have a strong affinity for water. For reproducible retention the amount of water in the mobile phase must therefore be carefully controlled. A simple way of doing this is to pre-saturate the mobile phase with water before use, but even so, since the mobile phase may absorb water from the atmosphere during use, retention changes may still occur. Another approach is to adjust the water content of the mobile phase so that the retention of a standard compound is always constant.

Silica can be produced in several different ways and the properties of the silica gel may vary considerably. Factors which will affect the chromatographic properties of the silica are surface area, pore size, and pore volume. The presence of metal ions on the silica may also have a profound effect on its properties. Since the silica surface is acidic, alkaline mobile phases ($pH > 7.5$) will slowly dissolve the silica (this applies also to the so-called bonded phases based on silica, to be described later). Although the batch-to-batch reproducibility of silicas is now quite good, differences will always be found between silicas produced by different suppliers, and this too applies to the bonded phases.

One particular form of adsorption chromatography widely used by lipid analysts and also used in HPLC is that of *argentation chromatography*, which is an example of *complexation chromatography*. Silica gel is impregnated with silver nitrate and the silver ions interact reversibly with double bonds to form polar complexes. *Cis* isomers interact more strongly than *trans* and retention increases with the number of double-bonds present in the molecule. Thus lipids can be separated by this technique according to the number and configurations of the double bonds in the acyl or alkyl moieties.

Alumina is not recommended as an adsorbent for lipid analysis as it may react with some solvents and lipid materials.

3.2 Partition chromatography

'True' partition is rarely carried out in modern liquid chromatography because of the problems of keeping the liquid coating 'stationary'. Instead, the liquid coating is chemically bonded to the surface of the stationary phase support material and this 'bonded liquid phase' has somewhat different properties to the 'bulk liquid phase', and this mode of chromatography is often referred to as 'bonded-phase partition chromatography'. Two forms of partition chromatography can be recognized according to the relative polarities of the stationary and mobile phases: *normal-phase* partition chromatography where the stationary phase is more polar than the mobile phase, and *reversed-phase* chromatography when the stationary phase is less polar than the mobile phase.

The majority of lipid separations carried out using the reversed-phase mode are for the separation of molecular species of lipids within a single lipid class, where the separation is dependent on the fatty acyl or alkyl chain-length and on the configuration at any double-bonds.

The most widely used stationary phase for this purpose is the octadecylsilyl–silica (ODS or C_{18}), which is prepared from silica by reacting the surface silanol groups with an organochlorosilane to give

$$\text{(surface Si)—O—Si—C}_{18}$$

The hydrocarbon chain may be modified to give other terminal groups (such as R–CN, R–NH_2) which change the nature of the stationary phase and allows these phases to be used in other forms of chromatography: for example, the nitrile (R–CN) phase can be used as a substitute for silica in adsorption chromatography, where it will give similar selectivities but shorter retention times than silica itself. It also has the advantage that its retention properties are not affected by water, and in common with all bonded phases it equilibrates very rapidly with the mobile phase and gives less peak-tailing; the amino (R–NH_2) phase may be used in both the normal- and reversed-phase mode as well as a weak ion-exchanger.

Bonded phases based on silica suffer from the same pH restrictions as silica itself, i.e. they should only be used in the range $2 < \text{pH} < 8$ for long life. Newer bonded phases based on a polystyrene-divinylbenzene matrix can tolerate a wider range of pH (1 to 13). The properties of bonded phases depend on the nature of the silica itself (particle diameter, shape, surface area, pore size and volume), the bonding method used, the nature of the bonded groups, the length of the carbon chain, the amount of carbon on the silica surface and whether the residual silanol groups on the silica surface have been treated to remove their polarity (called 'endcapping'). Because of these variations it is naïve to assume that all ODS columns, for example, will give the same retention independently of the manufacturer, and in following the published methods it is advisable to use the same stationary phase from the same manufacturer. In developing your own methodology it may be that a column from one manufacturer which does not give a good separation may with advantage be replaced by one from another manufacturer that will.

3.3 Ion-exchange and ion-pair chromatography

Ion-exchange packings are either silica- or polymer-based and both have been used in lipid analysis. The cellulose-based anion-exchangers diethylaminoethyl (DEAE)- and triethylaminoethyl (TEAE)-cellulose have been used to separate polar complex lipids using low-pressure column chromatography, and bonded phases have been used for phospholipid analysis. Ion-exchange involves the competitive substitution of ions on sites of opposite charge on the stationary phase and the technique is therefore only applicable to the ionic forms of molecules. Retention can be adjusted by the addition of different counter-ions to the mobile phase and by the use of pH control.

Ion-pair chromatography can either be considered as a form of partition chromatography or as a form of dynamic ion-exchange chromatography. The essential feature is that the stationary phase is non-polar (for example, ODS) and the mobile phase contains a large ionic counter-ion (for example, tetrabutyl ammonium [Bu_4N^+]). According to the partition mechanism, the counter-ion will form a neutral ion-pair with an ionic species and will then chromatograph on the non-polar stationary phase in the usual way. In the ion-exchange mechanism the counter-ion is sorbed on to the stationary phase (by interactions between the butyl and the ODS groups), forming an ion-exchange site on which separation may occur. The technique may have applications in the separation of phospholipids.

3.4 Exclusion chromatography

This technique, also called 'gel-permeation' or 'gel-filtration', depends on the exclusion of molecules from the pore structure of the stationary phase, which may be a porous polymer or a porous silica. Molecules above a certain size cannot permeate the pore structure and therefore are eluted first, whilst smaller molecules penetrate the pores to an amount depending on their size and are selectively retained. A difference in molecular weight of 10% is required before separation is observed, and therefore the method is usually used as a preliminary separation procedure. Elution times are short, the peaks symmetrical, and gradients are rarely used, but the peak capacity is low.

The technique has been used for the separation of fatty acids, mono-, di-, and tri-glyceride mixtures and in the separation of polymerized lipids and of lipoprotein complexes and apolipoproteins.

3.5 Chiral-phase chromatography

The introduction of chiral stationary phases which have a chiral molecule bonded chemically to a silica-based matrix, has made possible the separation of enantiomers by HPLC. The mechanism of retention is complex and is not yet fully understood.

Enantiomer separations of long-chain monoacyl-, monoalkyl-, and diacyl glycerols on *N*-(*S*)-2-(4-chlorophenyl)isovaleroyl-D-phenylglycine bonded silica and of diacyl- and dialkylglycerol enantiomers on *N*-(*R*)-1-(α-napthyl)ethylaminocarbonyl-(*S*)-valine chemically bonded to γ-aminopropyl silanized silica have been reported.

4. Equipment

No special equipment is needed for the HPLC of lipids and for a general discussion on instrumentation the reader is referred to ref. 2. However, it is pertinent to consider certain aspects of the practice of lipid analysis by HPLC.

4.1 Mobile phases

The actual choice of the mobile phase will be determined by the nature of the sample and the type of chromatography being employed, but for lipid analysis in general organic, rather than aqueous, solvents are usually used. The choice of mobile phase, and its effect on the optimization of the separation, is fully covered in the literature (3). Solvents such as hexane, chloroform, ethers, methanol, isopropanol, and toluene have been used extensively in lipid analysis. More specific solvents are used for some purposes; for example, acetone is a good solvent for phospholipids but not for glycolipids, and acetonitrile is used in the HPLC of lipids where a separation according to the chain-length of the fatty acyl groups is required.

An important property of the mobile phase is its UV cut-off (the wavelength at which the transmission is less than 10%). Most lipids absorb weakly in the UV range 200 to 210, but so do many of the solvents used in lipid analysis (especially chloroform). Even solvents which are transparent in this range (for example, methanol, water, acetonitrile, isopropanol, and hexane), often show considerable absorption, due to traces of impurities or additives such as anti-oxidants and plasticizers (plastic bottles should not be used for storage). Fractional distillation can be used for cleaning large amounts of solvents and adsorption column chromatography is suitable for smaller amounts, and can remove stabilizers, UV-absorbing compounds, peroxides, and water. For the purification of solvents which are less polar than ethyl acetate a column filled with a mix of silica and alumina is most effective. Up to 600 cm^3 of solvent can be cleaned on 100 g of column filling. 'HPLC grade' solvents are available commercially and are usually low in UV-absorbing compounds.

Isocratic elution (i.e. using a mobile phase of constant composition) can be used for simple separations, but for the analysis of samples containing a range of lipid types gradient elution (with a changing mobile phase composition) has to be used. However, some detectors used in lipid analysis (for example, the refractive index detector) cannot normally be used with gradient elution.

4.2 Columns

Most of the earlier work on the HPLC of lipids was done using silica columns, but more and more often the bonded-type phases are being used. Silica–C_{18} is the most commonly used stationary phase of this type and examples of its use will be given later.

Many lipid samples can be described as being 'dirty'; that is, they have a wide range of components many of which are not of interest and which may be irreversibly sorbed on to the column. These compounds could be removed in a preliminary clean-up step, but one of the advantages of HPLC is the possibility of injecting such samples directly on to the column, thus avoiding possible losses and contamination during the work-up procedure. The length of life of the column may be extended considerably when dealing with this type of sample by the introduction of a short 'guard' column prior to the analytical column. The guard column may be a short version of the main column packed with the same stationary phase. Since it is desirable that these guard columns can be changed frequently and easily they are best filled with a 'pellicular' type packing (with a diameter of about 40 μm) so that they can be repacked when spent by a simple tap-and-fill procedure.

The life of a column will also be extended if high pressures are avoided and if mobile phases containing water or inorganic ions are flushed out at the end of each period of use.

4.3 Detectors

The HPLC analysis of lipids is still hampered by the lack of a detector which satisfies the following major requirements:

(a) to respond to all lipid molecules and to be equally sensitive to all molecular types;

(b) not to be affected by changes in mobile phase composition (as in gradient elution) or by temperature;

(c) to be able to monitor small amounts of compounds (as in trace analysis);

(d) to be cheap enough to be available in any lipid laboratory.

Non-selective detectors react to the bulk property of the solution passing through, and tend to be less sensitive than the selective detectors which measure a particular property of the sample molecule, such as its UV absorbance or its fluorescence. Both types of detectors have been used in lipid analysis. For a discussion of detectors used in HPLC the reader is referred to refs. 2 and 3, but for lipid analysis the following will act as a guide.

4.3.1 Ultraviolet spectrophotometric detectors

These are the most common detectors used for HPLC. They can give great selectivity and sometimes sensitivity for the analysis of specific compounds, and they are relatively insensitive to changes in temperature, flow-rate, and mobile phase composition. The highest sensitivity is towards compounds with conjugated double bonds and aromatic ring systems, neither of which are common, however, in naturally occurring lipids, although some fatty acids found in seed oils contain conjugated double bonds and they are also common in lipids which have undergone hydroperoxidation. Carotenes and tocopherols can also be analysed with UV detection. Chemical derivatization to form compounds which do absorb in the UV region is a way of extending the use of this detector to other lipid molecules. For example, fatty acids can be converted to aromatic esters, glycolipids can be benzoylated, and diacylglycerols derived from phospholipids can be esterified with aromatic acid derivatives. Quantitative analysis with UV detection is easily carried out as long as the mobile phase composition and flow-rate are carefully controlled.

4.3.2 Fluorescence detectors

Fluorescing compounds can be detected with a high degree of sensitivity and selectivity. The sensitivity can be as much as 1000 times greater than with UV detection but the presence of oxygen and other compounds which quench fluorescence will reduce the sensitivity. Selectivity may be varied by changing both the excitation and the emission wavelengths.

Few lipids exhibit natural fluorescence but fluorescent derivatives can be prepared. Fatty acids have been derivatized using 9-anthryldiazomethane (ADAM), 9-aminophenanthranene (9-AP), and with 3-bromomethyl-6,7-dimethoxyl-1-methyl-2(1H)-quinoxalinone (Br-DMEQ).

4.3.3 Refractive index detectors

The refractive index (RI) detector is universal, i.e. it will respond to any molecule that has a different refractive index to that of the mobile phase. Its response is proportional to this difference so that, in theory at least, its sensitivity may be enhanced by changing the mobile phase. Its sensitivity in a favourable case (5×10^{-7} g) is some 1000 times less sensitive than the UV detector (5×10^{-10} g) for a molecule with a high absorption coefficient. The RI detector is very sensitive to uncompensated changes in ambient temperature, and it should be thermostatted to ± 0.01 °C for optimum sensitivity and stability. The detector is also sensitive to changes in mobile-phase composition and it is not really practical to use with gradient elution.

In spite of its drawbacks, refractive index detection has been widely used in lipid analysis, for the analysis of triglycerides in a range of vegetable and animal oils and fats, using non-aqueous reversed-phase chromatography (NARP), for fatty acid methyl esters, again using NARP, and in the size exclusion separation of fatty acids, mono-, di-, and tri-glycerides.

4.3.4 Infra-red spectrophotometric detectors

All organic molecules absorb infra-red light at some wavelength or other, and IR detectors have been used in the analysis of non-polar lipids containing the carbonyl function which absorbs at 5.75 μm. The detector is insensitive to temperature and flow-rate changes, but since most solvents absorb to some extent in the spectral regions of interest, background levels tend to be high. The sensitivity with a suitable molecule is similar to that of the refractive index detector, but IR detection has the advantage that gradient elution can be used. However, due to the changing mobile phase absorption with mobile-phase composition, solvent-pair matching is desirable to reduce the baseline drift (4).

4.3.5 Mass detectors

The introduction of a 'mass detector' or 'light-scattering detector' by Applied Chromatography Systems (ACS) Ltd., Macclesfield, UK, is beginning to have a profound affect on lipid analysis. The principle of the detector is that solvent from the column is evaporated in a stream of air or oxygen, but the solute molecules are not evaporated but are nebulized and pass as minute droplets through a light beam which is reflected and refracted. The amount of scattered light measured is related to the amount of material in the eluent. The sensitivity again is similar to the refractive index detector, but it is unaffected by temperature changes and by small changes in mobile-phase flow-rate. Furthermore, it can be used with gradient elution without any of the problems associated with the RI detector. The disadvantages are that the detector needs a supply of dry, filtered compressed air and the outlet containing the evaporated solvent must be vented outside the laboratory, either directly or via a fume chamber. The choice of mobile phase is limited to those which will evaporate in the heated chamber. If it is required to collect solute samples then a stream splitter must be introduced before the detector, since the sample cannot easily be collected. In order to obtain quantitative data the detector needs very careful calibration and strict adherence to the calibration conditions.

The separation and detection of a wide range of lipids (cholesteryl esters, triglycerides, fatty acids, and phospholipids) using gradient elution with the mass detector has been reported.

4.3.6 Flame ionization detectors

The demise of the Pye Unicam LC2 transport-flame ionization detector, which was introduced in the early 1970s as a detector for HPLC, was probably only mourned by lipid analysts, because its place was rapidly filled by other more reliable and easier-to-use detectors. In the LC2 detector the eluent was transported from the end of the column on a moving wire. After evaporation of the solvent in an oven the solute was burned to CO_2 and H_2O, and the CO_2 was converted to CH_4 over a nickel catalyst and detected in a standard FID detector. The choice of mobile phase was limited and quantitative results were poor since not all the eluted solute was detected. The Tracor 945 detector (Tracor Instruments, Austin, Texas, USA) introduced in 1983, also uses a transport system which utilizes a fibrous quartz belt at the periphery of a rotating disc which is enclosed and also contains the solvent applicator and a dual-flame ionization detector. All the eluted solute is detected, so that sensitivity is greater than that of the LC2 and is about the same as the mass detector (1 μg). Response differences for different compound types will be observed, but in most cases for lipids absolute, theoretical FID response factors can be used. Like the mass detector, the choice of solvents is limited to the more volatile ones and a stream-splitter would have to be introduced if it is desired to collect the solute.

The use of the Tracor 945 for lipid analysis has been the subject of two recent reviews (5, 6).

4.3.7 Radioactivity detectors

The use of lipids labelled with the beta-emitters ^{14}C, ^{3}H, and ^{32}P has been an important development in the study of lipid metabolism, giving high sensitivity and ease and accuracy of quantitation. Much of the earlier work was done using off-line detection with a scintillation counter. More recently, the use of on-line liquid scintillation counters has been applied to phospholipid analysis using ^{32}P-labelled compounds.

4.3.8 Mass spectrometry

The use of GC-MS is well-established in the analysis of lipids, especially for fatty acids and glycerides. The problems of interfacing HPLC with MS have now been largely solved and commercial systems are available, at a price. Since the mass spectrometer provides structural detail as well as serving as a detector it is particularly useful for the identification of molecular species within lipid classes. Reversed-phase HPLC coupled to a chemical-ionization mass spectrometer has been used to study sphingoid bases from sphingolipids after acid hydrolysis, and recent improvements in ionization techniques have allowed the spectral analysis of lipids without prior hydrolysis and/or derivatization.

5. Fractionation into lipid classes

With the introduction of HPLC into lipid analysis it has become possible to analyse individual simple and complex lipids in a single step. However, not all lipid mixtures are amenable to such analysis, or it may be desirable to use a different approach to the analysis. For example, since the HPLC analysis of lipids still suffers from problems with their detection, it may be desirable to use gas chromatography, when volatility permits, in order to make use of the sensitivity of the flame ionization detector. Impurities co-extracted with the lipid material may cause interference in the analysis of the lipid both because of co-elution and because of strong UV absorption. In such cases a preliminary separation into lipid classes may simplify the final analysis by removing both undesirable contaminants and unwanted lipid. The most commonly used technique for lipid fractionation is that of adsorption chromatography, but ion-exchange and exclusion chromatography have also been used.

5.1 Adsorption chromatography

The most useful adsorbent for lipid fractionation is silica gel. The use of a short column packed with about 1 g of silica allows about 30 mg of lipid to be loaded on to the column. Elution sequentially with chloroform or diethyl ether (10 ml), acetone (10 ml), and methanol (10 ml), will yield fractions containing simple lipids, glycolipids, and phospholipids respectively. The simple lipids may further be separated on silica gel or Florisil (TM) columns and eluted with hexane containing increasing amounts of diethyl ether (7).

The use of commercial columns has already been mentioned. Sep-Pak (TM) columns of silica gel (Waters Associates, Milford, USA) have been used to separate neutral, free fatty acid, and polar lipid classes using a novel solvent system containing methyltertiarybutylether (MTBE) (8).

The lipids may be extracted from human serum using the method of Bligh and Dyer (9) and stored under nitrogen at −20 °C.

Protocol 1. Separation of neutral lipids

1. Prior to use wash each Sep-Pak column with hexane/MTBE (96:4, 4 ml) followed by 12 ml hexane to remove interfering substances.
2. Evaporate to dryness under nitrogen the serum lipids from the extraction equivalent to 100 µl of human serum, dissolve in 2.0 ml hexane/MTBE (200:3) and apply to the Sep-Pak column.
3. Wash the vessel containing the extract with the same solvent (2 ml) and add the solvent to the column. Save the 2 ml eluted solvent.
4. Elute the cholesteryl esters with hexane/MTBE (200:3, 10 ml) in a total volume of 12 ml.
5. Elute the triglycerides using 12 ml hexane/MTBE (96:4).
6. Acidify the column with 12 ml hexane/acetic acid and discard the fraction which contains no lipid.
7. Elute the fatty acids with 12 ml hexane/acetic acid (100:2) and the cholesterol with 12 ml MTBE/acetic acid (100:0.2).

Protocol 2. Separation of polar lipids

After elution of the neutral lipids and fatty acids as in *Protocol 1*, polar lipids are eluted with combinations of MTBE/methanol/ammonium acetate (pH 8.6).

1. Elute approx. 50% of the phosphatidylinositol with 8.0 ml of MTBE/methanol/ammonium acetate (ph 8.6) (25:4:1) and the remaining 50% with MTBE/methanol/ammonium acetate (pH 8.6) (10:4:1).
2. Using 12 ml MTBE/methanol/ammonium acetate (pH 8.6) (5:4:1), elute 69% of the phosphatidylcholine, 50% sphingomyelin, and 2% lysophosphatidylcholine.

3. Increase the polarity of the solvent to MTBE/methanol/ammonium acetate (pH 8.6) (5:8:2) and elute an additional 22% phosphatidyl choline, 50% sphingomyelin, and 93% lysophosphatidylcholine. A bulk separation of neutral and polar lipids can be achieved on Sep-Pak columns eluting with MTBE/acetic acid (100:0.2). The lipids are dissolved in 2.0 ml of the same solvent, applied to the column and washed with a further 2.0 ml. The addition of a further 10 ml of solvent elutes the lipids. No polar lipids are eluted with the MTBE/acetic acid (100:0.2) solvent.

The use of aminopropyl bonded phase (Bond Elut) together with the Vac-Elut apparatus has been demonstrated (10) for the separation of polar and lipid classes including cholesteryl esters, triglyceride, diglyceride, monoglyceride, cholesterol, free fatty acids, and phospholipids with good recoveries. The method has been applied to the separation of lipids from bovine adipose tissues which contain large amounts of triglycerides and fatty acids.

5.2 Ion-exchange chromatography

Column chromatography using DEAE–cellulose in the acetate form has been widely used for simplifying complex lipid mixtures. Chloroform/methanol mixtures are usually used as the eluting solvent, but by extending the range of solvents used a more complete separation is possible. The fractionation of lipid classes from animal tissues may be carried out as follows:

Protocol 3. Ion-exchange fractionation of lipid classes

1. Convert the DEAE–cellulose to the acetate form before packing the column by washing sequentially with hydrochloric acid (1 mol.dm^{-3}), distilled water, potassium hydroxide (0.1 mol.dm^{-3}) and again with distilled water, and leave the packing material overnight in glacial acetic acid.
2. Pack into a column (30 × 2.5 cm) in a slurry of the same solvent.
3. Wash the column sequentially with methanol, chloroform/methanol (1:1 v/v), and finally chloroform.
4. Elute using the following elution scheme:

Fraction	*Lipids eluted*	*Solvents*	*Column volumes*
1	simple lipids	chloroform	10
2	phosphatidylcholine lysophosphatidylcholine sphingomyelin ceramide monohexoside	chloroform/methanol (9:1 v/v)	10

Protocol 3. *Continued*

Table (*cont.*)

Fraction	*Lipids eluted*	*Solvents*	*Column volumes*
3	phosphatidylethanolamine ceramide oligohexosides lysophosphatidylethanolamine	chloroform/methanol (1:1 v/v)	10
4	—	methanol	10
5	phosphatidylserine	glacial acetic acid	10
6	—	methanol	4
7	phosphatidic acid diphosphatidylglycerol phosphatidylglycerol phosphatidylinositol cerebroside sulfate	ammonium acetate solution[a]	10
8	—	methanol	10

[a] Chloroform/methanol (4:1 v/v) made 0.5 mol.dm^{-3} with respect to ammonium acetate and with 20 ml of 28% ammonium hydroxide added per litre.

Different lipid sources will, of course, give variations in the lipids eluted but the above will be typical for animal tissue.

5. Re-generate the column at the end of the analysis by introducing a reverse gradient, ending up with chloroform.

6. Separation of lipid classes

The simultaneous separation and quantitation of lipids by HPLC eluded the chromatographer for some time. The main problem lay in the use of the UV- and refractive index detectors. The absorbance due to the double bonds contained in the fatty acid chains, in the 200–215 nm region, is dependent on the concentration and degree of unsaturation of the species, and direct quantitation of a complex lipid mixture is not possible with UV detection (12). Traditionally, lipid class separations have been carried out by TLC, but many of the solvents used are not suitable for HPLC with UV detection because they are not transparent in the region where lipids absorb. With refractive index detection the gradient elution required for the elution of individual lipid classes cannot be used.

Two approaches to the problem have been made: (*i*) the development of a method using normal-phase chromatography with isocratic elution and an RI detector; and (*ii*) the use of detectors which would allow the use of gradient elution. Three detectors which have been used are the infra-red detector, the flame ionization detector, and the mass detector. Of these, the IR detector suffers from baseline shifts, and although solvent matching can be used to reduce the problem, little work has been done on this. The use of both the flame ionization detector and the mass detector places some limitations on the solvents used.

6.1 Analysis of free fatty acids and acylglycerols

Free fatty acids are not easily resolved by HPLC without the formation of derivatives or the addition of an organic acid (such as formic or acetic acid), as an ion suppressant. The formation of phenacyl esters of free fatty acids produces a species suitable for either UV or IR detection, whereas the use of an ion suppressant is suitable only for RI detection. The following method is suitable for the separation of free fatty acids and acylglycerols (18).

Column: Zorbax silica (250 × 4.6 mm i.d.) at 35 °C.

Mobile phase: isooctane/tetrahydrofuran/formic acid (80:20:0.5) at 2.0 ml min^{-1}.

Detector: Refractive index.

SAMPLE: Free fatty acids and acylglycerols of palmitic acid, methyl palmitate, cholesterol, cholesteryl oleate, and ricinoleic acid (internal standard) dissolved in mobile phase to reduce the solvent peak signal. Aliquots containing 0.2–5 mg total lipid are injected on to the column.

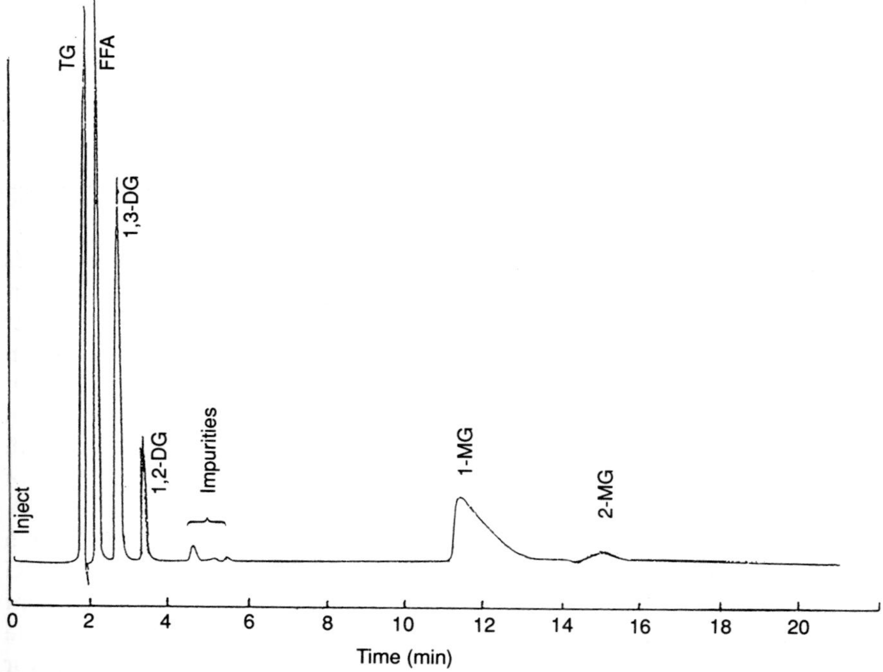

Figure 1. Separation of lipid classes containing only palmitic acid. TG = triglyceride; DG = diglyceride; MG = monoglyceride; FFA = free fatty acid. (Reprinted with permission from *Analytical Chemistry*.)

Table 1. Retention time of lipid classes

Lipid	**Typical retention time/min**
Triglyceride	1.73
Free fatty acid	2.20
1,3-Diglyceride	2.65
1,2-Diglyceride	3.37
1-Monoglyceride	11.30
2-Monoglyceride	14.98
Methyl ester	1.82
Cholesterol	4.00
Cholesterol ester	1.69
Ricinoleic acid	3.80

Table 2. Relative response factors of lipid types

Lipid	**Response factor**	
	relative to triglyceride	**relative to ricinoleic acid**
Palmitic lipids		
Tripalmitin	1.000	1.189
Dipalmitin	0.891	1.059
Monopalmitin	0.912	1.084
Palmitic acid	1.073	1.276
Stearic lipids		
Tristearin	1.000	1.076
Distearin	1.047	1.127
Monostearin	1.151	1.238
Stearic acid	1.152	1.240
Oleic lipids		
Triolein	1.000	1.175
Diolein	0.833	0.979
Monoolein	1.054	1.113
Oleic acid	0.971	1.141
Elaidic lipid		
Trielaidin	1.000	1.209

Results: A chromatogram of a separation of lipid classes containing only palmitic acid is shown in *Figure 1*; typical retention times are listed in *Table 1* and relative response factors of the lipid type in *Table 2*.

Notes: For qualitative work the analysis time may be reduced, and the peak shape of the monoglycerides improved, by increasing the flow rate after elution of the second diglyceride. The identity of the fatty acid may have a small effect on the retention times of some lipid classes. This was most noticeable with free fatty acids but was not observed with the triglycerides.

The response factors are reasonably consistent for all the lipid classes shown. The possibility of on-column intramolecular acyl migration would decrease the accuracy for the determination of individual positional isomers of the partial glycerides. The response factors of trielaidin (*trans*-isomer) and triolein (*cis*-isomer) indicate that the double-bond configuration has little effect on the detector response. Coefficients of variation for the method are generally better than 2.0%.

The method would not be suitable for samples containing polar lipids, which would probably block the column, unless a preliminary clean-up procedure was used.

The use of the mass detector enables gradient elution to be used and lipid samples with a wide range of polarity can then be analysed. A method suitable for lipid samples ranging in polarity from cholesterol esters to lysophosphatidylcholine is as follows (14, 15):

Column: 3 μm Spherisorb-silica (100 × 5 mm i.d.).

Mobile phase: Ternary gradient using: A, isooctane/tetrahydrofuran (99:1 v/v); B, isopropanol/chloroform (4:1 v/v); C, isopropanol/water containing 0.5 mM serine adjusted to pH 7.5 with ethylamine (1:1 v/v). The flow-rate was 2 ml min^{-1} throughout. The ternary gradient is given in *Table 3*.

Table 3. Ternary gradient system

	% Solvent		
Time/min	**A**	**B**	**C**
0	100		
1	100		
5	80	20	
5.1	42	52	6
20	32	52	16
20.1	30	70	
25	100		
30	100		

Note: The last 10 min of the programme is for column equilibration prior to injection of the next sample.

Detector: Mass selective detector Model 750/14. (Applied Chromatography Systems, Macclesfield, Cheshire, UK), with the following settings: internal air pressure 27 p.s.i., 'evaporator set' 40.

Sample: Extracts of rat brain and rat kidney containing cholesterol esters, triacylglycerols, and phospholipids. A sample of 0.4 mg of lipid in 5 μl of hexane/chloroform (1:1) was applied to the column, using a 10 μl sample loop.

Results: Chromatograms of the separations are shown in *Figures 2* and *3*. The rat-kidney extract (*Figure 2*) shows all the main simple and complex lipid classes separated, together with the more abundant phospholipid classes. The rat-brain tissue extract (*Figure 3*) shows the simple lipid free cholesterol, together with the main phospholipid classes including the acidic lipids.

The detector response was found to be linear over the concentration range 50 to 200 μg but fell off rapidly below 10 μg. Each compound in the sample required separate calibration. In order to obtain accurate quantitative results it is necessary to set up the detector parameters as well as the elution system in exactly the same way for both calibration and analysis from day to day. The manner of application of the sample was also found to be critical.

Notes: In order to accommodate the different lipid polarities it is necessary to elute the non-polar lipids with a hydrocarbon solvent and to use an aqueous solvent mixture to elute the phospholipids. The intermediate solvent, isopropanol, is required to maintain solvent miscibility between these two

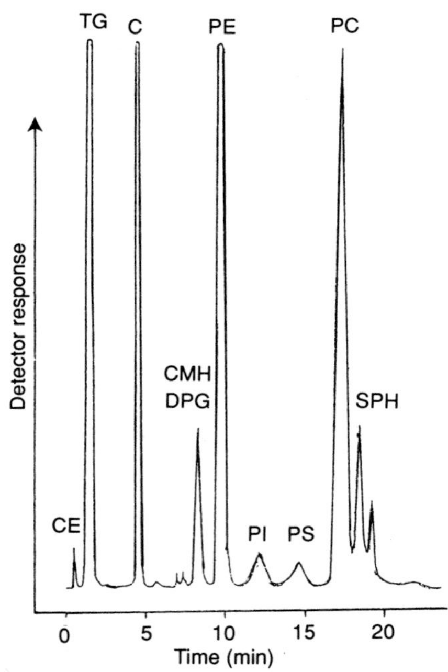

Figure 2. Separation of a lipid extract from rat kidney on a silica column with mass detection. CE = cholesterol esters; TG = triacylglycerols; C = cholesterol; DPG = diphosphatidylglycerol; CMH = ceramide monohexoside; PE = phosphatidylethanolamine; PI = phosphatidylinositol; PS = phosphatidylserine; PC = phosphatidylcholine; SPH = sphingomyelin (usually seen as a double peak). (Reprinted with permission from *Journal of Chromatography*.)

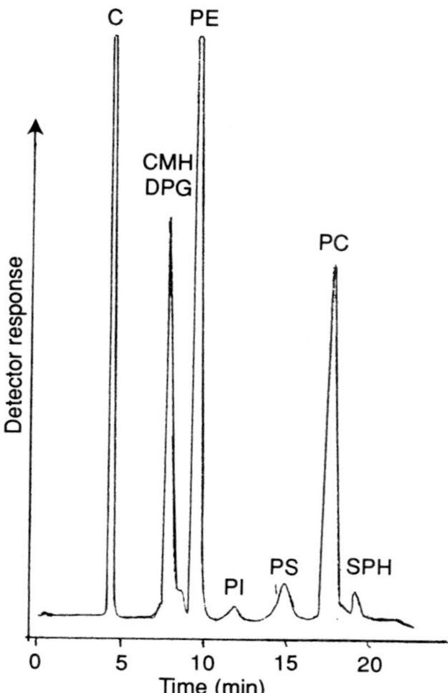

Figure 3. Separation of a lipid extract from rat-brain tissue. For abbreviations see *Figure 2*. (Reprinted with permission from *Journal of Chromatography*.)

extremes. The addition of the ion suppressant, serine, was found to improve resolution and column life without increasing the background noise from the detector.

Another approach to the separation and quantitation of fatty acids and acyl glycerols has been to use size exclusion chromatography. The method allows a simple and rapid separation of low-molecular-weight compounds with relatively small differences in molecular weight (16):

Column: LiChrogel PS4 and LiChrogel PS1 columns in series with the PS4 column placed first (EM Science, Gibbstown, NJ, USA). The columns (250 × 7 mm i.d.) were packed with 5 μm spherical styrene/divinylbenzene copolymer beads with an upper exclusion limit of 5×10^3 daltons for PS4 and 2×10^3 daltons for PS1; 100 daltons was the lower exclusion limit for both columns.

Mobile phase: Toluene at a flow-rate of 0.5 ml min^{-1}.

Detector: Refractive index detector.

Sample: Methyl esters, mono-, di-, and triglycerides containing $C_{12:0}$ to $C_{18:2}$ fatty acids, and a safflower oil. Sample concentrations were 25 mg ml^{-1} in either toluene or tetrahydrofuran; 0.5 mg injected on-column with a 20 µl sample loop. Monolaurin was used as internal standard.

Results: Chromatograms for the separation of the standard mixture and for a sample of lipolysed safflower oil are presented in *Figures 4* and *5*. Retention times are given in *Table 4*.

Notes: The method is only suitable for the separation of lipid classes, the various molecular species of individual lipid classes could not be separated. The presence of unsaturation in the fatty acid increases the retention by approx. 0.25 min compared to the retention of the saturated compound. It is necessary to esterify any free fatty acids in the sample otherwise they will overlap the monoglyceride peak.

For quantitation, calibration curves are linear up to 5% of each component which corresponds to 1 mg injected. As little as 0.05% (10 µg) of each

Table 4.

	Retention time/min			
Fatty acid	**TG**	**DG**	**ME**	**MG**
Lauric	17.98	19.26	21.88	24.25
Myristic	17.54	18.95	21.27	23.41
Palmitic	17.14	18.24	20.71	22.70
Stearic	16.86	17.88	20.20	22.15
Oleic	17.09	18.13	20.47	22.34
Linoleic	17.25	18.34	20.85	22.55

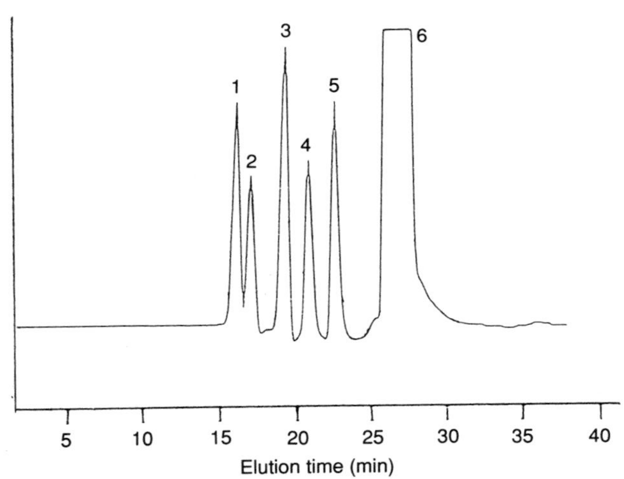

Figure 4. HPSEC separation of a test mixture containing; 1, simple triglycerides; 2, simple diglycerides; 3, methyl esters; 4, monoglycerides containing $C_{16:0}$, $C_{18:0}$, $C_{18:1}$, and $C_{18:2}$ fatty acids, 5, monolaurin (internal standard); 6, THF. (Reprinted with permission from *Journal of the American Oil Chemist's Society*.)

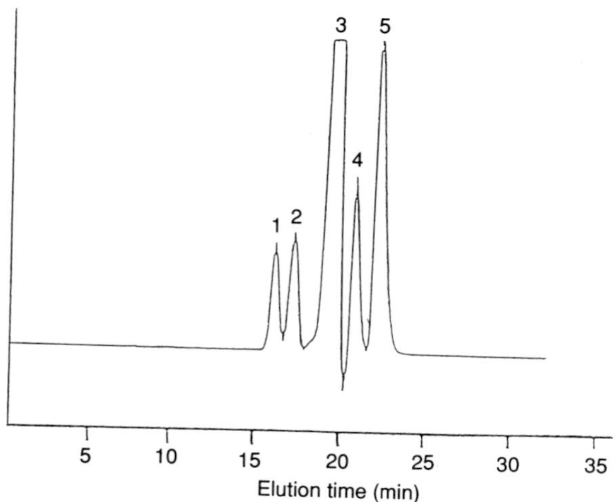

Figure 5. HPSEC separation of safflower oil lipolysis mixtures after diazomethane methylation of the free fatty acids; 1, triglycerides, 2, diglycerides; 3, methyl esters; 4, monoglycerides; 5, monolaurin. (Reprinted with permission from *Journal of the American Oil Chemists' Society*.)

component could be determined. An internal standard (monolaurin) was required for quantitative reliability. Correction factors for methyl esters, mono-, di-, and triglycerides containing $C_{16:0}$, $C_{18:0}$, $C_{18:1}$, and $C_{18:2}$ fatty acids were used to quantify individual components of these fatty acids. Correction factors in the same lipid class (for example, triglycerides) increased with unsaturation. For mixtures containing different fatty acids each peak may contain an entire group of lipids each differing in molecular weight and functional groups. Quantitation of such compounds will be unreliable unless individual standards are available. A method is presented for the calculation of weighted correction factors, based on the fatty acid composition of the sample as determined by GLC, which enables each lipid species to be determined with a precision of 1–5% relative standard deviation.

The method is applicable to a wide variety of samples, such as food emulsifiers, lipolysis mixtures, and in enzyme action studies.

7. Separation of fatty acids

The analysis of common saturated and unsaturated fatty acid methyl esters (FAMEs) is usually carried out by gas chromatography, and it is unlikely that HPLC will become the method of choice for these compounds. The greatest value of HPLC is for the analysis of components which are too involatile or too unstable for the high temperatures required in gas chromatography and for preparative scale separations for the isolation of particular fatty acids for structural analysis, or of isotopically labelled fatty acids for radioactivity measurements by liquid-scintillation counting. Positional and conformational isomers are often not easily separated by GC, and HPLC does provide an alternative approach.

The choice of detector for the HPLC analysis of fatty acids and FAMEs is important, and to some extent determines the nature of the mobile phase used. Refractive index detector response is similar for saturated and unsaturated fatty acids, whereas the UV extinction at a given wavelength depends on the number and configuration of double bonds. The RI detector is less sensitive than the UV detector and to dissolve enough FAME may require the use of a non-aqueous mobile phase. This will reduce retention times and also resolution so that longer columns may be required. Using reversed-phase HPLC it is possible to suppress ionization of the acids, and therefore free fatty acids may be resolved. Using phosphoric acid to buffer the mobile phase to pH 2.2, commonly occurring FFAs (including *cis* and *trans* isomers) can be separated using the conditions given in *Figure 6* in less than 13 min (17).

Many fatty acid separations are carried out on derivatized samples so that

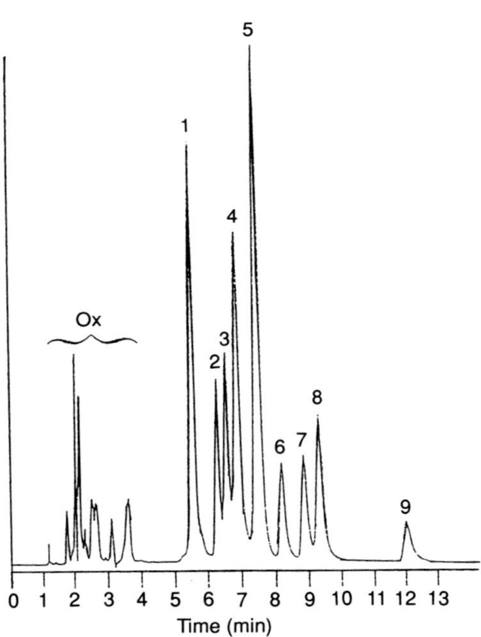

Figure 6. Separation of free fatty acids. *Conditions*: 5 μm Supelcosil LC-18 Column (250 × 44.6 mm) eluted with Ch_3CN/THF/0.1% H_3PO_4 (50.4:21.6:28) at 1.6 ml/min. with UV detection at 215 nm. (Ox) oxidation products; 1, 18:3c; 2, 16:1c; 3, 16:1t; 4, 18:2c; 5, 18:2t; 6, 16:0, 7, 18:1c; 8, 18:1t; 9, 18:0. (Reprinted with permission from *Supelco Reporter*.)

greater sensitivity can be achieved using the UV or fluorescence detectors. The common derivatives are the phenacyl esters for UV detection, and anthracene- and phenanthracene-containing moieties for fluorescence detection.

7.1 Analysis of fatty acids with UV detection

The usual method for the derivatization of fatty acids for UV detection is based on the method of Durst *et al.* (18) and involves the formation of the *p*-bromophenacyl esters of the fatty acids, using crown ethers as catalyst (see Chapter 2). The method (19) given in *Protocol 4* can be used for fatty acid standards and samples alike.

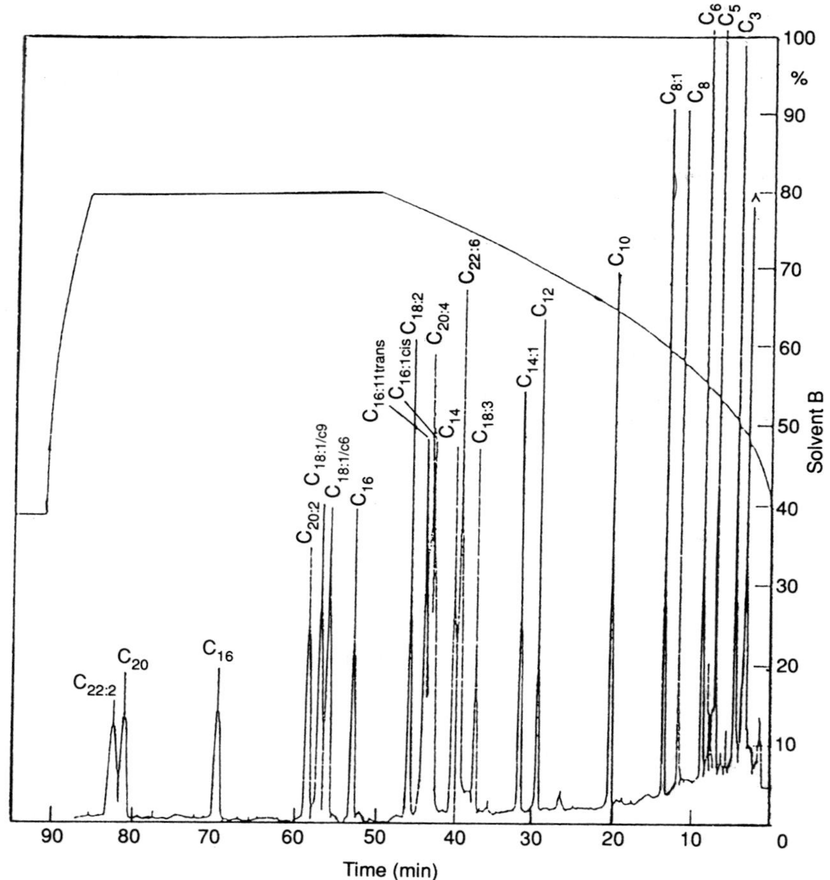

Figure 7. Separation of fatty acids as *p*-bromophenylesters. Conditions as in text. A = *p*-bromophenacyl bromide. (Reprinted with permission from *Chromatographia*.)

Protocol 4. Analysis of fatty acids

1. Make up fatty acid standards as 10 mM solutions in methanol and store in dark brown bottles in the refrigerator.
2. Neutralize 200 μl (2 μmol) of a fatty acid stock solution to a phenolphthalein end-point (pH 8.4) by the addition of about 40 μl of a methanolic solution of potassium hydroxide (0.05 M).
3. Add 100 μl of Kryptofix 222 (Merck: 4,7,13,16,21,24-hexaoxa-1,10-diazabicyclo-(8,8,8)-hexacosane, 2 mM in acetonitrile) and heat the solution gently for 10 min.
4. Add 110 μl (2.2 μmol) of *p*-bromophenacyl bromide (PBPB) and heat the solution for 15 min at 80 °C.
5. Dilute the solution with acetonitrile to give a volume of 10 ml (containing 200 μM ester).
6. Inject 20 μl of this solution (i.e. 4 nmol of one ester) into the chromatograph.
7. Calculate the amounts of reagents required for fatty acid samples of unknown concentration from the consumption of 0.05 M potassium hydroxide required for neutralization.
8. Analyse the derivatized fatty acids under the following conditions:

Column: Superspher C_8 Merck (250 × 4.0 mm i.d.)

Mobile phase: Solvent A acetonitrile/water (50:50). Solvent B acetonitrile.

Gradient: Solvent B 38% to solvent B 80% in 50 min, using an exponential gradient according to the equation $y = e^{0.5x}$). After 50 min solvent B 80% isocratic. Flow rate 1 ml min^{-1}.

Detector: Variable wavelength UV at 288 nm.

Sample: Standards. Ten saturated acids: (C_3, C_5, C_6, C_8, C_{10}, C_{12}, C_{14}, C_{16}, C_{18}, C_{20}). Seven mono-olefinic acids ($C_{8:1^{32}}$, *cis*-$C_{14:1^{9}}$, *trans*-$C_{16:1^{9}}$, *cis*-$C_{18:1^{9}}$, *cis*-$C_{18:1^{6}}$, $C_{22:1^{13}}$), and five polyolefinic acids (*cis,cis*-$C_{18:2^{9,12}}$. *cis,cis*-$C_{20:2^{11,14}}$, *cis*-$C_{18:3^{9,12,15}}$, *cis*-$C_{20:4^{5,8,11,14}}$, and $C_{22:6^{4,7,10,13,16,19}}$) – PBPB esters. 20 μl injected on column (4 nmol per ester).

Results: A chromatogram of the separation is shown in *Figure 7*.

All the derivatized fatty acids can be separated in one run, including the geometric isomers, palmitoleic and palmitelaidic acid, and the positional isomers, oleic and petroselinic acid. Standard deviations of less than 1% were achieved in the concentration range 4 nmol down to 4 pmol.

Notes: The resolution of *cis*-$C_{16:1}$, $C_{20:4}$, and *trans*-$C_{16:1}$ could be improved by a change in the gradient program or by using a silica column.

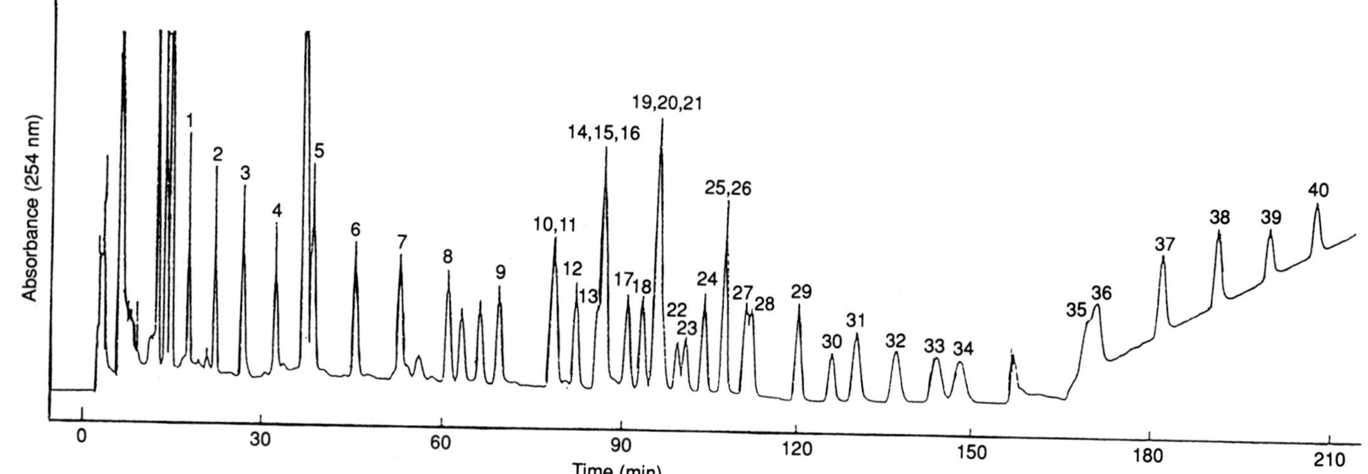

Figure 8. Separation of C_3 to C_{24} fatty acids as their *p*-bromophenacyl esters using a C_{30} bonded silica column. (Reprinted with permission from *Journal of Liquid Chromatography*.)

The use of a silica column to separate the *p*-bromophenacyl esters of fatty acids has been demonstrated by Takayama *et al.* (20) as follows:

Column: C_{30} bonded silica prepared as in ref. 20. Two columns in series, each being 300 × 3.9 mm i.d.

Mobile phase: Gradient elution using a convex gradient of acetonitrile/water (2:3 v/v) to 100% acetonitrile for 150 min followed by a linear gradient from acetonitrile to *p*-dioxin for 150 min at a flow rate of 1 ml min^{-1}.

Sample: PBA esters of 40 fatty acids (C_3 to C_{24}).

Results: The separation is shown in *Figure 8*. Not all the fatty acids were separated. However, the separation between the *cis*-$C_{16:1}$, *trans*-$C_{16:1}$ (palmitoleic acids) and the $C_{20:4}$ (arachidonic acid) is an improvement on the previous method and there is greater separation of the saturated and unsaturated fatty acid esters of the same chain-length. The method does give an alternative to the use of a C_8 column but retention times are longer. However, the separation of saturated C_{35} to C_{56} fatty acid esters from *Mycobacterium tuberculosis* using this column has also been demonstrated (21).

7.2 Analysis of fatty acids with fluorescence detection

For the derivatization and analysis of fatty acids with fluorescence detection the procedure in *Protocol 5* may be used (22).

Protocol 5. Preparation of BMP

2-Bromo-1-methylpyridinium iodide (BMP) is prepared as follows:

1. Dissolve 2-bromopyridine (5 ml) in 20 ml dried diethyl ether.
2. Add iodomethane (7 ml) and reflux the solution in a water bath for 1 h.
3. Wash the pale yellow precipitate with ether to give a white precipitate which is used at a concentration of 20 mg mol^{-1}.
4. Mix the fatty acid with 50 μl of a solution of 9-hydroxymethylanthracene in dichloromethane (2 mg/ml), BMP in dichloromethane (50 μl of suspension), and triethylamine (10 μl) and heat the mixture at 50 °C for 30 min.
5. Evaporate the excess reagent under nitrogen at 50 °C and dissolve the derivatized fatty acid in 1 ml of mobile phase ready for analysis, using the following conditions.

Column: Spherisorb C_8, 250 × 4.5 mm i.d., particle size 3 μm.

Mobile phase: Acetonitrile/water gradient: 93:7 (v/v) for 12 min, 86:14 (v/v) for 5 min, 100:0 (v/v) for 23 min. Flow rate 1 ml min^{-1}.

Detector: Fluorescence (excitation 360 nm, emission 420 nm).

Sample: Standard fatty acids. 1 = $C_{12:0}$, 2 = $C_{18:3}$, 3 = $C_{20:4}$, 4 = $C_{14:0}$, 5 = $C_{16:1}$, 6 = $C_{18:2}$, 7 = $C_{16:0}$, 8 = $C_{18:1}$, 9 = $C_{17:0}$, 10 = $C_{18:0}$, 11 = $C_{20:0}$, 12 = $C_{22:0}$, 13 = $C_{24:0}$ in heptane.

Results: The separation of the standard mixture is shown in *Figure 9*.

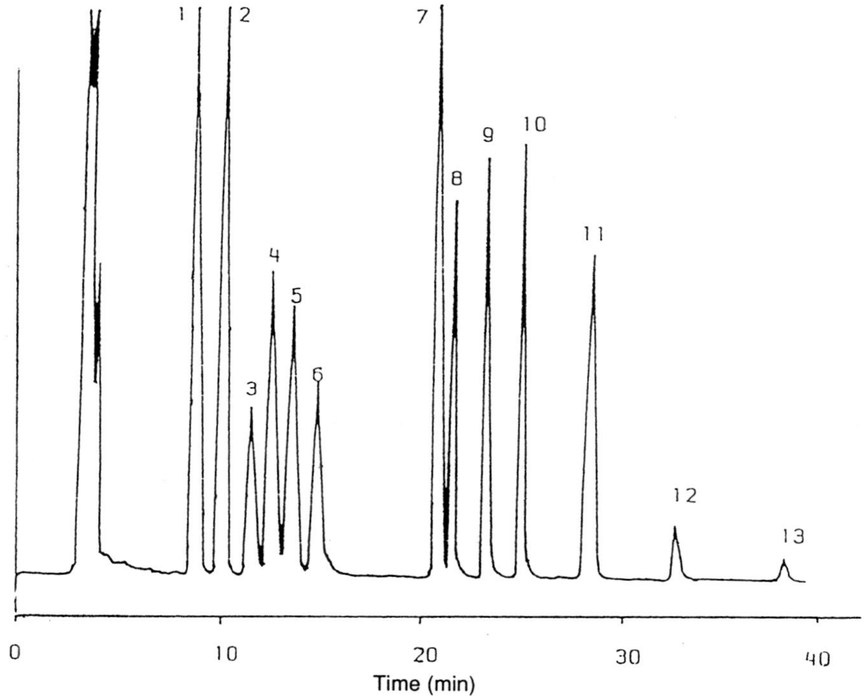

Figure 9. Separation of fatty acids as their anthrylmethyl derivatives and fluorescence detection. 1, $C_{12:0}$; 2, $C_{18:3}$; 3, $C_{20:4}$; 4, $C_{14:0}$; 5, $C_{16:1}$; 6, $C_{18:2}$; 7, $C_{16:0}$; 8, $C_{18:1}$; 9, $C_{17:0}$; 10, $C_{18:0}$; 11, $C_{20:0}$; 12, $C_{22:0}$; 13, $C_{24:0}$. (Reprinted with permission from *Journal of Chromatography*.)

All the standards could be separated in less than 40 min. The method was also applied to a plasma sample as follows:

(a) Extract aliquots of plasma (200 μl) with 3 ml of chloroform/*n*-hexane/methanol (28:21:1 v/v) containing 3.9 μg/ml of a $C_{17:0}$ internal standard.

(b) Vortex the mixture and centrifuge. Remove 2 ml of the organic layer and evaporate to dryness under nitrogen at room temperature.

(c) Derivatize the sample as described previously.

Fatty acid levels up to about 200 μmol/litre can be determined with a coefficient of variation (CV) of better than 11% although the CV increased as the age of the catalyst increased.

Plasma and serum samples contain lauric ($C_{12:0}$), myristoleic ($C_{14:1}$), and linolenic ($C_{18:3}$) acids at very low concentrations and many methods are unable to separate these compounds. The procedure given in *Protocol 6* (23) uses 3-bromomethyl-7,7-dimethoxy-1-methyl-2(1H)-quinoxalinone (Br-DMEQ) as a highly sensitive fluorescence reagent for free fatty acids in serum and separates these compounds.

Protocol 6. Preparation of fluorescent derivatives

1. Mix a 5 μl aliquot of serum with 200 μl of 0.5 M phosphate buffer (pH 6.5), 50 μl of 10 μM $C_{17:0}$ acid solution (internal standard) and 2.0 ml chloroform/*n*-heptane (1:1 v/v).
2. Vortex for 2 min and centrifuge for 5 min.
3. Evaporate the organic layer to dryness *in vacuo* and redissolve the residue in 200 μl acetonitrile.
4. Place 100 μl of this solution in a screw-capped 1.5 ml vial, add *c.* 20 mg of powdered K_2CO_3/Na_2SO_4 (1:1 w/w), 50 μl of 1.3 mM Br–DMEQ and 50 μl 5.7 mM 18-crown-6 solutions. Heat the closed vial at 50 °C for 20 min in the dark.
5. After cooling, inject a 10 μl aliquot into the chromatograph.

To prepare a calibration graph the same procedure is followed except that the internal standard is replaced with a solution containing 0.25 pmol to 5.0 nmol each of the fatty acids. The net peak–height ratios of the individual fatty acids and $C_{17:0}$ acid is plotted against concentration of the fatty acid spiked.

The analysis is then done using the following conditions:

Column: 5 μm YMC-Pack C8 (150 × 6 mm i.d.) (Yamamura Chemical Labs., Japan).

Mobile phase: Gradient elution using methanol/water (80:20 v/v) isocratically for 28 min then to 100% methanol in 18 min at 2.0 ml min^{-1}.

Detector: Fluorescence with a 20-μl flow cell. (λ-excitation 370 nm, λ-emission 455 nm).

Figure 10 shows a separation of a portion (5 μl) of standard fatty acids (2.0 nmol each/ml). The peak identification is: 1 = $C_{12:0}$, 2 = $C_{14:1}$, 3 = $C_{14:0}$, 4 =

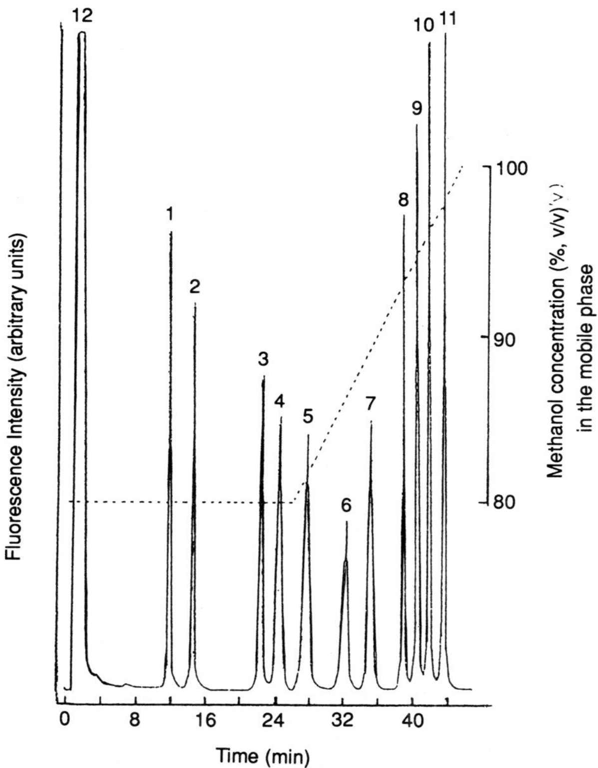

Figure 10. Separation of fatty acids with fluorescence detection. 1, $C_{12:0}$; 2, $C_{14:1}$; 3, $C_{14:0}$; 4, $C_{15:3}$; 5, $C_{16:1}$; 6, $C_{20:4}$; 7, $C_{15:2}$; 8, $C_{16:0}$; 9, $C_{15:1}$; 10, $C_{17:0}$; 11, $C_{15:0}$ acids; 12, Br-DMEQ. (Reprinted with permission from *Journal of Chromatography*.)

$C_{18:3}$, 5 = $C_{16:1}$, 6 = $C_{20:4}$, 7 = $C_{18:2}$, 8 = $C_{16:0}$, 9 = $C_{18:1}$, 10 = $C_{17:0}$, 11 = $C_{18:0}$, 12 = Br-DMEQ.

Recoveries for a normal serum sample were better than 95% with standard deviations between 3.0 and 4.5%.

7.3 Determination of double bonds in fatty acids from fats and oils

The routine determination of the number of double bonds in fatty acids from fats and oils may be achieved, using the pentafluorobenzyl (PFB) esters and normal phase chromatography on silica gel, when the separation depends primarily on the degree of unsaturation (24). The saponification of the fat or oil and the derivatization may be carried out in a single procedure as set out in *Protocol 7*.

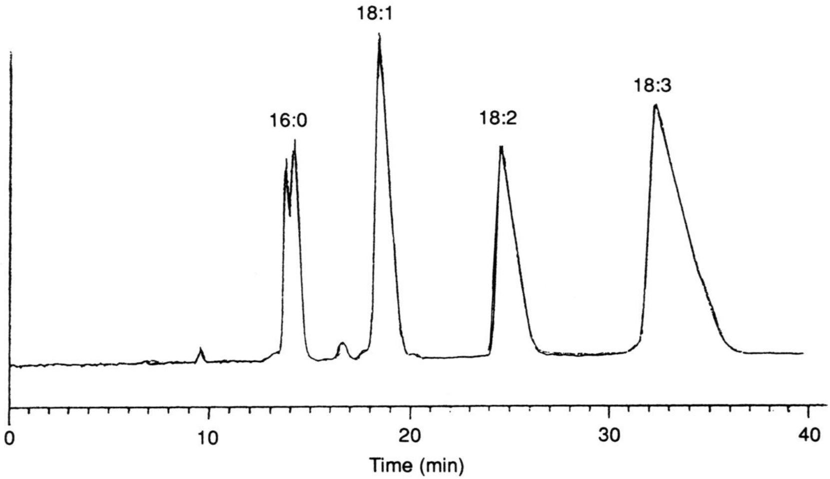

Figure 11. Separation of PFB esters from linseed oil. (1) PFB 16:0, (2) PFB 18:0, (3) PFB 18:2, (4) PFB 18:3. (Reprinted with permission from *Journal of the American Oil Chemists' Society*.)

Protocol 7. Saponification and derivatization of fatty acids

1. Add approximately 20 mg of the fat or oil to a screw-cap vial together with 1 ml of 0.2 M KOH in methanol and incubate at 90 °C for 25 min.
2. Evaporate off the methanol in a stream of nitrogen then add 1 ml of 0.1 M tetrabutylammonium hydrogen sulfate, 1 ml dichloromethane, and 20 μl pentafluorobenzyl bromide.
3. Incubate at room temperature with shaking for 40 min, when fatty acids longer than C_8 should be quantitatively derivatized.
4. Extract the dichloromethane layer and evaporate to dryness before taking up in hexane and apply to a Sep-Pak silica column. Elute with 10 ml hexane to remove excess Br-PFB and by-products, followed by 10 ml 15% dichloromethane/hexane and evaporate to give the PFB esters. Add 200 μl hexane and inject 10 μl into the HPLC using the following conditions:

Column: 5 μm silica (250 × 4.6 mm i.d.).

Mobile phase: dichloromethane/hexane/water-saturated hexane (10:45:45 v/v) at 2 cm^3 min^{-1}. Note: the water content must be kept constant for reproducible retention.

Detector: UV absorption at 254 nm or at both 263 and 216 nm. Although the esters have adequate absorption at 254 nm the use of two wavelengths allows ratio plots of the two wavelengths to be made. These are helpful in determining the number of double bonds in an unknown peak.

Results: A separation of the fatty acid esters from linseed oil is shown in *Figure 11*. The order of elution is $C_{18:0}$ and C_{16} (incomplete separation), $C_{18:1}$, $C_{18:2}$, and $C_{18:3}$. A plot of log (retention) against number of double bonds is linear and, as long as the carbon chain-length does not vary too much, can be used to determine the number of double bonds in the PFB esters.

Note: Typically with adsorption chromatography the peaks tend to be broad and the column efficiency low as in *Figure 11*. One of the reversed-phase systems given earlier could with advantage be used. The order of elution would then be reversed.

7.4 Separations using silver resin chromatography

Silver resin chromatography has been used (25) to isolate multigram quantities of the highly unsaturated omega-3 (ω3) components of fish oil fatty acids and esters. A 50% ω3 fish oil concentrate was esterified with sulfuric acid/methanol and Menhaden oil was transesterified (Na metal in methanol) to obtain the fatty acid methyl esters (FAMEs) and saponified (alcoholic

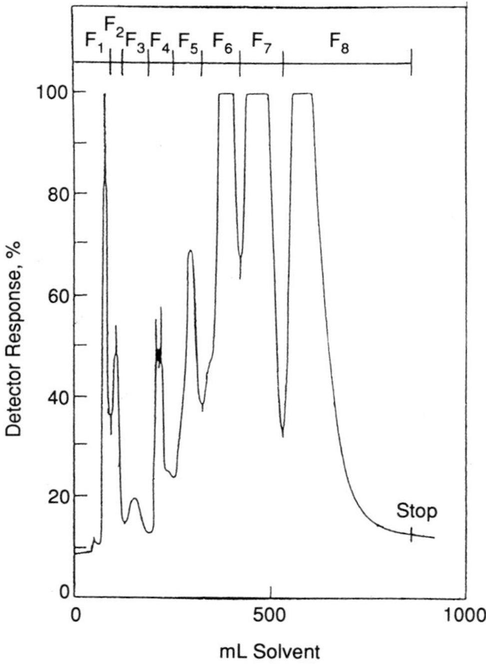

Figure 12. Fractionation of FOC methyl esters using silver resin chromatography. (Reprinted with permission from *Journal of the American Oil Chemists' Society*.)

potassium hydroxide) to prepare the fatty acids which were analysed using the following conditions:

Columns: (*i*) For sample sizes up to 500 mg of fatty acids or FAMEs: Slurry pack a column (60 × 2.7 cm) with 65 g of 200–270 mesh, 100% Ag^+/Na^+ XN1010 resin (i.e. one in which the sulfonic acid protons are first replaced by Na^+ ions and then 100% of these are replaced by Ag^+ ions. (*ii*) For sample sizes up to 10 g: Slurry pack a column (45 × 4.7 cm) with 250 g of 100/200 mesh 100% Ag^+/Na^+ XN1010 resin.

Mobile phases: Typically acetonitrile/acetone (35:65 v/v) at a flow-rate of 8.0 ml min^{-1} but depending on the actual sample. Isocratic or gradient elution can be used as required.

Detector: Refractive index.

Results: A typical separation is shown in *Figure 12*.

Notes: The composition of the fractions was determined by gas chromatography of the FAMEs. The FAMEs are fractionated into groups each composed of a group of fats with the same number of *cis* double bonds.

7.5 Separation of fatty acids with post-column ion-pair extraction and absorbance detection

This method allows the detection of fatty acids in nanogram quantities after separation by LC. It depends on the extraction of the fatty acids from aqueous solution as neutral ion-pairs with methylene blue dye and the detection of the dye by absorbance. For full details of the method the reader should consult ref. 26.

Column: 5 μm Spherisorb ODS-2 (150 × 4.6 mm i.d.).

Mobile phase: Gradient elution at 0.8 ml/min. Acetonitrile/water 79:21 (v/v) to 87:13 (v/v) from 0 to 10 min then to 99:1 (v/v) from 10 to 15 min and maintained isocratically to 20 min. The mobile phase is adjusted to pH 4.0 with 5% orthophosphoric acid.

Post-column ion-pair extractor: The effluent from the LC column is mixed with methylene blue solution (2 mg/ml in 0.02 M Na_2 HPO_4) pumped at 2.0 ml/min, and after passing through a mixing coil it is mixed with chloroform (1.0 ml/min) in a mixing tee and passes through an extraction coil before entering a phase separator where the fatty acids in chloroform are separated by gravity before being detected by a variable-wavelength detector set at 651 nm.

Results: The method can be used to determine fatty acids in orange juice at levels below 2.5 p.p.m., and in butter and margarine at levels between 34–586 and 34–256 p.p.m. respectively.

8. Separation of glycerolipids

In nature, each lipid class exists as a complex mixture of related compounds where the composition of the aliphatic residues varies from one molecule to the next. In the case of triacylglycerols (triglycerides) each of the three positions in the molecule may contain a different fatty acid. Therefore, for a complete structural analysis of a lipid it is necessary to separate it into *molecular species* according to the specific alkyl or acyl group (fatty acid, alcohol, ether) in the relevant positions in the molecule. For monoacylglycerols this is relatively easy, and it is not impossible for diacylglycerols. However, for triacylglycerols, although they may be simplified, it is not yet possible to obtain single species.

Because of the complex nature of the problem no single type of chromatography will give a satisfactory separation. The separation will depend on the properties of all the aliphatic residues in each molecule. Reversed-phase HPLC will separate molecules according to the combined chain-lengths of the fatty acids, with the retention being modified by the presence of double bonds (each double bond reduces the retention time by the equivalent of approximately two carbon atoms). Argentation chromatography will separate molecules according to their degree of unsaturation and adsorption chromatography can be used to separate species according to the polarity of the substituent groups.

The choice of solvent is an important consideration in the separation of glycerolipids. Reversed-phase chromatography is the preferred method of separation but the commonly used water/methanol mobile phases may lead to problems because of the low solubility of triacylglycerols in these solvents. The use of non-aqueous reversed phase (NARP) systems in the analysis of triacylglycerols has been reviewed (27). Acetone is the usual modifier used with refractive index detection, but with UV detection this is usually replaced by tetrahydrofuran.

8.1 Equivalent carbon number

The concept of the 'equivalent carbon number' (*ECN*) or 'partition number' (*PN*) has been used to rationalize the retention of triacylglycerols. The *ECN* is defined as the number of carbon atoms in the aliphatic residues (*CN*) minus twice the number of double bonds (*N*) per molecule, i.e.

$$ECN = CN - 2N.$$

Two components with the same *ECN* value are known as 'critical pairs' and present a special problem in the separation, as they will elute very close together. Thus, triacylglycerols with fatty acid combinations 16:0–16:0–16:0, 16:0–16:0–18:1, 16:0–18:1–18:1 and 18:1–18:1–18:1 have the same *ECN*.

In practice they will not have the same retention times, because the presence of a second or third double bond does not have the same effect as the first, but the concept can be used to define regions of the chromatogram in which compounds will elute.

8.2 Normal phase separation of triacylglycerols

Normal phase chromatography of triacylglycerols separates components depending on the degree of unsaturation and therefore gives a good separation of 'critical pairs'. There is also a small effect from chain-length, the longer chain-lengths being eluted first. The technique is therefore complementary to reversed-phase separations and can be used as follows (28):

Column: Silica Spheri-5 (250 × 4.6 mm i.d., 5 μm, Brownlee Labs., USA).

Mobile phase: 0.7% acetonitrile in half water-saturated hexane made up as follows; Dry 7 ml of acetone over molecular sieve and add it to 496.5 ml of water-saturated hexane. Shake the solution well and make up to 1 dm^3 with dry (molecular sieve) hexane. The flow-rate is 2 ml min^{-1}.

Detector: Hewlett-Packard diode array detector at 200 nm.

Sample: triacylglycerol standards; tristearoyl-, tripalmitoleoyl-, trioleoyl-, trilinoleoyl-, trilinolenyl-, and tri-11-eicosenoylglycerol.

Results: The separation depending on degree of unsaturation is shown in *Figure 13*, where the separation between triacylglycerols with 3, 6, and 9 double bonds is clearly shown. The dependence of retention on chain-length is shown in *Figure 14*.

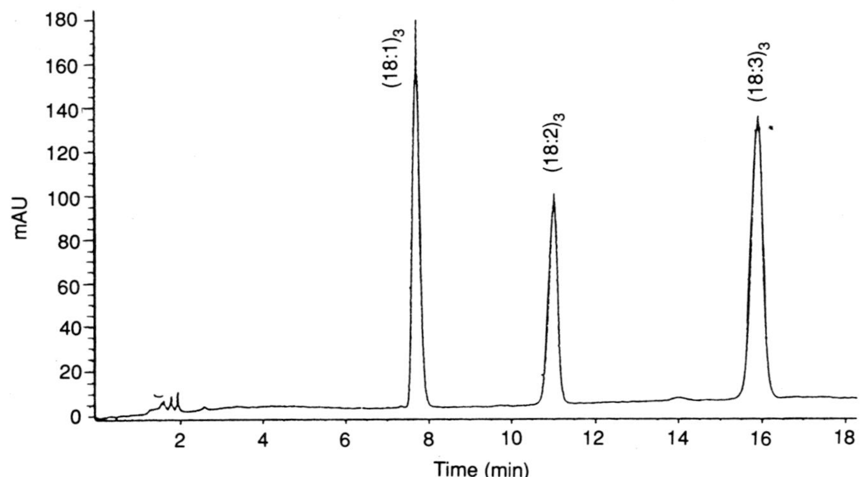

Figure 13. Separation of triacylglycerols according to degree of unsaturation. 1, trioleoylglycerol; 2, trilinoleolylglycerol; 3, trilinolenylglycerol. (Reprinted with permission from *Journal of Chromatography*.)

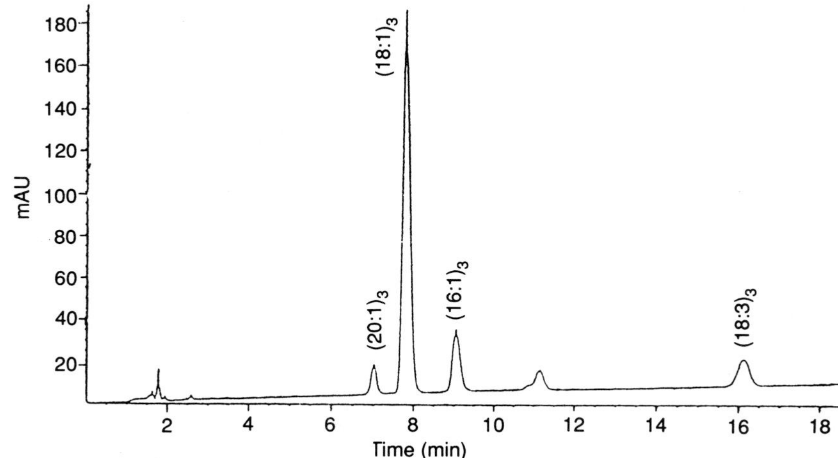

Figure 14. Separation of triacylglycerols according to chain-length. 1, tri-11-eicosenoylglycerol; 2, trioleoylglycerol; 3, tripalmitoleoylglycerol; 4, trilinolenoylglycerol. (Reprinted with permission from *Journal of Chromatography*.)

Notes: The method has been applied to the separation of triacylglycerols in sunflower oil, safflower oil, linseed oil, a margarine, and a flax seed extract.

8.3 Non-aqueous reversed-phase separations of triacylglycerols

There have been several studies on the separation of triacylglycerols using non-aqueous mobile phases with a variety of detector systems.

The following method (29) is suitable for the separation of high *ECN* triacylglycerols using gradient elution with methyltertiarybutylether (MTBE)/acetonitrile and UV detection.

Column: Spherisorb ODS-2 (15 cm × 4.6 mm i.d., 5 μm) at 30 °C.

Mobile phase: Gradient elution from 23% to 30% MTBE in ACN in 25 min at a flow-rate of 1 ml min^{-1}.

Detector: UV at 215 nm.

Sample: Standard triacylglycerols with *ECN*s from 12.00 to 54.00.

Results: Relative retention times (*RRT* min) and equivalent carbon numbers (*ECN*) are given in *Table 5* for a number of triacylglycerols and the chromatogram is given in *Figure 15*.

Notes: The pairs LPL–LnOP, LOP–LSL, LSP–POP, and LOS–OPO (Ln = $C_{18:3}$, L = $C_{18:2}$, O = $C_{18:1}$, S = $C_{18:0}$, P = $C_{16:0}$) could not be separated.

Table 5. Relative retention times (*RRT* min) and equivalent carbon numbers for simple triglycerides

TG	*CN*	*ND*	*ECN*	*RRT*
BuBuBu	12	0	12.00	0.72
ClClCl	24	0	24.00	1.75
CaCaCa	30	0	30.00	3.48
LnLnLn	54	9	34.74	6.49
LaLaLa	36	0	36.00	7.42
DeDeDe	39	0	39.00	10.54
LLL	54	6	41.16	12.90
PaPaPa	48	3	41.58	13.54
MMM	42	0	42.00	15.69
OOO	54	3	47.58	25.44
PPP	48	0	48.00	28.56
MaMaMa	51	0	51.00	35.73
SSS	54	0	54.00	46.84
NoNoNo	57	0	57.00	> 90
AAA	60	0	60.00	–
BeBeBe	66	0	66.00	–

Abbreviations: Bu = $C_{4:0}$, Cl = $C_{8:0}$, Ca = $C_{10:0}$, Ln = $C_{18:3}$, La = $C_{12:0}$, De = $C_{13:0}$, L = $C_{18:2}$, M = $C_{14:0}$, O = $C_{18:1}$, P = $C_{16:0}$, Ma = $C_{17:0}$, S = $C_{18:0}$, No = $C_{19:0}$, A = $C_{20:0}$, Be = $C_{22:0}$

The use of the UV detector reduces the choice of mobile-phase solvents because of the low wavelengths that have to be used. This restriction does not apply to the mass detector which has been evaluated for triacylglycerol analysis as follows (30):

Column: LiChrospher 100 RP-18 (250 × 4.6 mm i.d.).

Mobile phase: Solvent A, acetonitrile; solvent B, acetonitrile/ethanol/hexane (40:40:20 w/w/w). Linear gradient from 100%A to 100%B in 120 min. Flow-rate 1 ml min^{-1}.

Detector: Mass detector (ACS Ltd, Luton, UK) at 40 °C and inlet gas pressure 15 p.s.i.

Sample: Commercial standard (HPLC G-1, Nu-Check Prep, Inc., Minnesota, USA), and common refined oils. The standard was dissolved in hexane/isopropanol (1:1 w/w) and 12 μl were injected (approx. 150 μg total sample).

Results: The separation of the standard mixture is shown in *Figure 16*.

Notes: A comparison of the quantitative analysis of the standard mixture using the mass detector and the UV detector was made and is given in *Table 6*. The results indicate a small non-linearity of response for the mass detector but the values agree reasonably well with the nominal values. The values obtained with the UV detector show the effect of unsaturation in the molecule on detector response and it is essential to calibrate the detector for these molecules if meaningful quantitative values are to be obtained.

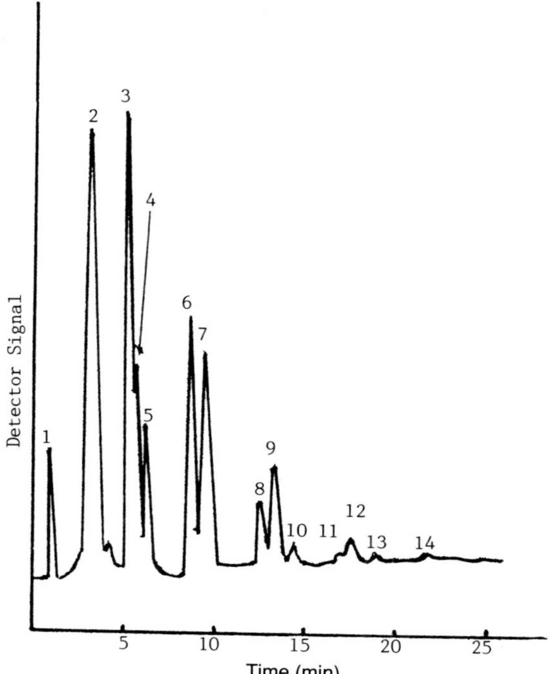

Figure 15. Separation of soybean oil trigylcerides using non-aqueous reversed phase HPLC. 1, LnLLn; 2, LLnL; 3, LLL; 4, LnLO; 5, LnLP; 6, LOL; 7, LPL + LnOP; 8, OLO; 9, LOP + LSL; 10, PLP; 11, OOO; 12, LOS + OPO; 13, LSP + POP; 14, OSO. *Abbreviations*: Ln = $C_{18:3}$, L = $C_{18:2}$, O = $C_{18:1}$, S = $C_{18:0}$, P = $C_{16:0}$. (Redrawn and reprinted with permission from *Chromatographia*.)

Table 6. Comparison of the quantitative analysis of a triglyceride mixture using the mass detector and the UV detector

Triglyceride	Nominal value (weight %)	Mass detector (area %)	UV (254 nm) detector (area %)
24:0	9.4	9.8	6.1
27:0	6.3	6.1	4.5
30:0	9.4	10.1	5.8
33:0	6.3	6.0	2.8
54:9	3.1	2.5	44.2
36:0	9.4	10.3	4.6
39:0	6.3	6.6	2.6
54:6	3.1	2.9	15.1
48:3	3.1	2.7	2.0
42:0	9.4	11.3	3.3
45:0	6.3	7.2	2.0
54:3	3.1	3.2	1.6
48.0	9.4	11.0	2.8
51:0	6.3	5.7	1.4
54:0	9.4	4.4	1.3

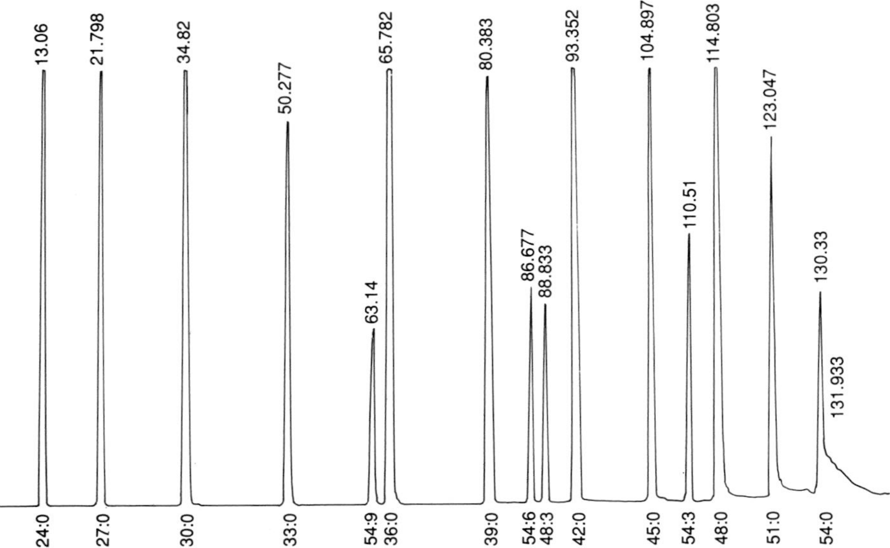

Figure 16. Separation of a standard triglyceride mixture $C_{24:0}$ to $C_{54:0}$ with mass detection. (Reprinted with permission from *Lipids*.)

Other detector systems which have been used for triacylglycerol analysis are the flame ionization detector (31) and a post-column reactor detector (32).

The flame ionization detector (Tracor Model 945, Austin, Texas, USA) was used in combination with two Hibar LiChrospher 100 CH-18/2 columns (250 × 4.0 mm i.d., 5 μm) in series at 40 °C with non-linear acetonitrile/acetone gradient elution at a flow-rate of 1 ml min^{-1}. The analysis of butter–oil triglycerides is demonstrated.

The use of a post-column reactor detector to monitor triglycerides separated on a 3 μm Hitachi gel 3057 (ODS) column, either isocratically or by gradient elution, using an ethanol/acetonitrile mobile phase has been demonstrated (32). The same detector system has also been used (33) to monitor triacylglycerols, which were first separated using argentation HPLC on two stainless steel columns (150 × 4.6 mm i.d.) packed with 10% silver nitrate impregnated Develosil 60–3 eluted with benzene and an infra-red detector (1745 cm^{-1}), after which trapped fractions were analysed by NARP–HPLC using two Hitachi gel 3057 columns (150 × 4.6 mm i.d.) eluted with ethanol/acetonitrile (60:40), followed by post-column reactor detection. The post-column reactor depends on the hydrolysis of the triacylglycerols with potassium hydroxide and oxidation of the resulted glycerin with periodic acid to formaldehyde. The formaldehyde is then reacted with acetylacetone in the presence of ammonium acetate to form 3,5-diacetyl-1,4-dihydrolutidine which is detected at 410 nm. The limit of detection if 0.1 mmol of trilaurin and response is linear between 0.3 and 60 nmol trilaurin.

The combination of HPLC and GLC has been applied to the determination of the distribution of saturated and unsaturated fatty acids amongst the different triacylglycerols in adipose tissue (34). Following a preliminary fractionation of the triacylglycerol fraction by TLC on silica gel 60 eluted with hexane/diethyl ether/acetic acid (85:25:1), HPLC was carried out on a 5-μm Spherisorb ODS column (250 × 4.6 mm i.d.) with gradient elution using 2-propanol/acetonitrile (45:55) for 30 min then to 2-propanol/acetonitrile (60:40) in 15 min at a flow-rate of 1 ml min^{-1}. Detection was by UV spectrophotometry at 213 nm. Fractions from the HPLC were collected, transesterified to fatty acid methyl esters, and analysed by packed column GLC.

8.4 Separation of diacylglycerols

The class separation of diacylglycerols has been described in Section 6. For the separation of molecular species of diacylglycerols reversed-phase chromatography of the diacetate derivatives (35) or of the dinitrobenzoate derivatives (36) has been used. The chromatographic conditions are similar to those given for the separation of triacylglycerols; for example, a silica-ODS column (250 × 4.6 mm i.d.) eluted with acetonitrile/isopropanol (4:1 v/v).

Attention has been given to the resolution of enantiomeric diacylglycerols and the following method is given as an example of the use of a chiral stationary phase for this purpose (37):

Column: Sumipax OA-4100 (Sumiitomo Chem., Japan). (250 × 4.0 mm i.d., 5-μm). This is *N*-(*R*)-1-(α-napthyl)ethylaminocarbonyl-(*S*)-valine chemically bonded to γ-aminopropylsilanized silica.

Mobile phase: Hexane/ethylene dichloride/ethanol (80:20:1) at 1 ml min^{-1}.

Detector: UV spectrophotometer at 254 nm.

Sample: Synthetic 1,2-diacyl-*rac*-, 1,2-diacyl-*sn*-, 1,3-diacyl-*sn*-, 2,3-diacyl-*sn*- and 1,2-dialkyl-*rac*-glycerols. The 3,5-dinitro phenylurethane derivatives were prepared as follows:

(a) React 1 mg of sample with 2 mg 3,5-dinitrophenyl isocyanate in 4 ml dry toluene in the presence of 40 μl dry pyridine at room temperature for 1 h.

(b) Purify the crude extract by TLC on silica gel GF plates (20 × 20 cm, 0.25 mm) eluted with hexane/ethylene dichloride/ethanol (40:10:3)

For HPLC 1 μl of the samples in chloroform was injected.

Results: *Figure 17* shows the separation of the optically active 1,2-dipalmitoyl-*rac*-glycerol together with the inactive compounds. Similar separations were obtained for other racemic acyl- and alkyl-glycerols.

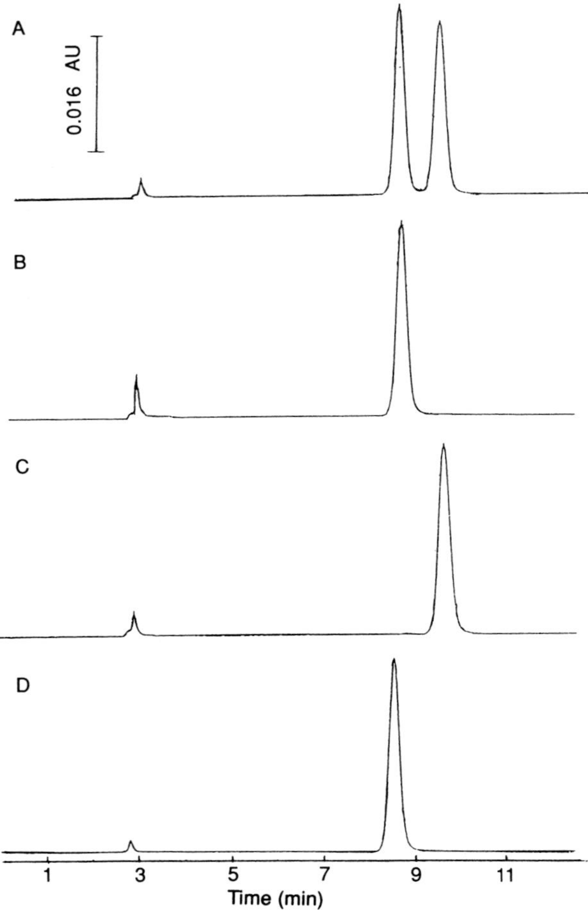

Figure 17. Separation of diacylglycerol isomers on a chiral stationary phase. A = 1,2-dipalmitoyl-*rac*-glycerol; B = 1,2-dipalmitoyl-*sn*-glycerol; C = 2,3-dipalmitoyl-*sn*-glycerol; D = 1,3-dipalmitoyl-*sn*-glycerol. (Reprinted with permission from *Lipids*.)

Notes: As well as the enantiomer separation the OA-4100 column gives the carbon number separation of the acyl and alkyl groups and the separation of saturated and unsaturated species due to the silica gel support.

8.5 Separation of monoacylglycerols

Monoacylglycerols have been commonly determined by the periodate oxidation method. This method, however, only gives the total amount of monoglyceride; the values for individual component monoglycerides are not obtained. For the GLC analysis of monoglycerides the trimethylsilylether derivatives are usually used. The following reversed-phase HPLC method

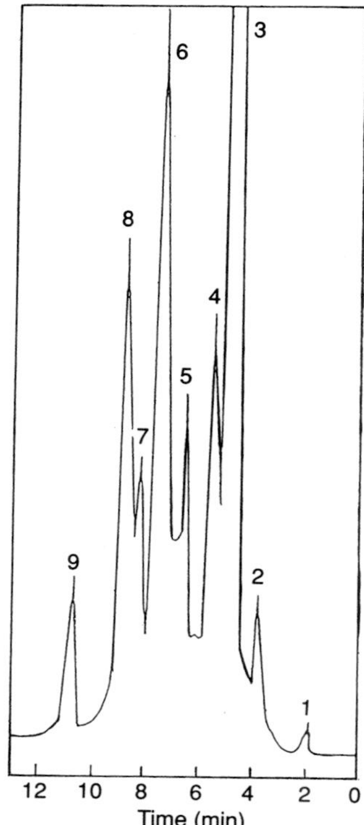

Figure 18. Separation of saturated and unsaturated monoacylglycerols using reversed phase HPLC. 1 + 2, impurities; 3, chloroform + diethyl ether; 4, C_{12}-MG; 5, C_{14}-MG; 6, $C_{18:2}$-MG; 7, C_{16}-MG; 8, C_{18}-MG; 9, C_{20}-MG. (Reprinted with permission from *Journal of American Oil Chemists' Society*.)

(38) gives the separation and quantitation of various monoglycerides without derivatization.

Column: Unisil Q-C8. (250 × 4.6 mm i.d., 5 μm) (Gasukuro Kogyo Co., Tokyo, Japan.)

Mobile phase: Isocratic elution with CH_3CN/H_2O (7:3 v/v) at 1 ml min^{-1}.

Detector: UV detector at 210 nm.

Sample: Saturated and unsaturated monoglycerides in the range C_6 to C_{18}.

Results: A separation of saturated and unsaturated monoacylglycerols is shown in *Figure 18*.

Notes: Quantitative determinations of the C_{12}, C_{14}, and C_{16} saturated monoacylglycerols gave standard deviations of less than 1.5 and coefficients of variation (%) of between 1.0 and 2.8 using the UV detector. These values were somewhat better than those obtained with a refractive index detector.

9. Separation of phospholipids

Phospholipids are one of the principal components of biological tissues, and the physical and chemical properties of cell membranes are critically dependent on the nature of the head group and the fatty acyl composition of the phospholipid. However, there is a lack of quick and easy methods for the characterization of phospholipids. Both TLC and GLC are used, but usually require multistep sample preparation procedures which are laborious and time-consuming. One method used is to determine the total fatty acid composition, by hydrolysing the acyl ester linkage followed by methylation of the fatty acid methyl esters, by GLC. To obtain more information, selective partial hydrolysis with phospholipases, followed by determination of the fatty acid composition at the *sn*-1 and *sn*-2 positions of the phospholipid molecule, can be used. However, the identity of the two fatty acids in individual molecular species is not obtained. HPLC on silica columns can be used to separate phospholipid classes, and a single phospholipid class can be separated into species according to differences in the fatty acid composition by reversed-phase HPLC.

Ultraviolet absorption at 200 nm is the most common method of detection but has several disadvantages. The mobile phase is limited to those which are UV transparent such as acetonitrile/methanol/water and hexane/isopropanol/water. The absorbance is mainly due to the double bonds in the fatty acyl moieties and functional groups such as carbonyl, carboxyl, phosphate, and amino groups. The degree of unsaturation differs between acyl groups and between phospholipid classes, so that peak areas are not a measure of weight per cent or molar quantities of the lipids, and it is impossible to obtain direct quantitation of phospholipids if they have an unknown degree of unsaturation. Quantitation of phospholipids can be done following HPLC separation using a modified Fiske and Subbarow procedure (39) for lipid phosphorus or by using a detector system which does not rely on UV absorption, such as mass spectrometry (40) or the mass detector (41, 42).

9.1 Separation of individual lipid classes

9.1.1

Phospholipids can be separated into different classes by normal-phase HPLC under the following conditions (43).

Column: Zorbax Sil column (250 × 4.6 mm i.d.).

Mobile phase: Solvent A, hexane/2-propanol (3:2 v/v); solvent B, hexane/2-propanol (3:2 v/v) with 5.5% water.

Gradient—for phospholipid classes except plasmologens:

Time/min	0	11	16	22	25
% B	50	100	100	50	50

for lyso compounds (see below):

Time/min	0	17	25	38	41
% B	50	100	100	50	50

Flow-rate in both cases was 1.5 ml min^{-1}.

Detector: UV at 205 nm. Absorbance range 0.64.

The separated phospholipids were determined by the method of Rouser *et al.* (44).

Sample: Beef steak, liver, heart, and kidney extracted by the Folch method. 20 μl extract was injected on column. Phospholipid standards were used to identify individual classes.

Results: The paper gives no retention data, only quantitative results, but the method presented separates the major phospholipid classes efficiently.

9.1.2 Phospholipid classes in human milk

The major phospholipid classes in human milk can be separated by isocratic elution from a silica column as follows (45):

Column: Alltech Econosphere 5 μm Silica (250 × 4.6 mm i.d.).

Mobile phase: Acetonitrile/methanol/sulfuric acid (100:3:0.5 v/v/v) at 2.5 ml min^{-1}.

Sample: Standards of phosphatidylinositol (PI), phosphatidylserine (PS), lysophosphatidylethanolamine (LPE), phosphatidylcholine (PC), lysophosphatidylcholine (LPC), and sphingomyelin (SM). Stock phosphorus solution containing 1.0 mg P/ml. Milk from mothers delivering full-term infants. Total phospholipid samples were made up in chloroform/diethyl ether (1.2 v/v) at 1.5 mg/ml.

Detector: UV detector at 202 nm to monitor HPLC effluent. Recovered phospholipids were analysed by a modified Fiske and Subbarow method (39).

Results: A separation of the phospholipids is shown in *Figure 19*.

Detection limits and reproducibility are based on UV absorption, although the actual phospholipid values are based on phosphorus. Because of this, detection limits vary with lipid class. Typical values are: L-α-PE, 0.25 μg; L-α-PC, 0.50 μg; SM, 2.50 μg, with a mean coefficient of variation of 3.7%.

Notes: Total lipids are extracted from milk samples by the modified Folch procedure. Phospholipids and non-polar lipids were separated by column chromatography on Unisil silica.

Phospholipids were determined as set out in *Protocol 8*.

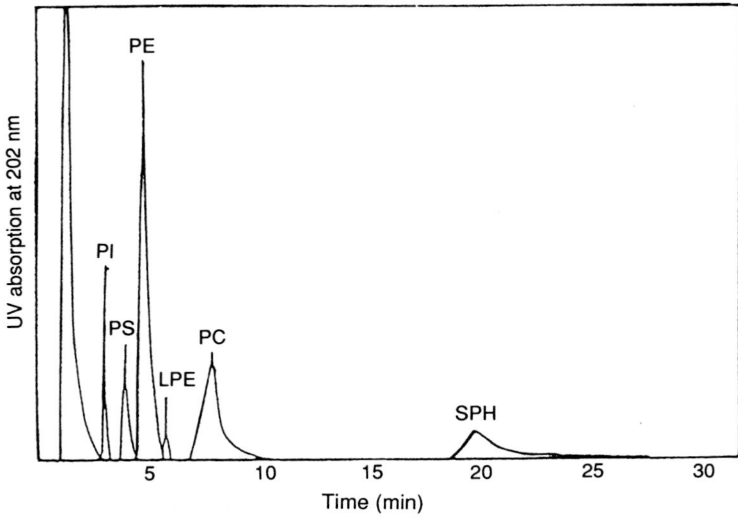

Figure 19. Separation of phosopholipid classes from an extract of human milk. PI = phosphatidylinositol, PS = phosphatidylserine, PE = phosphatidylethanolamine, LPE = lysophosphatidylethanolamine, PC = phosphatidylcholine, SPH = sphingomyelin. (Reprinted with permission from *Journal of Dairy Science*.)

Protocol 8. Determination of phospholipids

1. Collect the separated phospholipids (monitored by UV absorption) in a series of test-tubes using a similar set of test-tubes containing eluent as blank.
2. Evaporate off solvent at 100 °C under nitrogen.
3. Prepare phosphorus standards containing 0.1 to 10 μg of phosphorus from the stock phosphorus solution.
4. Add 200 μl sulfuric acid (2.5 N) to the sample, blank, and standard tubes.
5. Reflux tubes at 200 °C, cool to 150 °C and add one drop of concentrated hydrogen peroxide to each tube and reheat to 200 °C. If any samples are not colourless add more H_2O_2 and reheat the sample.
6. Prepare ammonium molybdate solutions as follows: (a) dissolve 2.5 g ammonium molybdate in 50 ml sulfuric acid (10 N) and dilute to 100 ml with distilled water; (b) dissolve 25 g ammonium molybdate in 300 ml sulfuric acid (10 N) and dilute to 1000 ml with distilled water.
7. Add 250 μl of (a) to tubes containing the standards and 250 μl of (b) to the tubes containing the blank and sample tubes. Dilute all tubes with 1.5 ml of distilled water.
8. Use a Fiske and Subbarow reducer (Sigma Chemicals, St Louis, Missouri,

USA) containing a 15% solution of 0.8% 1-amino-2-napthol-4-sulfonic acid, sodium sulfite and sodium bisulfite. Add 100 μl of this solution to each tube, mix the tubes by inversion and allow them to react for 20 min.

9. Measure the absorbance of all the tubes at 660 nm and convert the absorbance values to micrograms phosphorus from a standard curve prepared from the standard phosphorus solutions.

The combination of HPLC and colorimetric measurement of the phosphorus provides a precise and accurate method for the determination of human milk phospholipids.

Similar chromatographic conditions have been used (46) for the separation of plant phospho- and glyco-glycerolipids.

9.1.3 Separation with the mass detector

The use of the mass detector allows the direct determination of phospholipids, since it does not suffer the drawbacks of UV detection. The following method (42) is suitable for the analysis of phospholipid (and other lipid) classs in blood serum.

Column: 10 μm Silica-100 (250 × 4.6 mm i.d.).

Mobile phase: 20 min gradient from chloroform (100%) to methanol/28% ammonia in water/chloroform (92:7:1). (Flow-rate not given).

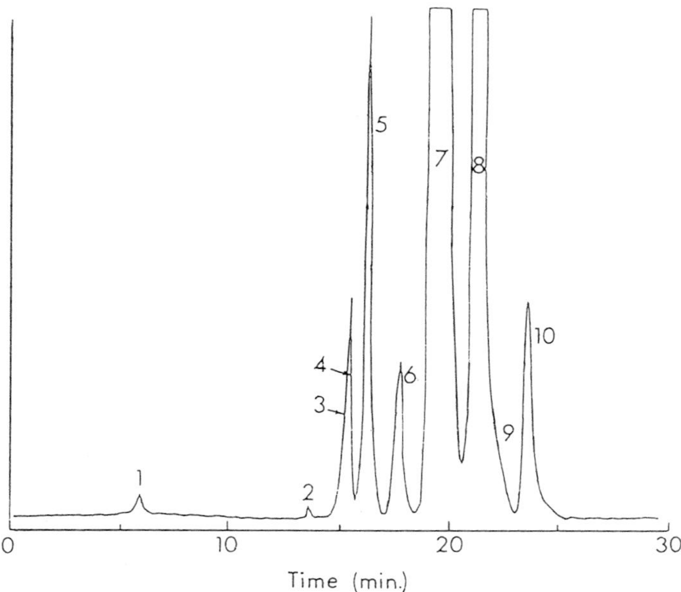

Figure 20. Separation of blood serum lipids with a mass detector. See text for peak identification. (Reprinted with permission from *Journal of Liquid Chromatography*.)

Sample: Standard sample of blood serum lipids: 1. cardiolipin, 2. phosphatidyl acid, 3. phosphatidyl glycerol, 4. diphosphatidyl diglycerol, 5. phosphatidyl ethanolamine, 6. phosphatidyl inositol, 7. phosphatidyl choline, 8. sphingomyelin, 9. phosphatidyl serine, 10. lysophosphatidylcholine. Sample size: 100 μg in 5 μl chloroform.

Results: The separation is shown in *Figure 20*.

Notes: If the separation of non-polar lipids is required a mobile-phase gradient from diethyl ether/chloroform or acetonitrile/chloroform to 100% chloroform is first run.

The response factor for fatty acid methyl esters, for di- and tri-glycerides, and for cholesterol is the same, but for polar lipids it is about four times larger. Within each of these two classes quantitation may be made directly from the peak areas.

9.2 Separation of individual lipid classes

A single phospholipid class can be separated into species according to differences in the fatty acid composition by reversed-phase HPLC. Buffers and mobile-phase additives (for example, choline chloride) have been used to improve resolution, but this may complicate the identification of the separated species. The following method (41) uses the mass detector and has been used to separate species of egg PC and PE as well as rat-liver PC.

Column: Nucleosil-5 C_{18} column (250 × 4.6 mm i.d.) at 30 °C.

Mobile phase: Methanol/water/acetonitrile (8:1:1 w/w/w) at 1.5 ml min^{-1}.

Samples: purified egg PC and PE and rat-liver PC. The rat-liver PC was extracted from homogenized rat liver with hexane/isopropanol (3:2) and the total lipid extract was fractionated by normal-phase HPLC and the PC fraction collected for subsequent analysis. The amount of sample injected was 200 μg.

Results: A chromatogram of the egg PC is given in *Figure 21*.

Notes: The chromatogram shows the difference in detection using the mass detector and a UV detector. Because the UV detector has a high sensitivity it detects more components. However, the mass detector provides the more useful quantitative information because the peak size is largely a function of the sample concentration. The maximum amount of information is obtained by using both detectors in series.

10. Separation of cholesterol and cholesterol esters

Cholesterol is found in both the free form and in the esterified form. The free cholesterol in tissues can often be determined by HPLC of a simple lipid extract, while the total cholesterol (free plus esterified) can be determined

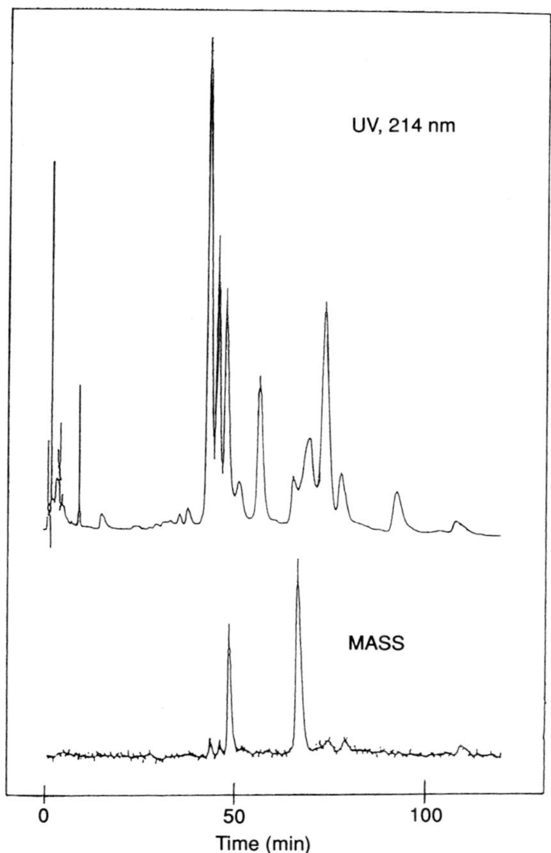

Figure 21. Separation of phosphatidylcholine from egg. Comparison of the response from a UV detector and a mass detector. (Reprinted with permission from *Journal of Chromatography*.)

after hydrolysis. Since the esters only contain one fatty acid constituent in each molecule the separation of individual species is the same as for simple fatty acid derivatives. However, if esters of sterols other than cholesterol are present the elution pattern will be more complex. The majority of separations are carried out on ODS columns with acetonitrile/isopropanol mobile phase. The following method (47) is suitable for the analysis of α-tocopherol, free cholesterol, esterified cholesterols, and triacylglycerols in human plasma and in fractions containing individual lipoproteins.

Column: 3 μm Hitachi Gel No. 3057, a silica-ODS column, (150 × 4 mm i.d.) at 50 °C.

Mobile phase: Isocratic, acetonitrile-isopropanol (75:25 v/v) at 1 ml min^{-1}.

Sample: Standards as listed (together with their capacity factors on the Hitachi Gel No. 3057 column) in *Table 7*. Pig-liver triglycerides. Plasma samples from healthy adult men.

Lipoprotein fractions were prepared by the standard method of flotation ultracentrifugation using KBr to adjust the density of the plasma. Fractions were collected at densities of 1.006, 1.063, 1.125, and 1.21.

Table 7. Capacity factors (k') of main fat-soluble compounds on the column of Hitachi gel No. 3057

No.	Compound	k'
1	α-Tocopherol	1.25
2	Cholesterol	3.01
3	Ch-benzoate	3.22
4	Ch-caprylate ($C_{8:0}$)	4.16
5	Ch-linolenate ($C_{18:3}$)	6.18
6	Ch-arachidonate ($C_{20:4}$)	6.51
7	Ch-linoleate ($C_{18:2}$)	7.72
8	Ch-palmitoleate ($C_{16:1}$)	7.72
9	Ch-myristate ($C_{14:0}$)	8.34
10	Ch-oleate ($C_{18:1}$)	10.06
11	Ch-palmitate ($C_{16:0}$)	10.68
12	Ch-heptadecanoate ($C_{17:0}$)	12.08
13	Ch-stearate ($C_{18:0}$)	13.65
23	Trilinolein (LLL)	4.35
39	Triolein (OOO)	8.59
41	1,2 or 1,3-Di-palmitoylolein (PPO or POP)	9.59
43	Tripalmitin (PPP)	10.61
44	1,2-Dioleoylstearin (OOS)	11.44

Protocol 9. Preparation of sample extracts

1. Pipette 100 μl of plasma or lipoprotein into a glass test-tube (23 × 115 mm).
2. Add 1 ml water and 1 ml ethanol, and vortex for 2 min.
3. Add 5 ml *n*-hexane containing 15 μg cholesteryl benzoate or cholesteryl heptadecanoate as internal standard and shake the tube for 10 min.
4. Pipette a 4-ml aliquot of the hexane layer into a test-tube (16 × 110 mm) and evaporate at 45 °C under reduced pressure.
5. Dissolve the residue in 100 μl ethanol ready for injection on to the chromatograph. Volume usually 20 μl).

Detector: UV absorption at 205 nm and 0.04 a.u.f.s. Fluorescence detection at $\lambda_{ex} = 225$ and $\lambda_{em} = 320$ was also used for cholesterol.

Results: A chromatogram for the standard compounds is shown in *Figure 22*. α-tocopherol (145 ng), cholesterol (2.23 μg), and cholesterol esters (1.5–6.85 μg) are shown as the bold line. The standard pig-liver triglyceride (10 μg) is shown as the dotted line. Peaks are identified in *Table 7*.

Notes: Cholesteryl esters with saturated fatty acids ($C_{8:0}$, $C_{14:0}$, $C_{16:0}$, $C_{17:0}$, and $C_{18:0}$) are separated according to their fatty chain-length. Those with unsaturated fatty acids are separated according to both chain-length and number of double bonds, and elute in the order $C_{18:3}$, $C_{20:4}$, $C_{18:2}$, and $C_{18:1}$. The peak for $C_{16:1}$ is eluted between $C_{14:0}$ and $C_{18:2}$. As shown by the dotted line, the pig-liver triglyceride is chromatographed under the same conditions as the cholesteryl esters and contains about 30 individual triacylglycerol species.

Although the chromatographic behaviour of the triacylglycerols is complex their elution order depends on the structure of the fatty acids as for the cholesteryl esters. The reference also reports results for the lipoprotein fractions together with the concentrations of the components found. Standard deviations were generally less than ± 2.0.

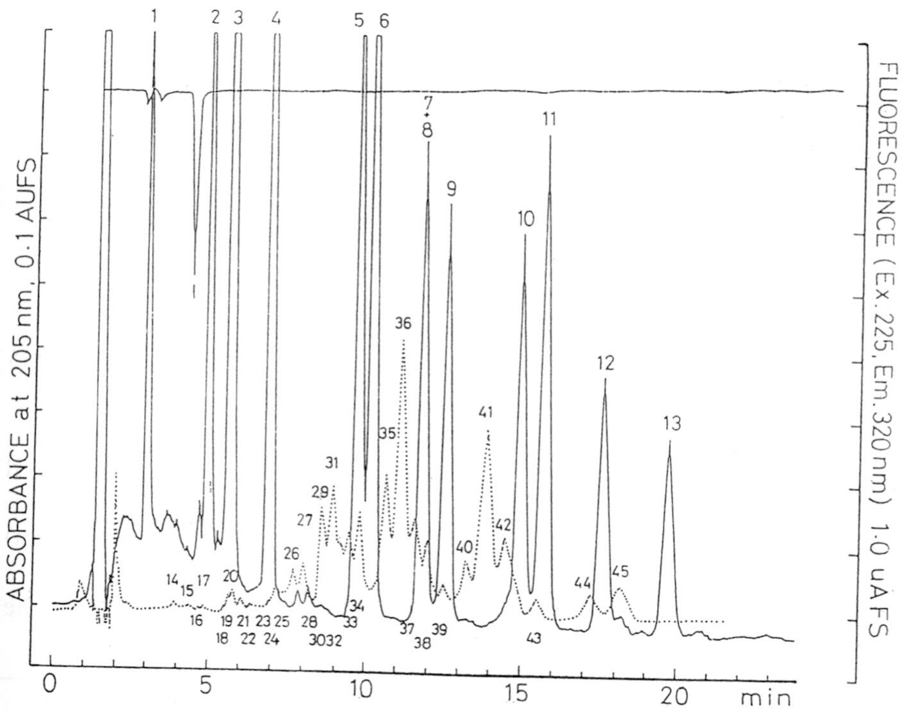

Figure 22. Separation of cholesterol and cholesterol esters. Peak identification in *Table 5*. (Reprinted with permission from *Journal of Chromatography*.)

11. Conclusions

HPLC offers a real alternative to TLC and GC in lipid analysis but until mass detectors, flame ionization detectors, or mass spectrometers are standard in any lipid chemists' laboratory detection will continue to be an overriding factor in the separation.

Not all classes of lipids have been included in the examples (for example, sphingolipids) but the techniques described can be adapted to most lipid mixtures. For a more complete survey of the HPLC of lipids the reader is referred to the book by W. H. Christie (48). There are also useful review articles on the profiling of carbohydrates, glycoproteins, and glycolipids (49), the separation of glycolipids and phospholipids (50), and lipids and their constituents (51). There is also a chapter on recent developments in lipid analysis in *Food constituents and food residues, their chromatographic analysis* (52). The biannual review on food published in *Analytical Chemistry* (53) is a further source of references.

References

1. *Analytical chemistry by open learning (ACOL) texts*. John Wiley, Chichester.
2. Hamilton, R. J. and Sewell, P. A. (1982). *Introduction to high performance liquid chromatography* (2nd edn). Chapman & Hall, London.
3. Meyer, V. R. (1988). *Practical high performance liquid chromatography*. John Wiley, New York and Chichester.
4. Atkin, D. S. J., Hamilton, R. J., Mitchell, S. F., and Sewell, P. A. (1982). *Chromatographia*, **15**, 97.
5. Hammond, E. W. (1988). *Chromatography & Analysis*, **10**, 7.
6. Maxwell, R. J., Nungesser, E. H., Marmer, W. M., and Foglia, T. A. (1988). LC-GC Int. 1 (0), 56.
7. Christie, W. W. (1982). *Lipid analysis* (2nd edn). Pergamon Press, Oxford.
8. Hamilton, J. G. and Comai, K. (1988). *Lipids*, **23** (12), 1146.
9. Bligh, E. G. and Dyer, W. J. (1959). *Can. J. Biochem. Physiol.*, **37**, 911.
10. Kaluzny, M. A., Duncan, L. A., Merrit, M. V., and Epps, D. E. (1959). *J. Lipid Res.*, **26**, 135.
11. Galanos, D. D. and Kapoulas, V. M. (1962). *J. Lipid Res.*, **3**, 134.
12. Christie, W. W. (1985). *Z. Lebensm.-Unters. Forsch.*, **181**, 171.
13. Ritchie, A. S. and Gee, M. H. (1985). *J. Chromat.*, **329**, 273.
14. Christie, W. W. (1985), *J. Lipid Res.*, **26**, 507.
15. Christie, W. W. (1986). *J. Chromat.*, **361**, 396.
16. Christopoulou, C. N. and Perkins, E. G. (1986). *J. Am. Oil Chem. Soc.*, **63** (5), 679.
17. In *Supelco Reporter* (1984). **3** (3), 6.
18. Durst, H. D., Milano, M., Kikta, E. J., Connelly, S. A., and Grushka, E. (1975). *Anal. Chem.*, **47**, 1797.
19. Osterroht, C. (1987). *Chromatographia*, **23** (6), 419.

20. Takayama, K., Jordi, H. C., and Benson, F. (1980). *J. Liquid Chromat.*, **3** (1), 61.
21. Takayama, K., Quereshi, N., Jordi, H. C., and Schnoes, H. K. (1979). *J. Liquid Chromat.*, **2** (6), 861.
22. Baty, J. D., Pazouki, S., and Dolphin, J. (1987). *J. Chromat.*, **395**, 403.
23. Yamaguchi, M., Matsunaga, R., Hara, S., and Nakamura, M. (1986). *J. Chromat.*, **375**, 27.
24. Netting, A. G. (1986). *J. Am. Oil Chem. Soc.*, **63** (9), 1197.
25. Adlof, R. O. and Emken, E. A. (1985). *J. Am. Oil Chem. Soc.*, **62** (11), 1592.
26. Lawrence, J. F. and Charbonneau, C. F. (1988). *J. Chromat.*, **445**, 189.
27. Barron, L. J. R. and Santa-Maria, G. (1987). *Chromatographia*, **23** (3), 209.
28. Rhodes, S. H. and Netting, A. G. (1988). *J. Chromat.*, **448**, 135.
29. Barron, L. J. R. and Santa-Maria, G. (1989). *Chromatographia*, **28** (3/4), 183.
30. Herslof, B. and Kindmark, G. (1985). *Lipids*, **20** (11), 783.
31. Nurmela, K. V. V. and Satama, L. T. (1988). *J. Chromat.*, **435**, 139.
32. Kondoh, Y. and Takano, S. (1986). *Anal. Chem.*, **58**, 2380.
33. Takano, S. and Kondoh, Y. (1987). *J. Am. Oil Chem. Soc.*, **64** (3), 380.
34. Baty, J. D. and Rawle, N. W. (1987). *J. Chromat.*, **395**, 395.
35. Nakagawa, Y. and Horrocks, L. A. (1983). *J. Lipid Res.*, **24**, 1268.
36. Takamura, H., Narita, H., Urade, R., and Kito, M. (1986). *Lipids*, **21**, 356.
37. Takagi, T. and Itabashi, Y. (1987). *Lipids*, **22** (8), 596.
38. Maruyama, K. and Yonese, C. (1986). *J. Am. Oil Chem. Soc.*, **63** (7), 902.
39. Fiske, C. H. and Subbarow, Y. (1925). *J. Biol. Chem.*, **66**, 375.
40. Kim, H.-Y. and Salem, N. (Jr) (1987). *Anal. Chem.*, **59**, 722.
41. Sothiros, N., Thorngren, C., and Herslof, B. (1985). *J. Chromat.*, **331**, 313.
42. Stolyhwo, A., Martin, M., and Guiochon, G. (1987). *J. Liquid Chromat.*, **10** (6), 1237.
43. Yeo, Y. K. and Horrocks, A. L. (1988). *Food Chemistry*, **29**, 1.
44. Rouser, G., Fleischer, S., and Yamamoto, A. (1970). *Lipids*, **5**, 495.
45. Hundrieser, K. and Clark, R. M. (1988). *J. Dairy Sci.*, **71**, 61.
46. Marion, D., Douillard, R., and Gandemer, G. (1988). *Rev. Franç. Corps Gras*, **35**, 229.
47. Seta, K. and Okuyama, T. (1990). *J. Chromat.*, Tokyo Symposium Volume (in press).
48. Christie, W. W. (1987). *High performance liquid chromatography and lipids*. Pergamon Press, Oxford.
49. Kakehi, K. and Honda, S. (1986). *J. Chromat.*, **379**, 27.
50. Eberendu, A. R. N., Venables, V. J., and Daugherty, K. E. (1985). *Liquid Chromat.*, **3** (5), 425.
51. Kuksis, A. and Myher, J. J. (1986). *J. Chromat.*, **379**, 57.
52. Christie, W. W. and Noble, R. C. (1984). In *Food constituents and food residues, their chromatographic analysis* (ed. J. F. Lawrence). Marcel Dekker, New York.
53. Athnasios, A. K., Gross, F. G., and Given, P. S. (Jr) (1989). *Anal. Chem.*, **61**, 45R.

6

Radiotracers in lipid analysis

COLIN G. TAYLOR

1. Introduction

Radiolabelled organic compounds may be detected at very low levels and quantified easily and accurately; the most common radionuclides used for such labelling being ^{14}C, tritium (^{3}H), ^{32}P, and ^{35}S. Species labelled with these nuclides have found extensive use in studies of chemical and biochemical pathways including those involving lipids.

Forty years ago, radiochemical techniques played an important part in the elucidation of the tricarboxylic acid cycle, the common biochemical path for the oxidation of fuel molecules. By labelling carboxylic acids and their esters with ^{14}C in specific positions, steps in the cycle were defined and it was also demonstrated that certain symmetric molecules (prochiral ones) such as citric acid, when bound to a substrate, may react asymmetrically to form optically active (chiral) molecules (1).

In one early application of radiotracers to lipid analysis, microgram amounts of fatty acids were first separated by paper chromatography and then converted to their radiocobalt salts by treatment of the chromatogram with a solution containing ^{60}Co. The acids were located and measured by counting separate pieces of the chromatogram and the separated fractions were finally treated with Wij's iodine containing ^{131}I to obtain iodine values by measurement of activity (2).

In another early application, [^{14}C]methyl stearate was used to determine, by isotope dilution analysis, the proportion of methyl stearate in a mixture with tristearin, and to study the interchange of methyl groups between stearate and oleate (3).

Over recent years, numerous lipids and related compounds have been synthesized with a radioactive label: many of these are now available commercially at high specific activity. Furthermore, radionuclides have been introduced into the cellular lipids of micro-organisms, higher plants, cell cultures and animals in order to identify lipid and phospholipid components by separative/radiochemical techniques(4).

This chapter is designed to acquaint the potential user with techniques which are available for the radiochemical analysis of lipids. Methods of measurement, the theory of isotope dilution analysis, and assessment of radiochemical purity are described and illustrated by practical examples. Also information on the availability of labelled lipids and on their safe handling is given.

The experimental procedures have been chosen for the ease with which they can be performed in a basic nuclear laboratory. These procedures apply to radiolabelled compounds in general and do not necessarily involve the use of radiolipids. Some of the procedures are applicable to fractions which can be obtained by separative techniques described elsewhere in this book.

2. Measurement of labelled compounds

The choice of technique for the measurement of a particular radionuclide is governed by the type and energy of the radiation emitted by that nuclide. The four most commonly used nuclides employed as tracers in lipid chemistry are all pure beta emitters. Maximum energies of their emissions are listed in *Table 1*.

Tritium, because of the very low energy of its radiation, is measured at the poorest efficiencies by all counting methods. Some methods (for example, conventional Geiger or solid scintillation counting), cannot be used at all for the measurement of tritium, because the beta-particles do not have enough energy to penetrate the detector. Thus, the measurement of tritium presents special problems, a satisfactory solution to which has been achieved only by the use of liquid scintillation counting (LSC).

^{14}C and ^{35}S are also low-energy beta-emitters. These nuclides can be measured by most counting techniques, but high efficiencies can only be attained by proportional counting or by LSC. By contrast, ^{32}P, a high-energy beta-emitter, can be measured at reasonably high efficiencies by almost any counting technique. By use of Cerenkov counting (a type of LSC), ^{32}P may be easily and specifically measured when other beta-emitters of lower energy are present.

The need to measure radionuclides at the highest possible efficiency is

Table 1. Data for some nuclides commonly used as radiotracers in lipid chemistry

	Maximum beta energy/MeV	Half-life
^{3}H	0.019	12.3 years
^{14}C	0.158	5760 years
^{35}S	0.167	87 days
^{32}P	1.71	14.3 days

1 eV = 1.60×10^{-19} joules

important for reasons of economy (less radionuclide being required to attain a given sensitivity); and of safety (a lower level of associated radiation).

2.1 Autoradiography

This is the earliest and simplest technique for the detection of radioactivity and for examining the distribution of radioactivity throughout a solid medium. Radioactive spots on a surface such as a paper chromatogram or a thin-layer chromatographic (TLC) plate may be located by contacting an X-ray film with the surface for an appropriate time and developing the film in the usual way. The procedure should be carried out in a photographic dark room.

To prepare an autoradiograph from a chromatogram, the film should be attached firmly and the exposure should take place in an envelope of light-proof paper. The origin and the solvent front of the chromatogram should be marked on the film after development, or they can be treated with a radioactive 'ink' (such as a solution containing ^{32}P) before contacting, so that their images appear on the film, as in Chapter 3.

Exposure times will depend upon the radionuclide used and the quantity of it which is present. The modern unit of radioactivity is the becquerel (Bq): 1 Bq = 1 disintegration per sec = 27 picocuries (pCi): the curie, which is the older unit of radioactivity, was defined as the total rate of disintegration of 1 g of radium in equilibrium with its decay products ($3.7 \times 10^{10} s^{-1}$). A spot containing tritium at a surface density of 20 Bq mm^{-2} would require an exposure time of about 24 h. The times for similar activities of ^{14}C (or ^{35}S) and ^{32}P are shorter, about 8 h and 4 h respectively, because of the higher energies and longer ranges of the beta-radiations from these nuclides. Chromatograms containing ^{32}P should be kept separate from other chromatograms during exposure, otherwise false spots may develop. This precaution is not necessary for tritium, ^{14}C, or ^{35}S.

It is difficult to quantify the labelled species present in a spot from the intensity of the image produced on an autoradiograph. For this reason, autoradiography should be regarded as a technique of location only. Located spots may be cut out of a paper chromatogram or scraped off a TLC plate for measurement of their activities by an instrumental method.

The image from a radiochromatogram can be enhanced by spraying the chromatogram before exposure with a solution containing a phosphor such as PPO (see Section 2.4.1). This is particularly advantageous for tritium spots because the exposure time can then be substantially reduced (4).

High-performance autoradiography film (Hyperfilm), which is now available from Amersham International, may lead to the reinstatement of autoradiography as a cheap and rapid alternative to scintillation counting for radio-TLC (Section 3.1.) Also, radiographs of compounds labelled with ^{32}P, ^{35}S, ^{14}C, or ^{125}I may now be imaged on a phosphor screen (the PhosphorImager, marketed by Molecular Dynamics). Such screens are said to be 100 times more sensitive than film. They can be used at room temperature, require no chemical processing, and can be re-used indefinitely. Quantification is carried out from the image on the screen by means of a laser.

2.2 Geiger counting

The Geiger–Müller (GM) counter fitted with an end-window tube is the simplest, cheapest and most robust instrument for the accurate measurement of radioactivity (5, 6). The GM counter is a gas ionization counter: it is particularly suited to the measurement of beta-emitters, but it suffers two disadvantages in this respect. First, the efficiency of counting tends to be low, especially for the measurement of nuclides which emit low energy radiation. Thus, ^{14}C is counted with an efficiency of less than 5%. Tritium cannot be measured at all using an end-window GM counter but windowless gas-flow GM counters have been developed for this purpose. Second, it is not possible to distinguish electronically between radiations of different energies or types, because the pulses produced in a GM tube are all of approximately the same size. Discrimination can only be achieved by interposing thin sheets of various materials (absorbers) between the radioactive source and the tube, to cut out lower energy radiations.

Nevertheless, the GM counter has been extensively used for the measurement of labelled organic compounds. Although it has been largely superseded by LSC in research and industry, the GM counter should not be overlooked as a rapid means of locating radioactivity when sensitivity is not the prime consideration.

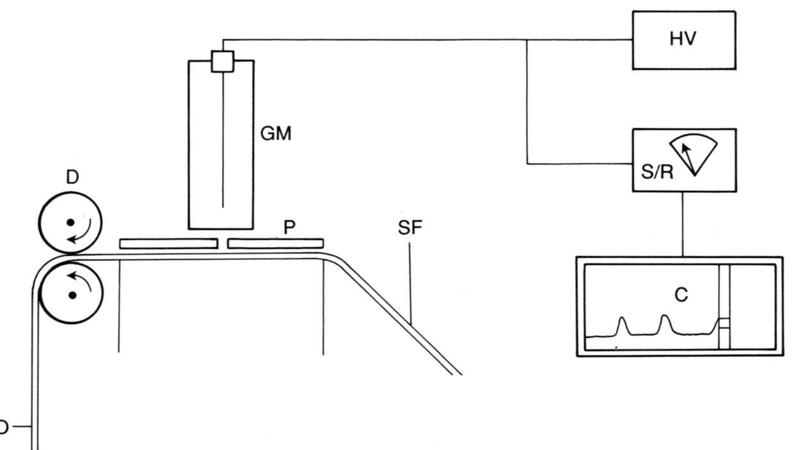

Figure 1. Geiger counting assembly for the scanning of radiochromatograms on paper strip. C = chart recorder, D = drive (rollers), GM = Geiger–Müller counting tube, HV = high voltage unit, O = origin of chromatogram, P = slotted lead plate, SF = solvent front of chromatogram, S/R = scaler/rate-meter.

A schematic diagram of a GM counter is part of *Figure 1*. The use of GM counting for the scanning of radiochromatograms is described in Section 3.1.

2.3 Proportional counting

There are three regions of operation for gas ionization counters according to whether the voltage applied to the counting tube is low, medium, or high (5, 6). These regions are designated 'ionization chamber', 'proportional', and 'GM'. Counters based upon the GM region have been discussed in Section 2.2. Ionization chambers find use mainly in health physics instruments and for the measurement of alpha activity. In the proportional region the applied voltage is sufficiently high to produce secondary ionizations in the counting gas (usually argon-methane) but not high enough to produce total ionization in the region of the anode and subsequent massive discharge ('the avalanche') which is characteristic of the GM counter. A pulse is created, the size of which is proportional to the energy of the particle producing it, and to the applied voltage. Thus, one radionuclide can be counted in the presence of another.

The instrumentation for a proportional counter is similar to that for a Geiger counter (*Figure 1*) with the inclusion of an amplifier between the counting tube and the rate-meter. Also the high-voltage supply must be stabilized so that a constant voltage to the tube is precisely maintained. The amplifier gain is set so that the rather small pulses from the tube are amplified sufficiently to cause the rate-meter and recorder to respond. Discrimination is achieved by suitable choice of gain so that pulses arising from a nuclide emitting high-energy radiation may be detected whilst those arising from a nuclide emitting low-energy radiation are excluded. In this way for example, ^{32}P may be counted in the presence of ^{14}C.

In gas-flow proportional counters the end-window counting tube is replaced by a chamber into which is placed the source to be measured. Air is flushed from the chamber by the counting gas and the source is counted under conditions of minimal absorption. Even tritium can be counted at high efficiency by this method. Gas-flow counting is a rather lengthy procedure however and it has been almost completely replaced by LSC for the measurement of organic compounds. Gas-flow counting still finds use for the calibration of standards, because the efficiency of counting can be very close to 100%.

2.4 Scintillation counting

This technique is based upon the interaction of radiation with certain materials (phosphors or scintillators) leading to the emission of light quanta which are converted to electronic pulses. Scintillation counting is now the preferred technique for the measurement of radiation because of the high counting efficiencies which are attainable and because one radionuclide can readily be measured in the presence of others by this means.

2.4.1 Liquid scintillation counting (LSC)

This is the method most commonly used for the radiochemical determination of lipids, for which purpose it is usually necessary to measure tritium or ^{14}C. The poor efficiencies of measurement of these low-energy beta-emitting nuclides by gas ionization counting have already been discussed. With LSC, efficiencies of up to 60 and 96% for tritium and ^{14}C respectively can be attained.

i. Principles of LSC

The sample is dissolved or suspended in a solvent (such as toluene) in which an organic phosphor [usually diphenyloxazole (PPO)] is also dissolved. A secondary scintillator such as 1,4-bis-2-(5-phenyloxazolyl)-benzene(POPOP) may also be present. The energy of radiation emitted in such an environment is efficiently transmitted through the solvent to phosphor molecules, which then undergo the scintillation process. Because the sample and phosphor are in intimate mixture, the probability of an emission leading to a scintillation is very high.

The solvent system is contained in a vial in close proximity to a photomultiplier (PM) tube so that the scintillations can give rise to electronic pulses which are amplified and fed as voltage pulses to a counter. The pulse sizes are proportional to the energies of the emissions which produce them. Thus, when a beta emitter is being measured the output consists of pulses, the energy spectrum of which is distributed similarly to the beta continuum. (*Figure 2a*) By use of a simple discriminator (an electronic device which allows only pulses lying within a set range to be registered) the count rates due to emissions from samples (signal) can be optimised with respect to the background count rate (low energy pulses arising mainly from electronic noise, and cosmic radiation).

If two nuclides, such as tritium and ^{14}C, are present in a sample, the discriminator may be set at a voltage corresponding to the maximum energy of the less energetic nuclide, enabling the specific measurement of the more energetic nuclide.

ii. Quenching

The principal problem in liquid scintillation counting is that of quenching (reduction of the signal, and displacement of the spectrum to regions of lower energy; *Figure 2b*). Quenching is caused by the presence of foreign species in the sample and may arise in two ways (*Figure 3*).

Chemical quenching is due to interference with the transfer of energy through solvent to the phosphor. Most species can cause chemical quenching to some extent, but certain species quench heavily: these include halogenated or nitrogen-containing compounds, metal salts, and dissolved oxygen.

Colour quenching results from the presence of species which reduce the transmittance, from the phosphor to the PM tube, of the emitted light. The

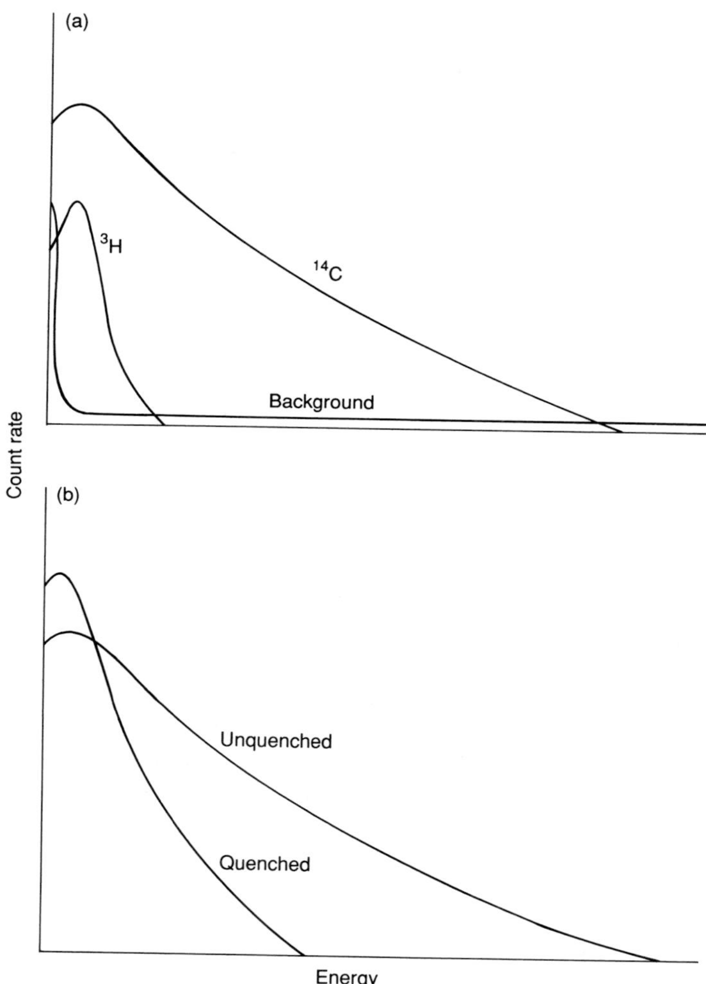

Figure 2. Pulse distributions in liquid scintillation counting.
(a) Background, ^3H and ^{14}C spectra.
(b) Effect of quenching on the ^{14}C spectrum.

wavelength of this light lies in the region of 400 nm: therefore serious colour quenching occurs when the liquid is yellow. The purpose of the secondary scintillant is to reduce colour quenching, by absorbing light from the primary scintillant and re-emitting at a longer wavelength at which the absorption by any coloured species present is less.

Because of quenching, it is not usually possible to determine the quantity of radioactive analyte in a sample by simply measuring the count rate and comparing it with count rates obtained from standards. Only if the samples

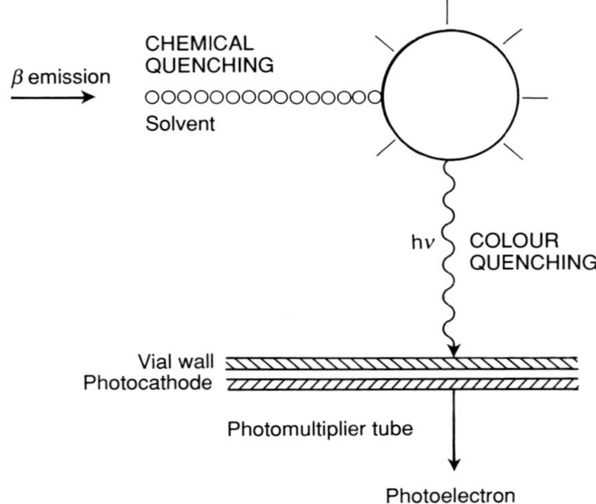

Figure 3. Origins of quenching in liquid scintillation counting.

and the standards are present in exactly the same liquid environment (i.e. quenched to the same extent) are the count rates strictly comparable. It is therefore usually necessary to measure not only the count rate of a sample but also the efficiency with which it is being counted. From these two parameters, the absolute disintegration rate of the sample is obtained.

Since the inception of LSC nearly forty years ago much effort has been expended to develop methods of measuring counting efficiency in order to correct for quenching.

iii. Methods of quench correction

The earliest, simplest and most accurate method of quench correction is internal standardization. For tritium or ^{14}C counting the internal standards most widely used are ^{3}H- or ^{14}C-labelled hexadecane which can be readily purchased at known specific activity. A method for determining the activity of a sample of a labelled organic compound by internal standardization is described in *Protocol 1*. Also described is a method of studying the effect of a quenching agent on the measured count rate.

Protocol 1. Determination of the activity of a sample of a ^{14}C-labelled compound, and study of quenching

Select a compound which is lipophilic, for example a ^{14}C-labelled lipid. For practice, [^{14}C]hexadecane itself could be used.

1. Prepare a solution in toluene containing a known concentration of the sample compound. The volume specific activity should be in the range 0.1–0.5 μCi(4–19 kBq)/ml.

2. Place 0.100 ml of the sample solution in a LSC vial. Add 10 ml of a 0.4% solution of PPO in toluene, and mix.
3. Measure the count rate in a liquid scintillation counter with gain and discriminator settings optimized for ^{14}C, for a period of 1 min.
4. Remove the vial from the instrument and add to it by dropping pipette a known weight (about 30 mg) of [^{14}C]hexadecane standard of known specific activity (about 1 μCi(37 kBq)/g).
5. Mix, and recount the sample to which standard has been added.
6. From the increase in count rate and the weight of standard added calculate the % efficiency with which the sample and standard have been counted.
7. From the % efficiency and the count rate of the sample (corrected for background) calculate the absolute activity of the sample in disintegrations per minute (d.p.m.), Bq and Ci.
8. From the weight of the compound taken and the absolute activity of the sample calculate the specific activity of the labelled compound in Ci(Bq)/g. If the relative molecular mass of the compound is known, obtain the specific activity in Ci(Bq)/mol. Compare this value with the stated value if this is available.
9. To the vial add 0.1 ml portions of CCl_4, mixing and recounting after each addition.
10. Plot a graph of quantity (or concentration) of CCl_4 vs. counting efficiency to illustrate the quenching effect of CCl_4.

The experiment could also be carried out with tritiated compounds.

Example

Specimen data:

Count rate of sample =52349 c.p.m.
Specific activity of standard =37.4 kBq g^{-1}
Weight of standard added =28.6 mg
Count rate of sample plus standard=92001 c.p.m.

Calculation:

Count rate due to standard added = 92001 − 52349 = 39652 c.p.m.

Activity of standard added $= \dfrac{37.4 \times 28.6 \times 1000 \times 60}{1000} = 64178$ d.p.m.

Efficiency of counting $= \dfrac{39652}{64178} \times 100 = 61.8\%$

$$\text{Absolute activity of sample} = \frac{52349}{61.8} \times 100 = 84707 \text{ d.p.m.}$$

$$\text{or} \quad \frac{84707}{60} = 1412 \text{ Bq} = 1.41 \text{ kBq}$$

$$\text{or} \quad \frac{1412}{3.7 \times 10^{10}} = 38.2 \times 10^{-9} \text{ Ci} = 38.2 \text{ nCi}$$

(Specific activities and weights of added standards are usually measured to three significant figures. Absolute activities should therefore usually be quoted to only three figures.)

Although the internal standard method of quench correction is accurate, it is time-consuming and expensive. Today, the principal use of internal standards is for the preparation of sets of quenched standards. Members of each set contain the same amount of standard but different amounts of a quenching agent such as nitromethane or an interfering substance which is present in the samples.

Quenched standards are used to calibrate instrumental methods of quench correction. The use of these methods has led to the development of the LS counter from a relatively simple single-channel instrument to the modern two- or three-channel instrument with external standards and computer interfacing. Performance has also been improved by the use of two photomultiplier tubes for coincidence counting to reduce the background count, and by pulse summation, to improve counting efficiency, especially for tritium (*Figure 4*) (7).

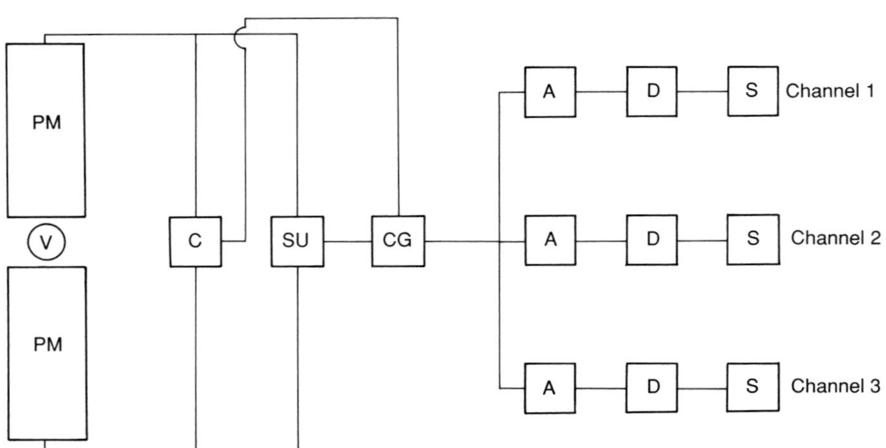

Figure 4. Circuit diagram of a three-channel liquid scintillation spectrometer. A = amplifier, C = coincidence unit, CG = coincidence gate, D = discriminator, PM = photomultiplier tube, S = scaler, SU = summation unit, V = vial.

All instrumental methods of quench correction involve the measurement of a quench index parameter (QIP) derived from either the sample spectrum, or from the spectrum of an external standard (a gamma-emitter such as ^{226}Ra) which can be moved automatically to a reproducible position close to the sample vial. By measuring the QIP and the counting efficiency of a series of quenched standards, a calibration graph is constructed. The counting efficiency of a sample is obtained by measuring its appropriate QIP and interpolating on the graph. The absolute activity of the sample is then obtained from the measured count rate. The various QIPs which can be measured on modern LS counters are listed and described in *Table 2*.

Table 2. Quench index parameters for liquid scintillation counting

(a) *Spectral index of the sample* (SIS)
The sample is counted once. The average energy of the pulses forming the sample spectrum is measured by means of an interfaced computer with a large number of memories (up to 1000) covering the energy-range of the spectrum. The SIS is the numerical value of the memory corresponding to the average energy (the memory in which the centre of gravity of the spectrum lies). Only one channel is required.

(b) *Sample channels ratio* (SCR)
The sample is counted once through two channels, one being set to cover a wider energy-range than the other. The ratio of the two count rates (SCR) changes with the extent of quenching.

(c) *External standard counts* (ESC)
The sample is counted twice, the first time through a channel optimized for sample pulses with the external standard in its remote position (out), and the second time through another channel set to exclude sample pulses, with the external standard in. The count rate recorded in the second channel is due to pulses generated in the vial by gamma radiation from the external standard. These pulses are subject to quenching just as the sample pulses are, so the external standard count constitutes a QIP.

(d) *External standard ratio* (ESR)
The sample is counted twice as for ESC but the second count is recorded through two other channels (2 and 3) which are set to measure different regions of the external standard spectrum alone. The ESR is the ratio of the counts in channels 2 and 3.

(e) *Inflection point Compton edge* (Beckman H number)
The external standard spectrum is analysed to obtain the coordinate of the point of inflection at the high energy edge. This point shifts according to the extent of quenching.

(f) *Spectral quench parameter of the external standard* SQP(E))
Similar to (e) but the coordinate of the maximum energy of the external standard spectrum is estimated.

(g) *Transformed spectral index calculations*
Packard have devised this QIP based upon the spectrum of a ^{133}Ba external standard. A mathematical transform is applied to the energy distribution to correct for spectral distortions due to wall effects, volume variation, and colour quenching. The QIP, expressed in energy units, is calculated from the transformed spectrum and is said to provide high statistical precision.

In *Protocol 2* a method is described for the measurement of the activity of a quantity of a ^{14}C-labelled compound by means of a QIP.

Protocol 2. Measurement of the activity of a ^{14}C-labelled compound by means of the external standard count quench index parameter

For practice, it is convenient to use the same compound chosen for the internal standard method described in *Protocol 1*.

1. Prepare a standard solution of the compound and from this prepare a test sample by mixing a portion of the solution with toluene/PPO scintillant.
2. Set channel 1 of the LS counter to cover the maximum energy range of ^{14}C (0.155 keV). Set channel 2 to cover the energy range 155–2000 keV.
3. Count the sample with and without the external standard present.
4. Count a series of quenched ^{14}C standards (each about 10^5 d.p.m., usually supplied with the instrument) with and without external standard present.
5. Construct a calibration graph from the quenched standard counts, plotting counting efficiency (channel 1) vs. external standard count (channel 2).
6. Interpolate the external standard count from the sample (channel 2) on the calibration graph, to obtain the sample counting efficiency. From this efficiency obtain a value for the activity of the sample (d.p.m., Bq or Ci) from the sample count rate (channel 1). See example in *Protocol 1*.

   The experiment could be extended by counting a series of test samples each containing the same quantity of labelled compound but quenched to different extents. Calculated activity values for each sample should agree within limits of experimental error.

iv. Dual labelling

Radioactive lipids labelled with two different nuclides (usually tritium and ^{14}C) have been employed in many biochemical investigations. QIP methods for the quantitation of dual-labelled compounds are available. One such method is described in *Protocol 3*. These methods are, of course, also applicable to the analysis of binary mixtures of ^3H- and ^{14}C-labelled compounds.

Protocol 3. Measurement of tritium and ^{14}C in dual-labelled compounds or binary mixtures using the external standard count quench index parameter

A three-channel LS counter is required for this method. For practice purposes, a standard solution containing $[^3H]C_{16}H_{34}$ in the range 0.2–1.0 μCi(7–37 kBq)/ml and $(^{14}C]C_{16}H_{34}$ in the range 0.1–0.5 μCi(4–19 kBq)/ml could be used.

1. Prepare a standard solution of the compound or a solution containing the mixture in known concentration, and mix a portion of this solution with toluene/PPO scintillant as described in *Protocol 1*.
2. Set channel 1 to cover the maximum energy range of ^{3}H(0–19 keV). Set channel 2 to cover the range of ^{14}C which excludes ^{3}H energies (19–155 keV). Set channel 3 to exclude both ^{3}H and ^{14}C (155–2000 keV).
3. Count the test solution with and without the external standard present. Count a series of quenched ^{14}C-standards 1.67 kBq (10^5 d.p.m.) and a series of quenched ^{3}H standards 16.7 kBq (10^6 d.p.m.) similarly.
4. Plot calibration graphs as follows.
   (a) ^{14}C efficiency in channel 2 vs. ^{14}C external standard count in channel 3;
   (b) ^{14}C efficiency in channel 1 vs. ^{14}C external standard count in channel 3;
   (c) ^{3}H efficiency in channel 1 vs. ^{3}H external standard count in channel 3.
5. Obtain the ^{14}C activity of the sample from the sample count in channel 2 and graph (a).

   Obtain the contribution of ^{14}C to the total sample count rate in channel 1, and hence the sample count rate in channel 1 due to ^{3}H, from the ^{14}C activity of the sample, the external standard count rate for the sample in channel 3, and graph (b). Obtain the ^{3}H activity in the sample from the sample count rate in channel 1 due to ^{3}H and graph (c).

Example

Specimen data: Count rates for a dual-labelled sample

	Channel		
	1	2	3
Without external standard/c.p.m.	104269	58946	0
With external standard/c.p.m.	High	High	E

(The counts in channels 1 and 2 with external standard are high, but are ignored.)

Calibration graphs generated from quenched ^{14}C and ^{3}H standards counted with and without external standard, showing sample counting efficiencies obtained by interpolation of the external standard count *E*.

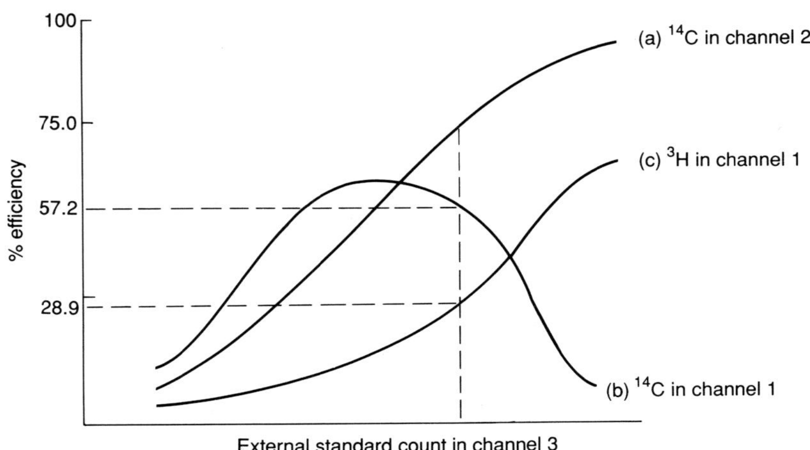

Calculation

Count rate of sample in channel 2	= 58946 c.p.m.
Counting efficiency of sample ^{14}C in channel 2 [from curve (a)]	= 76.0%
Activity of ^{14}C in sample	$= \dfrac{58946}{76.0} \times 100 = 77561$ d.p.m.
Counting efficiency of sample ^{14}C in channel 1 [from Curve (b)]	= 51.2%
Count rate in channel 1 due to ^{14}C	$= \dfrac{77561 \times 51.2}{100} = 39711$ c.p.m.
Total count rate in channel 1	= 104269 c.p.m.
Count rate due to ^{3}H	$= 104269 - 39711 = 64558$ c.p.m.
Counting Efficiency of sample ^{3}H [from curve (c)]	= 28.9 %
Activity of ^{3}H in sample	$= \dfrac{64558 \times 100}{28.9} = 223384$ d.p.m.

The activities of ^{14}C and ^{3}H in the sample can be expressed as follows.

	^{14}C	^{3}H
d.p.m.	77561	223384
KBq	1.29	3.72
nanocurie (nCi)	34.9	101

v. Cerenkov counting

When energetic beta-particles pass through a liquid medium, they excite molecules to produce visible scintillations in the range 350–600 nm. In order for this phenomenon (the Cerenkov effect) to occur the particles must be travelling at a velocity which exceeds that of light in the medium. The threshold energy of a beta-particle to produce the effect in water is 263 keV. Thus, any nuclide which emits beta-radiation with a maximum energy greater than this can be counted in aqueous solution by LSC without the addition of a phosphor.

Cerenkov counting is a convenient method for measuring ^{32}P- labelled compounds and has been applied to the determination of phospholipids (8). Counting efficiencies of up to 50% can be attained. The effect of substances which normally cause heavy chemical quenching is minimal, but colour quenching takes place to the same extent as in conventional LSC.

Quench corrections in Cerenkov counting may be performed by the usual means (for example, internal standard, SIS or SCR), but if sample solutions are colourless, quench corrections need not be carried out.

vi. Sample preparation

If the sample to be counted is a lipid or a mixture of lipids (for example, glycerides) it can usually be readily dissolved in toluene/PPO or other liquid scintillant. To count aqueous samples in a single phase system, a number of liquid scintillation 'cocktails' have been developed. One of the earliest of these is Bray's scintillant (naphthalene 60 g, PPO 4 g, POPOP 0.2 g, methanol 100 g, ethylene glycol 20 ml, 1,4-dioxane to 1 litre). This scintillant forms a single phase with up to 30% of water, and yields counting efficiencies of up to 22% for tritium and 86% for ^{14}C.

Tissue samples may be dissolved prior to LSC by treatment with hyamine hydroxide. Other, proprietary, solubilizing agents containing quaternary ammonium bases are available from Amersham International, New England Nuclear, E. Merck, and Packard. These agents may also be used to elute radioactive materials from paper chromatograms and TLC plates.

Some samples, such as culture media and acid aqueous solutions, are more conveniently measured in heterogeneous systems. A detergent (for example, Triton X-100), may be added to form a colloidal suspension of the sample in the liquid scintillant.

The presence of a base or a detergent in a system leads to a reduction in counting efficiency, and errors may arise if a simple QIP is used to evaluate the activity in a heterogeneous system.

2.4.2 Gamma scintillation counting

A gamma scintillation counter consists of a solid phosphor detector (usually sodium iodide activated with thallium) coupled to a PM tube; the detector assembly being connected through an amplifier and discriminator to a scaler and/or a recorder. As with liquid scintillation counting, pulse size is proportional to radiation energy. Thus, by pulse height analysis one radionuclide can be counted specifically in the presence of others.

Gamma radiation, unlike beta radiation, is monoenergetic. Hence a gamma spectrum contains a peak (the photopeak) corresponding to the total transfer of gamma energy to an electron (a photoelectron). The pulse spectrum also includes a continuum of lower-energy pulses resulting (a) from other processes (Compton scatter and pair production) which take place in the phosphor, and (b) from background (cosmic radiation, noise from the PM tube, etc.)

When setting up a scintillation counter for measurement of a particular radionuclide, it is desirable to optimise the instrument so that the maximum sample count (signal) is recorded with the minimum background (noise). The procedure for doing this is described in *Protocol 4*.

Protocol 4. Optimization of a gamma scintillation counter to measure a particular radionuclide

1. Obtain or prepare a source (about 0.1 μCi (4 kBq)) of the nuclide of interest. For practice, sealed sources of ^{137}Cs(E = 660 keV) or ^{60}Co(E = 1.17 and 1.33 MeV) could be used.
2. Adjust the upper discriminator setting to its maximum and the lower discriminator setting to zero.
3. With the source in the counter, increase the voltage on the PM tube (the HV) to the threshold at which counting begins (usually between 800 and 1000 V). Set the HV to 50 V above the threshold and count the source for 1 min.
4. Carry out similar counts, increasing the lower discriminator setting between each count in steps of 5% of the total discriminator range. Continue these counts until the count rate has fallen to about 10% of the first count rate.
5. Repeat the series of measurements several times, increasing the HV between each series by 50 V steps up to 1200 V.
6. Remove the source from the counter and measure the background at the same HV and discriminator conditions as those used with the source.
7. For each set of conditions, subtract the background count rate (B) from the gross source count rate to give the nett source count rate (R) and hence obtain values for R^2/B. The optimum HV and discriminator settings for measurement of the nuclide are those at which R^2/B is a maximum.

Notes

[a] The parameter R^2/B, rather than R/B, is chosen to minimize the effect of the poor precision associated with measuring the background for only 1 min.

[b] Optimization may be improved by making measurements of source and background counts with narrower intervals between settings in the region of R^2/B.
[c] The procedure can be considerably shortened by the use of a Simplex optimization computer programme (9).

The most important gamma-emitting radionuclides in lipid chemistry are the isotopes of iodine, ^{131}I(360 and 640 keV) and ^{125}I(36 and 28 keV (X-ray)]. These find use for radioimmunoassay (Section 4.3.) and for the iodination of unsaturated lipids (2). Scintillation counters are available with discriminator windows pre-set for the measurement of these isotopes.

3. Radiochromatography

Chromatography coupled with radiation detection is a powerful technique for the separation, identification, and measurement of labelled organic compounds which has been developed considerably over recent years (10). This section is an account of the two main types of radiochromatography, both of which have been applied to lipid analysis.

3.1 Paper and thin-layer chromatograms

Radiochromatogram scanners incorporating GM counters have been marketed by many manufacturers of nucleonic equipment. A schematic diagram of a typical scanner for paper chromatograms is shown in *Figure 1*.

The paper strip of drawn at constant speed beneath the window of a GM tube, and scanned from the origin to the solvent front through the slit in a lead plate placed between strip and tube. The count rate from the tube is displayed on a rate-meter with output to a pen recorder. Spots on the strip are thereby recorded as peaks on the chart.

An experiment in which a radiochromatogram is prepared on paper strip and then scanned manually using a GM counter is described in *Protocol 5*.

Protocol 5. Preparation and scanning of a radiochromatogram

1. Mix 80 ml of *t*-butyl alcohol with 10 ml of formic acid and 40 ml of water. Place a dish containing 20 ml of this solution in a chromatographic tank fitted with a glass trough and glass rods. Cover the tank and allow it to stand for 1 h so that the atmosphere in it equilibrates with the solution.
2. Prepare a solution of [^{32}P]phosphoric acid by mixing 1 ml of 1 μCi/ml carrier-free [^{32}P]sodium phosphate with 1 ml of 10% phosphoric acid.
3. Cut a 40 cm length of 3-cm-wide Whatman No. 4 paper strip. Mark origins in pencil at a distance of 10 cm either side of the mid-point of the strip.

Protocol 5. *Continued*

4. Place 10 μl of [^{32}P]phosphoric acid on one origin. (Identify this origin with a mark.) Place 10 μl of 10% phosphoric acid on the other origin.
5. Remove the lid from the tank and introduce the strip with its mid-point resting in the trough, secured by a rod, and its two parts hanging vertically over supporting rods.
6. Pour the remainder of the aqueous butyl alcohol-formic acid into the trough and replace the lid.
7. Allow the chromatograms to run until the solvent fronts are within a few centimetres of the ends of the strip. (About 12 h.)
8. Prepare the spray reagent by dissolving 5 g of ammonium molybdate in 100 ml of water and pouring the solution with constant mixing into 40 ml of 30% HNO_3.
9. Remove the strip from the tank and mark the positions of the solvent fronts. Hang the strip in air to dry for a few minutes. Divide the strip at the mid-point.
10. Spray the inactive half of the strip and mark the limits of the yellow spot which appears. Estimate from these limits an R_f value for the inactive phosphate species.
11. Mark the active half of the strip into numbered sections, 1 cm in width, from the origin to the solvent front. Cut the strip into these sections.
12. Measure the activity of each section by Geiger counting. (Place each section in a fixed position close to the window of the GM tube and count for a 10-sec period.)
13. Construct a histogram of section number vs. count rate and from this estimate the R_f value of the active phosphate species. Compose the R_f values of inactive phosphate and active phosphate.
14. Select the sections which contain all the radioactivity. (The one with maximum activity and one or two each side of it.) Count these as a single source. The counting time should be sufficient to allow at least 10^4 counts to be recorded and the count rate should be corrected for lost counts (obtained from tables) and background (previously measured).
15. Calculate a value for the counting efficiency of ^{32}P under the conditions of this experiment, from the corrected count rate and the absolute disintegration rate of the activity taken.
16. Perform further counts on the composite source interposing various calibrated aluminium absorbers between the source and the GM tube. Absorbers should be chosen to cover the range 0–120 mg/cm^2 in five or six approximately equal intervals. At least 10^4 counts should be accumulated for each measurement and count rates should be corrected as before.

17. Plot a graph of absorber thickness (mg/cm^2) vs. log corrected count rate (c.p.m.). From the graph (which should be nearly linear) estimate the 'half-thickness' ($x_{½}$ = the absorber thickness required to reduce the count rate to one-half of the value with no absorber present). Calculate a value for the maximum beta-energy of $^{32}P(E_{max})$ from the empirical expression $x_{½} = 46\ E_{max}^{1.5}$. Compare the value obtained to the accepted one (1.71 MeV).

The labelled species involved in this experiment, ([^{32}P]phosphate), has found wide use in lipid chemistry. It is inexpensive and is safe and easy to handle. The experiment includes measurements of counting efficiency and of the maximum energy of the beta radiation emitted from ^{32}P.

The most convenient method for examining radio-TLC plates is direct scanning with radiation detectors. If the nuclides of interest are tritium and/or ^{14}C, windowless flow-through GM or proportional detectors are necessary. Scanning is performed by mounting the plate on a motor-driven platform and moving the plate in one or two dimensions beneath the detector. Such equipment is expensive, however, and can only be justified if the throughput is high.

The most sensitive method for examining TLC plates is LSC. For this, the stationary phase must be removed in sections and suspended in a scintillation cocktail for counting. If performed manually, this procedure is time-consuming and requires some skill. Equipment is commercially available however for dividing a chromatogram automatically into a series of narrow zones.

TLC plates may be imaged by positioning a windowless proportional detector over the whole plate with the anode wire running the length of the chromatogram (*Figure 5*). Radiation from a spot on the plate causes local ionisation of the gas (Ar/CH_4) in the chamber of the detector. A pulse is thereby produced at a point on the wire immediately above the spot. The

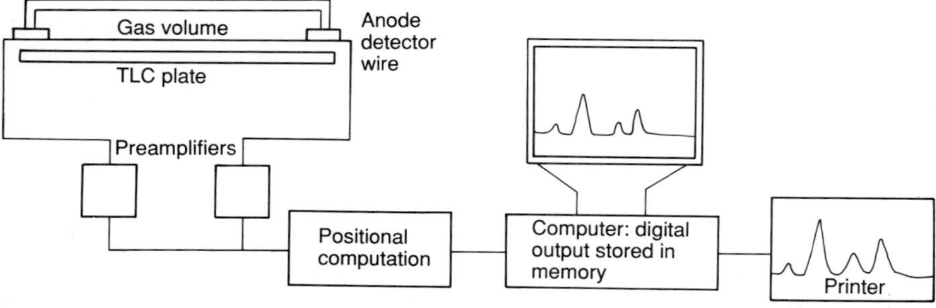

Figure 5. Proportional counting assembly for the simultaneous imaging of radio-TLC plates.

pulse divides and flows to pre-amplifiers at each end of the wire where the ratio of the pulse sizes is governed by the resistance of the portion of the wire from the source of the event to each pre-amplifier. Measurement of this ratio enables the position of the source to be fixed. An image of the chromatogram is displayed on a monitor and counts are allowed to accumulate until the required sensitivity has been attained. Chromatograms may be stored and hard copies printed (11).

3.2 Column chromatography

The majority of developments in this field have taken place with high-performance liquid chromatography (HPLC). For radio-HPLC it is necessary to add to a conventional HPLC system (pump, injector, mass detector, and controller/integrator) a radiochemical detector, usually in series with the mass detector. For the measurement of compounds labelled with a gamma-emitting nuclide the detector can consist simply of a loop of Teflon tubing in the well of a NaI(Tl) crystal coupled to a scintillation counter and recorder. For the associated measurement of mass, either a UV/vis or an electrochemical detector is used.

For the measurement of beta-emitters, flow-through cells with solid scintillants (anthracene, plastic, glass, or CaF_2) can be used. An early application of such cells was to the determination of tritium and ^{14}C in aqueous solution (12).

High sensitivity for on-line quantitation of eluent containing tritium and ^{14}C has been achieved by means of the homogeneous flow cell. The eluent is merged with a liquid scintillator and the merged solution is passed through a flow-through cell mounted in a LSC. By means of this technique, steroids resulting from the incubation of tissues with radioactive precursors have been quantitated (13) and studies have been made of the metabolism of platelets and plasma (14), and of phospholipid methylation in myogenic cells (15). Radio-HPLC has also been applied to the determination of the following radiolipids of pharmaceutical interest:

α-CH_3[1-^{11}C]heptadecanoic acid (16), 15-(*p*-[^{75}Br]bromophenyl) pentadecanoic acid (17), 17-[^{123}I]heptadecanoic acid (18), [^{131}I]-11-*N*(*p*-iodophenyl sulfonamide undecanoic acid (19), [1-^{11}C]palmitic acid (20, 21), ω-(*p*-^{123}I-iodophenyl)-pentadecanoic acid (22) and [^{11}C]valproic acid (21).

[^{11}C]valproic acid has been synthesized in order to study the behaviour of sodium valproate, an anticonvulsant drug used in the treatment of epilepsy. Examination of this labelled acid by HPLC, using a radiometric detector and a refractive index detector simultaneously, has revealed the presence of two unidentified radioactive impurities in the crude product (*Figure 6a*). The purified product, obtained by acidification of a solution in aqueous sodium bicarbonate, followed by extraction into ether, exhibits high radiochemical purity (*Figure 6b*).

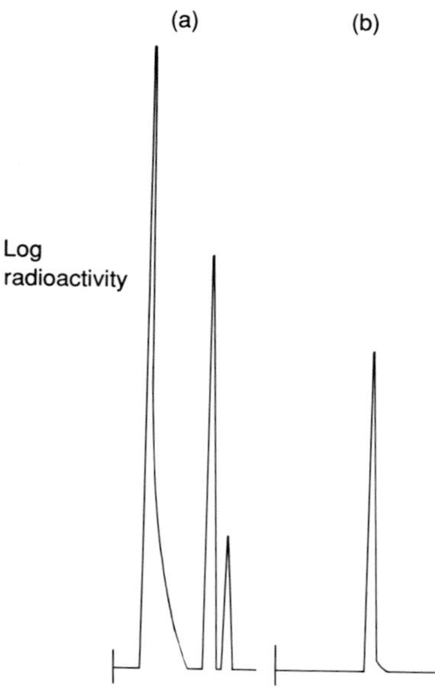

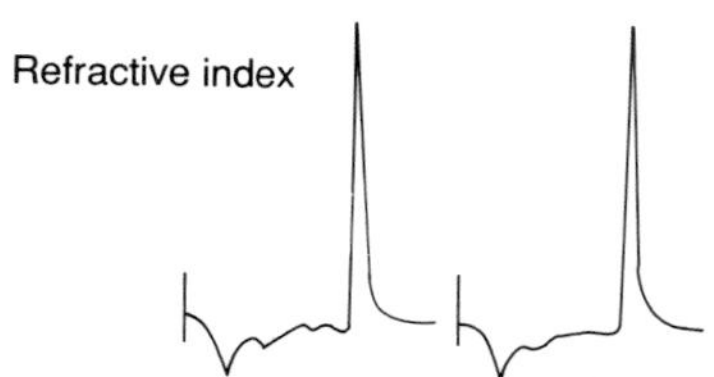

Figure 6. HPLC chromatograms of [^{14}C]valproic acid, (a) crude product; (b) purified acid. Column; Partisil 10-ODS-3; eluent CH_3CN:H_2O:HOAc 35:64:1; flow-rate 0.9 ml/min.

4. Radioisotope dilution analysis

The term covers a number of analytical procedures involving the measurement of specific activity (count rate per unit mass or volume), which have found particular application to the determination of components in mixtures of organic compounds, such as reaction mixtures produced during the synthesis of both inactive and labelled compounds. Radioimmunoassay (RIA), a special example of substoichiometric isotope dilution analysis (SIDA), has found wide application to the determination of trace quantities of proteins, complex lipids, and drugs.

4.1 Single dilution analysis

To a portion of a sample, containing an unknown quantity (X g) of an analyte, is added a standard quantity (M g) of the labelled analyte with measured activity (C c.p.m.). Some pure analyte or analyte derivative is separated from the mixture and its specific activity is measured (S c.p.m./g). The amount of inactive analyte which has diluted the radioactive standard is found from the expression $X = C/S - M$.

An early reported analysis of this type is a determination of methyl stearate in admixture with tristearin (3). To 2.96 g of the mixture was added 0.498 g of ^{14}C-labelled methyl stearate with a total activity of 374 arbitrary units. By recrystallization, 0.250 g of pure methyl stearate was isolated, having an activity of 41.5 units. Thus, the percentage of methyl stearate in the mixture was

$$\frac{\dfrac{374}{41.5/0.250} - 0.498}{2.96} \times 100 = 59\%.$$

An inorganic isotope dilution analysis which can be readily carried out in a radiochemical laboratory is described in *Protocol 6*.

Protocol 6. Determination of phosphate by isotope dilution analysis

For practice, a sample solution 0.03 M in Na_2HPO_4 and containing 0.4% NaCl may be used.

1. To 10.0 ml of sample solution add about 0.2 μCi (8 kBq) of carrier-free ^{32}P as aqueous $^{32}PO_4^{3-}$. Dilute the mixture to 25.0 ml.
2. Measure the activity of a 10.0 ml aliquot of the diluted solution by liquid Geiger or Cerenkov counting (S c.p.m.)
3. Transfer the solution from the counting vessel to a conical flask (the transfer need not be quantitative) and dissolve 12 g of NH_4NO_3 in the transferred solution.
4. Prepare a solution of ammonium molybdate by dissolving 100 g of MoO_3 in a mixture of 400 ml of water and 80 ml of NH_4OH(s.g. 0.88). Mix this solution (filtered if necessary) slowly, with stirring beneath the surace of a mixture of 400 ml of conc. HNO_3 and 600 ml of water. Store the final solution in a warm place for several days and decant from any sediment.
5. Run 50 ml of the ammonium molybdate solution, heated to 40–45 °C, into the flask, with constant swirling. Allow the flask to stand for 30 min.

6. Filter most of the precipitated ammonium phosphomolybdate $(NH_4)_3PO_4.12MoO_3$ on a sintered glass crucible. Wash the filtered precipitate with 1% KNO_3 solution until the filtrate no longer gives an acid reaction with methyl orange.
7. Wash the precipitate several times with acetone and allow it to dry in an oven at 105 °C for 15 min.
8. Accurately weigh a portion (about 200 mg) of the precipitate (*W* g) and dissolve the portion in 5 ml of 4 M HNO_3. Dilute this solution with water to 25.0 ml.
9. Measure the activity of a 10.0 ml aliquot of the diluted solution (*P* c.p.m.).
10. Obtain the concentration of phosphate in the sample solution from the expression

$$\text{Concentration (g/litre)} = \frac{W \times 95 \times S \times 100}{1877 \times P}.$$

(The relatively molecular masses of PO_4^{3-} and $(NH_4)_3PO_4.12MoO_3$ are 95 and 1877 respectively.)

An advantage of isotope dilution analysis over other methods involving chemical separations is that it is not necessary to separate the analyte quantitatively. The separated analyte must, however, be in a pure form so that its specific activity can be measured.

If the analyte is radioactive, it can be determined by the related technique of reverse dilution analysis using an inactive standard. The expression $X = C/S - M$ can again be applied, where *C* is now the activity of the portion of sample taken. Reverse dilution is applicable when the analyte is the only radioactive component of the sample (for example, for the calibration of a standard radioactive solution).

Reverse dilution analysis can also be used to measure the radiochemical purity (*R*) of a labelled compound. The compound is diluted with a large amount of its unlabelled form. After thorough mixing, the specific activity (A_1) is measured. Part of the mixture is then exhaustively purified and the specific activity (A_2) of the pure product is measured. Radiochemical purity is obtained from the expression

$$R = \frac{A_2}{A_1} \times 100.$$

The radiochemical purity of [^{3}H]xylocaine, prepared by exchange labelling with the boron trifluoride complex of tritiated phosphoric acid, has been measured in this way (23). The labelled product is diluted with pure, unlabelled xylocaine and the mixture is successively recrystallised from petroleum ether, until the specific activity falls to a constant value. In one such measurement, about 10 mg of product was mixed with about 100 mg of unlabelled xylocaine. The specific activity of the mixture was found to be 7.14×10^6 d.p.m. mg^{-1}. After recrystallization, the specific activity of the pure product was found to be 6.93×10^6 d.p.m. mg^{-1}, and the radiochemical purity was calculated as 97%.

4.2 Double dilution analysis

When radioactive species other than the analyte are present, the activity of the analyte cannot be measured directly. It is then necessary to take two (usually equal) portions of sample, to add a known quantity (M_1) of pure inactive analyte to one portion, and a different quantity (M_2) to the other. Separations of pure analyte are carried out from each portion and the specific activities of the separated analytes (S_1 and S_2 c.p.m./g) are measured. The quantity of analyte (X) in the portions taken, the activity (A c.p.m.) associated with this quantity, and the specific activity of the analyte in the sample (A/X c.p.m./g) can be obtained by simultaneously solving the equations

$$A/(X + M_1) = S_1 \quad \text{and} \quad A/(X + M_2) = S_2$$

from which $X = \dfrac{S_1M_1 - S_2M_2}{(SS_2 - S_1)}$ and $A = \dfrac{S_1S_2\,(M_1 - M_2)}{(S_2 - S_1)}$.

4.3 Substoichiometric isotope dilution analysis (SIDA) and radioimmunoassay (RIA)

In dilution analysis the limit of detection is governed by the sensitivity of the method used to quantify the separated analyte. But in SIDA methods, including RIA, measurement of this quantity is avoided and only the separated activity is measured. This allows extremely low limits of detection to be achieved.

For SIDA, the sample (X g of analyte) is added to the standard (M g of labelled analyte with activity C c.p.m.). To the mixture (sample + standard) and to a further M g of standard alone is added a reagent which forms a separable derivative with the analyte. Exactly the same quantity of reagent is added to both (sample + standard) and to standard alone and this quantity must be substoichiometric to the quantity (M) of standard analyte. (An equivalence ratio of about 1:2 is usually employed.) Under these conditions of excess analyte the quantities of derivitized analyte separating from the (sample + standard) and from the standard should be exactly the same. The activities, A_{S+St} and A_{St}, (but not the weight (M_s)) of the two equal amounts of separated analyte are measured.

For (sample + standard) $C/(X + M) = A_{S+St}/M_s$

For standard alone $C/M = A_{St}/M_s$.

These expressions can be combined to give

$$X = M(A_{St}/A_{S+St} - 1) \quad [1]$$

Thus, the analytical result is obtained simply from a ratio of activities and the weight of labelled standard taken.

The first recorded application of RIA, for the determination of insulin (24), pre-dates the codification of SIDA (25) by several years. RIA is used for the determination of molecules of physiological importance which exhibit immunoresponse, especially large proteins. In order for smaller molecules which do not exhibit immunoresponse to be determined by RIA these molecules must be capable of acting as haptens, i.e. they must be bound to an antigenic carrier molecule such as human serum albumin. Haptens are usually compounds with relative molecular mass less than 1000: they include phospholipids and triglycerides.

The label for the standard analyte in RIA is sometimes tritium, but is more usually ^{125}I, which can be introduced into a large molecule without affecting its biochemical behaviour. The reagent is an antibody specific to the analyte, raised in an animal into which the analyte(antigen) has been injected. The antibody is added substoichiometrically to the labelled standard to form a complex which is separated by centrifugation and counted.

An RIA calibration is carried out by taking a series of inactive analyte standards through the procedure. Samples (for example, serum or plasma), are analysed by subjecting them to the same procedure and interpolating the measured activities on the calibration curve. The analyte axis of the curve corresponds to the X term in the SIDA equation [1] and the activity axis corresponds to the term A_{S+St}. The relationship is therefore a hyperbolic one (*Figure 7*).

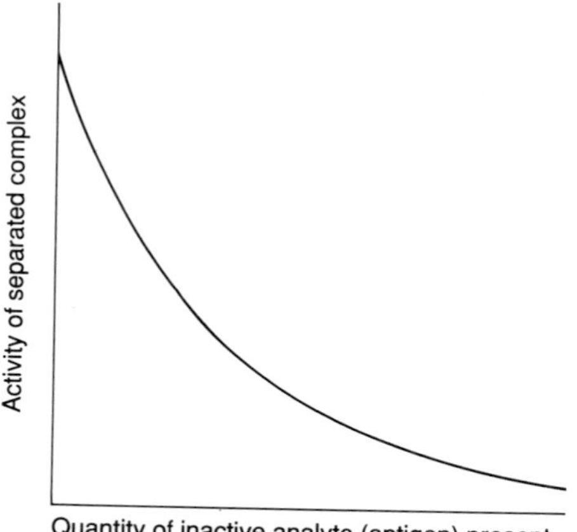

Figure 7. The general form of RIA calibration curves.

RIA methods tend to be lengthy and require the use of fairly expensive reagents, usually purchased in the form of kits. In *Protocol 7*, such a method for the determination of insulin in blood serum is described. In *Protocol 8*, a simulated RIA method is described which employs inexpensive inorganic reagents but which incorporates all the important features of the technique.

Protocol 7. Determination of insulin in blood serum by radioimmunoassay (published by permission of Kodak Clinical Diagnostics Ltd)

For this analysis, a commercial radioimmunoassay kit marketed by Kodak Clinical Diagnostics Ltd, is used.

The following are supplied in this kit.

(*i*) 2 ml of[^{125}I] insulin solution in stabilized phosphate buffer, containing not more than 10 μCi (370 kBq) of ^{125}I.

(*ii*) Insulin binding reagent (two vials).

(*iii*) Human insulin standard (about 2000 μunits—the exact value is stated on the vial).

(*iv*) Solid buffer (two containers).

Items (*ii*), (*iii*), and (*iv*) are freeze-dried.

The quantities are sufficient for the determination of 68 duplicate unknowns on one occasion or 28 on each of two occasions, together with calibrations.

If this protocol is being used for group teaching purposes, it is instructive to carry out replicate analyses on the blood of a single subject sampled twice: first, after a period of starvation, and second, about 30 min after a carbohydrate-rich meal.

1. A medical practitioner or qualified phlebotomist should sample the subject's blood (5–10 ml) into a centrifuge tube.
2. Centrifuge the sample at a low speed in order to separate the red cells intact from the serum.
3. Remove the straw-coloured serum into a glass or plastic tube. Unless the analysis is to be carried out immediately, stopper the tube and store the serum in a deep freeze. The water used in subsequent steps should be freshly distilled. All glassware should be scrupulously clean.
4. Dissolve the solid buffer (the contents of one container) in 30 ml of water and dilute the solution to 100 ml.
5. Dissolve the insulin binding reagent (the contents of one vial) in 8 ml of water using a vortex mixer.
6. Dilute 1 ml of the [^{125}I] insulin solution with 7 ml of buffer solution.
7. Prepare a stock standard solution of insulin by adding to the human

insulin standard sufficient buffer to produce a concentration of exactly 100 μunits ml^{-1}.

8. Prepare a 160 μU ml^{-1} working standard solution of insulin by mixing 2.1 ml of buffer with 400 μl of stock standard insulin.
9. Prepare other working insulin standards by mixing volumes of buffer and 160 μU ml^{-1} standard insulin as follows:

Working insulin standard concentration/μU ml^{-1}	10	20	40	80
buffer/μl	1500	700	600	500
160 μU ml^{-1} standard/μl	100	100	200	500

10. Into plastic, round-bottomed tubes, 70 mm in length and 10 mm in diameter, pipette solutions according to the following scheme (all volumes are in microlitres). (Pipette tips should first be wetted by drawing up and discharging buffer solution several times before pipetting insulin solutions.)

	Total counts	Standards μU ml^{-1}						Blank	Sample 1	Sample 2
Tube	1–3	4–6	7–9	10–12	13–15	16–18	19–21	22–24	25, 26	27, 28
Buffer	–	100	–	–	–	–	–	100	–	–
Insulin standard	–	–	100	100	100	100	100	–	–	–
Serum sample	–	–	–	–	–	–	–	–	100	100
Binding reagent	–	100	100	100	100	100	100	–	100	100

11. Vortex mix the contents of each tube. Store all tubes at 2–4 °C for 45 min, then add 100 μl of diluted [^{125}I] insulin solution to each tube. Set aside tubes 1–3 for subsequent counting (see step 16).
12. Vortex mix the contents of the other tubes again and store these tubes at 2–4 °C for at least 2.25 h. Then add 700 μl of buffer.
13. Vortex mix a third time and centrifuge the tubes with a force of at least 1500 *g* for 25 ± 2 min. At the end of this period, allow the tubes to come to rest gradually (do not apply a brake).
14. Decant the supernatant liquid from each tube by inverting the tube gently in one continuous movement (avoid tapping or shaking). Place the inverted tubes on paper tissue and allow them to drain for at least 15 min. Remove any surplus liquid from the neck of each tube with paper tissue before returning the tube to an upright position.
15. Wipe the bottom and the rim of each tube with damp paper tissue.
16. Measure the radioactivity in each tube using a gamma scintillation counter set to measure ^{125}I (see *Protocol 4*). Count each tube either for 1

Protocol 7. *Continued*

min or, if constant precision is required, over a period of time sufficient for 10^4 counts to be recorded.

17. Construct a calibration curve by plotting counts per minute versus standard concentration, for the data from tubes 4–21. Draw the best curve through the mean points of the replicate counts, ignoring any outlying results. The curve should be hyperbolic in form.

18. Evaluate the concentration of insulin in each sample by interpolation of sample count rates (tubes 25 and above) on the calibration curve.

19. Check the performance of the assay by comparing
   (a) replicate results;
   (b) blank count rates (tubes 22–24) with total count rates (tubes 1–3), (the former should not exceed 5% of the latter);
   (c) count rates with no insulin standard added (tubes 4–6) with total count rates, (the former should lie between 20 and 35% of the latter).

Protocol 8. Determination of phosphate by substoichiometric isotope dilution analysis: an experiment to illustrate the principles of radioimmunoassay

An approximately 0.2 M aqueous phosphate solution (Na_2HPO_4 or NaH_2PO_4) may be used as the sample.

1. Prepare an aqueous magnesia solution which contains 25 g of $MgCl_2.6H_2O$ and 50 g of NH_4Cl/litre (0.125 M Mg).

2. Place 1.0 ml of ^{32}P-labelled 0.2 M, 0.2 μCi(8 kBq)/ml sodium hydrogen phosphate solution into each of eight plastic centrifuge tubes.

3. Into tubes 1–6 place 0.0, 0.5, 1.0, 1.5, 2.0, and 2.5 ml respectively of 0.20 M sodium hydrogen phosphate solution. Into tubes 7 and 8 place 1.0 ml portions of the sample solution.

4. Dilute the contents of each tube to 5 ml and mix thoroughly.

5. Add 1.0 ml of magnesia solution to each tube followed by 5 drops of NH_4OH (s.g. 0.88). Mix the contents and allow the tubes to stand until the precipitates coagulate.

6. Centrifuge the tubes and discard the supernatant liquid.

7. Wash the precipitates with 5 ml of 5% ammonia. Centrifuge again and discard the washings.

8. Dissolve the washed precipitates in 10 ml of 2 M HCl.

9. Transfer the solutions to counting vials and count each vial for 1 min in a liquid scintillation counter set for Cerenkov counting (tritium settings are suitable).
10. Plot a calibration graph of quantity of inactive phosphate present in tubes 1–6 vs. the count rates recorded for the precipitates in each of these tubes. The graph should be a hyperbolic curve.
11. From the sample count rates and the graph obtain the concentration of phosphate in the sample solution. (Sample count rates should lie near the centre of the calibration range and should agree closely with each other.)

In *Table 3*, substances which have been measured by RIA are listed: the list is not exhaustive.

Table 3. Some analytes which have been determined by radioimmunoassay

Nonpeptide hormones
- Aldosterone
- Cortisone
- Estrogens
- Testosterone
- Prostaglandins
- Thyroxine
- Triiodothyronine

Peptide hormones
- Adrenocorticotropic hormone (ACTH)
- Angiotensins
- Bradykinins
- Calcitonin (CT)
- Cholecystokinin-pancreozymin (CCK-PZ)
- C-peptide
- Enteroglucagon
- Follicle-stimulating hormone (FSH)
- Gastrin
- Glucagon
- Growth hormone
- Human chorionic gonadotropin (HCG)
- Human chorionic somatomammotropin (HCS)
- Insulin
- Lipotropin (LPH)
- Luteinizing hormone (LM)
- Melanocyte-stimulating hormone (MSH)
- Oxytocin
- Parathyroid hormone (PTH)
- Proinsulin
- Prolactin
- Secretin
- Somatomedins
- Thyroid-stimulating hormone (TSH)
- Thyrotropin releasing factor (TRF)
- Vasopressin

Table 3. *Continued*

Non-hormonal substances
- Albumin
- Anti-haemophilic factor (AHF)
- Australia antigen (HAA)
- Carcinoembryonic antigen
- C_1-esterase
- Cyclic nucleotides
- Digitoxin
- Digoxin
- α-Fetoprotein
- Folin acid
- Fructose 1,6-diphosphatase
- Immunoglobulin G(IgG)
- Intrinsic factor
- LSD
- Morphine
- Neurophysin
- Rheumatoid factor
- Thyroxine-binding globulin

5. Radiochemical purity

When a labelled organic compound is used as a tracer it is necessary to ensure that it is radiochemically pure, i.e. that none of the radioactivity present is in the form of other compounds such as synthesis by-products, decomposition products, or homologues. Labelled compounds can decompose on storage due to self-radiolysis. Data on the stability of some labelled lipids (*Table 4*) (26) indicate that substantial decomposition of tritiated glycerides takes place over periods of a few months, whereas ^{14}C-labelled glycerides remain stable for much longer.

The radiochemical purity of labelled compounds is usually monitored by chromatography, the simplest procedures being paper chromatography and TLC. A small portion of the compound is chromatographed under conditions which lead to the separation of any possible impurities. The chromatogram, which may be one- or two-dimensional, is scanned radiometrically, and the positions of the activity spots are located. If the compound is pure there should be only one spot. The percentage purity is obtained by quantitative

Table 4. Stability of some labelled glycerides in benzene solution

Glyceryl ester	**S** (mCi/mM)	**C** (mCi/ml)	**T** (°C)	**t** (months)	**D** (%)
Trioleate-9,10-^3H	354	28.4	20	12	17
Tristearate-9,10-^3H	4000	57	20	4.5	10
Tristearate-9,10-^3H	4300	30	20	9	8
Tri(oleate-1-^{14}C)	9.8	na	20	16	nd
Tri(stearate-1-^{14}C)	14.5	na	20	55	nd
Tri(stearate-2-^{14}C)	1.36	na	−40	24	2

S = specific activity, C = concentration T = temperature, t = storage time, D = decomposition, na = not available, nd = not detected measurement of all the active spots, and impurities may be identified by comparison of the chromatogram with another produced by running authentic samples of possible labelled impurities under similar conditions.

Column chromatography may be used to check radiochemical purity. An HPLC instrument fitted with both UV and radiation detectors can indicate with high resolution the presence of labelled and unlabelled impurities. The percentage purity can be quantified as before, the impurities can be identified by means of authentic unlabelled compounds, and methods of further purification by preparative chromatography can sometimes be devised. A simple experiment for monitoring the radiochemical purity of a labelled iodine compound is described in *Protocol 9*.

Protocol 9. Estimation of the radiochemical purity of carrier-free $[^{131}I]$sodium iodide

1. Cut three 40-cm lengths of Whatman No. 4 paper strip, 3 cm wide.
2. Apply 10 μl portions of aqueous solutions to marked points in the centre of the strips, 10 cm from one end, as follows:

   strip 1, 0.5% KI, 1% KIO_3 and 0.5 Ci/ml $Na^{131}I$
   strip 2, 1% KI
   strip 3, 2% KIO_3
3. Suspend the strips as described in *Protocol 5* in a chromatographic tank, the atmosphere in which has been equilibrated with methanol/water (9:3).
4. Elute with methanol/water for about 3 h.
5. Remove the strips from the tank, mark the positions of the solvent fronts and allow the strips to dry for a few minutes.
6. Spray strips 2 and 3 with the following solutions:

   strip 2, 1% KIO_3 in 2 M acetic acid
   strip 3, 1% KI in 2 M acetic acid.

   The position of the species on each strip will be indicated by the immediate appearance of an I_2 stain. With a pen, indicate the position of each stain before it fades. Estimate from these positions R_f values for iodate and iodide.
7. Following the procedure described in *Protocol 5*, cut strip 1 into sections and count each section by GM or scintillation counting, and thereby locate the region or regions of highest activity.

Compare the R_f values of iodate, iodide, and radioiodine species and deduce:

(a) the form or forms of the radioiodine and

(b) whether any iodide/iodate exchange has occurred.

---

6. Availability of radiolipids and related compounds

The practical procedures described in this chapter can be carried out using radiotracers which are relatively inexpensive, in order to illustrate radiochemical methods which are relevant to lipid analysis. If these procedures are adapted to particular needs in lipid chemistry, more expensive radiolabelled compounds are required. In *Table 5* is a list of some of the simpler radiolipids together with some labelled precursors which are currently available from Amersham International one of the world's largest suppliers of radiolabelled compounds. A similar range is available from New England Nuclear (NEN).

Tritiated compounds are usually supplied at higher specific activities than

Table 5. Some radiolabelled lipids and other related labelled organic compounds available from Amersham International

	Specific activity (mCi/mmol)	**Unit price (£/100 μCi)**
[U-^{14}C]acetic acid, sodium salt	50–60	106
[1-^{14}C]acetic acid, sodium salt	50–60	80
[2-^{14}C]acetic acid, sodium salt	50–60	94
[^{3}H]acetic acid, sodium salt	2000–5000	
cholesteryl[1-^{14}C]linoleate	50–60	156
[1,2 (*n*)-^{3}H]cholesteryl linoleate	40 000–60 000	77
[1,2 (*n*)-^{3}H]cholesteryl oleate	40 000–60 000	77
cholesteryl[1-^{14}C]oleate	50–60	174
[1,5-^{14}C]citric acid	100–120	120
[1-^{14}C]eicosa-5,8,11,14,17-pentaenoic acid	50–60	956
[2-^{14}C]eicosa-8,11,14-trienoic acid	50–60	388
[U-^{14}C]glycerol	140–180	160
[1(3)-^{3}H]glycerol	1000–3000	8
[2-^{3}H]glycerol	500–1000	6
L-[U-^{14}C]glycerol 3-phosphate, NH_4salt	<100	228
glycerol tri[1-^{14}C]oleate	50–60	206
glycerol tri[9,10(*n*)-^{3}H]oleate	500–1000	8
glycerol tri[1-^{14}C]palmitate	50–60	106
[^{14}C]hexadecane, standard	10^{-4}	7400
[^{3}H]hexadecane, standard	4×10^{-4}	2000
DL-[1-^{14}C]lactic acid, sodium salt	10–30	120
L-[U-^{14}C]lactic acid, sodium salt	50–150	184
[1-^{14}C]lauric acid	10–30	127
[1-^{14}C]linoleic acid	50–60	106
[1-^{14}C]linolenic acid	50–60	174
[U-^{14}C]palmitic acid	400–800	174
[1-^{14}C]palmitic acid	50–60	94
[9,10(*n*)-^{3}H]palmitic acid	40 000–60 000	5
[^{32}P]orthophosphate	carrier-free	4
[1-^{14}C]pyruvic acid	10–30	120
[1-^{14}C]pyruvic acid, sodium salt	10–30	106
[1-^{14}C]stearic acid	50–60	74
1-stearoyl-2-[1-^{14}C]arachidonyl-*sn*-glycerol	50–60	648
[^{35}S]sulfate	carrier-free	5

U = uniformly labelled

^{14}C-labelled compounds because of the shorter half-life of tritium and the poorer efficiencies with which it is counted. Labelled compounds vary considerably in cost, the more easily prepared compounds usually being the cheapest. Price is also affected by demand.

Labelled water-soluble precursors can be incorporated into lipids in cell-free systems, in the presence of the appropriate enzymes and co-factors. Some precursors which have been synthesised are given in *Table 6*. Many other labelled lipids substrates have been synthesized. Some of these are listed in *Table 7*.

Table 6. Some labelled water-soluble precursors for which synthetic methods are available

Precursor	**Ref.**
[γ-^{32}P]adenosine triphosphate	27
cytidine diphosphate[^{14}C]choline	28
cytidine diphosphate[^{14}C]ethanolamine	28
deoxycytidine diphosphate[^{14}C]choline	28
deoxycytidine diphosphate[^{14}C]ethanolamine	28
[1,3-^{14}C] or [2-^{14}C]glycerol	29
[2-^{3}H] or [1,3-^{14}C]-*sn*-3-glycerophosphate	30
[^{32}P]*rac*-3-glycerophosphate	31, 32
[^{32}P]*sn*-3-glycerophosphate	30
[1-^{14}C]isopentenyl pyrophosphate	33
[Me-^{14}C]phosphorylcholine	34
[^{14}C]phosphorylethanolamine	28
[^{32}P]phosphorylethanolamine	35

Table 7. Some labelled lipids and substrates for which synthetic methods are available

Compound	**Ref.**
[^{3}H] or [1-^{14}C]alkylglyceryl ethers	36
cytidine diphosphate-*sn*-1,2-[U-^{14}C]diacylglycerol	37
sn-1,2[U-^{14}C]diacylglycerol	38
1,2-[^{3}H]diolein	39
2-[^{14}C]linoleoyl phosphatidylethanolamine	40
[1-^{14}C]long-chain alcohols	41
1-[^{3}H]palmitoyl phosphatidylethanolamine	40
[1-^{14}C]oleic acid	42
[^{3}H]oleic acid	43
sn-3[U-^{14}C]phosphatidic acid	44
phosphatidylcholines	
1,2-di-[^{14}C]eicosatrienoyl	45
1,2-di-[^{14}C]linoleoyl	45
2-[^{14}C]eicosatrienoyl	46
2-[^{14}C]linoleoyl	40
1,2-di-[^{14}C]oleoyl	45
1-[^{3}H]palmitoyl	40
1-[^{14}C]stearoyl-2-[^{14}C]linoleoyl	46
[1-^{14}C] or [^{3}H]polyunsaturated fatty acids	47
[1-^{14}C]saturated fatty acids	48, 49, 50 51, 52
[^{3}H]triolein	43

7. Health and safety

Absorption of radiation by the human body can cause two types of effect, somatic and genetic. Somatic effects are those of radiation upon the individual and may be acute or long-term; for example, carcinogenisis. Genetic effects are those which affect subsequent generations; for example, mutations leading to haemophilia, mental deficiency, or diabetes.

The basic unit of radiation dose is the gray (Gy), the dose received by tissue which has absorbed 1 joule of energy per kilogram. The practical unit of radiation dose is the sievert (Sv). This is the dose in Gy times a quality factor which depends upon the type of radiation received. Factors range from unity for electrons and X-rays to 20 for the heavier ionizing particles.

The quantities of radionuclides used in radiochemical analyses are usually so small that the radiation dose received by the analyst is insignificant, provided that suitable precautions are taken to limit exposure and to avoid ingestion. With good working practices the absorbed doses of radiation should be only a small fraction of the maximum permissible dose to radiation workers of 1 mSv per week.

7.1 Protection from external radiation

(a) *Shielding*. Beta radiation should be shielded by light materials, such as aluminium or plastic, in order to minimize the production of bremsstrahlung (X-rays produced by the stopping of beta-particles). Gamma radiation is absorbed exponentially and can best be attenuated by heavy materials such as lead. The half-thickness of lead for gamma-radiation from ^{131}I (0.36 MeV) is 3 mm.

(b) *Distance*. Radiation obeys the inverse square law. Strong sources should therefore be handled with tongs or forceps. Although beta-radiation is substantially absorbed by air, neglect of this precaution could lead to significant finger doses when handling high-energy beta-emitters such as ^{32}P.

(c) *Time*. The period spent close to a radiation source should be kept to a minimum.

Shielding, distance, and time should be optimized to give the safest working conditions.

7.2 Protection from ingestion

Ingestion of radionuclides can take place through the mouth, the nose, skin, or open wounds. Thus, in a radioactive area:

(a) there should be no eating, drinking, smoking, or other mouth operations;

(b) operations leading to the production of radioactive vapours should be carried out in a ventilated fume hood;

(c) surgical rubber gloves or thin plastic gloves should be worn and hands, and any part of the body on which a spill has occurred, should be washed before leaving the area;

(d) before working with radioactivity, any cuts should be covered with waterproof dressing for protection against spillage.

Although the external hazard from the low-energy radiation of tritium and ^{14}C is minimal, these nuclides are a serious hazard if ingested, because of the large amount of energy deposited over the short range of their beta-radiations.

7.3 Legislation and administration

Legislation exists in many countries to control the use of radionuclides. In the UK this legislation is embodied in The Radioactive Substances Act (1960) and The Ionising Radiations Regulations (1985), and is enforced by the Health and Safety Executive via the Factory Inspectorate.

Laboratories in which only small quantities (< 1 mCi) of the common radionuclides are used are normally exempt from these regulations. For laboratories subject to such regulations it may be necessary to appoint a radiation protection adviser and to arrange for medical checks to be carried out. Laboratories in the UK where the dose rate is likely to exceed 7.5 μSv/h are designated as controlled areas where high standards of contamination monitoring and of washing and changing facilities are required. Records should be kept in all laboratories of the quantities of radioactive materials acquired and disposed of, with dates.

7.4 Monitoring

Film badges should be worn when working with radioactive substances. There are approved dosimetry services which supply and develop these badges (in the UK, the Radiation Protection Service). Records of received radiation doses provided by such services should be archived (in the UK, for 50 years).

Regular contamination monitoring should be carried out. Portable monitors (GM for beta-radiation, scintillation for gamma-radiation) are available for this purpose. Monitoring for tritium and other low-energy beta contamination is best carried out by wipe sampling in which the surface to be checked is wiped with a tissue moistened with acetone/water. Any activity on the tissue is then measured by LSC.

7.5 Waste disposal

Small quantities (less than 1 mCi) of the radionuclides commonly used in lipid analysis may be disposed of as ordinary liquid or solid waste. The quantities generated in research or other experimental work would not normally exceed this amount. A problem arises with the disposal of liquid waste from LSC. In the past, such waste has been locally incinerated, but this has caused concern, not principally because of release of radioactivity to the atmosphere, which is inevitable, but because of the toxicity and flammability of the components of a scintillant. It is now usual to arrange for a waste disposal contractor to collect waste from LSC so that it can be destroyed under carefully controlled conditions.

Certain scintillants are claimed to be biodegradable. If these scintillants are water-miscible, they could be diposed of via the laboratory sink, but agreement with the sewage authority should obtained before doing this.

References

1. Krebs, H. A. (1970). *Perspect. Biol. Med.*, **14**, 154.
2. Kaufmann, H. P. and Budwig, J. (1951). *Fette u. Seifen*, **53**, 253.
3. Rushman, D. F. (1953). *J. Oil & Colour Chemists Assoc.*, **36**, 352.
4. Kates, M. (1986). In *Techniques of lipidology*, (2nd edn.). Elsevier, Amsterdam.
5. Geary, W. (1986). *Radiochemical methods*. John Wiley, Chichester.
6. Keller, C. (1988). *Radiochemistry*. Ellis Horwood, Chichester.
7. Packard Instrument Co. (1987). *Liquid scintillation analysis, science and technology*. Publ. No. 169–3052, Downers Grove, Illinois.
8. Satyaswaroop, P. G. (1972). *Indian J. Biochem. Biophys.*, **9**, 101.
9. Burton, K. (1989). *Anal. Proc.*, **26**, 285.
10. Wieland, D. M., Tobes, M. C., and Mangner, T. J. (ed.) (1986). *Analytical and chromatographic techniques in radiopharmaceutical chemistry*. Springer-Verlag, New York.
11. Shulman, S. D. (1983). *J. Liquid Chromat.*, **6**, 35.
12. Schram, E. and Lombaert, R. (1962). *Anal. Biochem.*, **3**, 68.
13. Kessler, M. J. (1982). *J. Liquid Chromat.*, **5**, 313.
14. Alan, I., Smith, J. B., and Silver, M. J. (1983). *Lipids*, 18, 534.
15. Koch, T. K. (1983). *Biochem. Biophys. Res. Commun.*, **114**, 339.
16. Livni, E., Elmaleh, D. R., Levy, S., Brownell, G. L., and Strauss, W. H. (1982). *J. Nucl. Med.*, **23**, 169.
17. Coenan, H. H., Harmand, M. F., Kloster, G., and Stocklin, G. (1981). *J. Nucl. Med.*, **22**, 981.
18. Dudczak, R., Klelter, K., Frischauf, J., Losert, U., Angelberger, P., and Schmoliner, R. (1984). *Eur. J. Nucl. Med.*, **9**, 81.
19. Fritzberg, A. R. and Eshima, D. (1982). *Int. J. Appl. Radiat. Isotopes*, **33**, 451.
20. Padgett, H. C., Robinson, G. D., and Barrio, J. R., (1982). *Int. J. Appl. Radiat. Isotopes*, **33**, 1471.
21. Schmall, B., Conti, P. S., Sundaro-Wu, B., Dahl, J. R., Jacobsen, J. K., and Lee, R. (1982). *J. Labelled Compd. Radiopharm.*, XIX, 1278.
22. Reidel, G., Bauer, T. and Palst, H. W. (1983). *Int. J. Appl. Radiat. Isotopes.*, **34**, 1642.
23. Telč, A., Brunfelter, B. and Gosztonyi, T., (1972). *J. Labelled Compd.*, **8**(1), 13.
24. Berson, S. A. and Yalow, R. S. (1954). *Nature*, 184, 1648.
25. Ruzicka, J. and Stary, J. (1968). *Substoichiometry in radiochemical analysis*. Pergamon Press, Oxford.

26. Bayley, R. J. and Evans, E. A. (1968). In *Storage and stability of compounds labelled with radioisotopes*. The Radiochemical Centre, Amersham, UK, Review 7, 43, and 61.
27. Glynn, I. M. and Chappell, J. B. (1964). *Biochem. J.*, **90**, 147.
28. Schneider, W. C. (1969). In *Methods in enzymology*, Vol. 14 (ed. J. Loewenstein), p. 684. Academic Press, New York.
29. Gidez, L. I. and Karnovsky, M. L. (1952). *J. Am. Chem. Soc.*, **74**, 2413.
30. Chang, Y. Y. and Kennedy, E. P. (1967). *J. Lipid Res.*, **8**, 447.
31. McMurray, W. C., Strickland, K. P., Berry, J. F., and Rossiter, R. J. (1957). *Biochem. J.*, **66**, 634.
32. Sastry, P. S. and Kates, M. (1966). *Can. J. Biochem.*, **44**, 459.
33. Cornforth, R. H. and Popjack, G. (1969). In *Methods in enzymology*, Vol. 15 (ed. R. B. Clayton), p. 359. Academic Press, New York.
34. Gal, A. E. and Fash, F. J. (1976). *Lipids*, **12**, 314.
35. Artom, C. (1957). In *Methods in enzymology*, Vol. 4 (ed. S. P. Colowick and N. O. Koplan), p. 809. Academic Press, New York.
36. Ostwald, E. O., Piantadosi, C., Anderson, C. E., and Snyder, F. (1966), *Lipids*, **1**, 241.
37. Agranoff, B. W. and Suomi, W. D. (1963). *Biochem. Preparations*, **10**, 47.
38. Hanahan, D. J. and Vercamer, R. (1954). *J. Am. Chem. Soc.*, **76**, 1804.
39. Krabisch, L. and Borgstrom, B. (1965). *J. Lipid Res.*, **6**, 156.
40. Waite, M. and VanDeenan, L. L. M. (1967). *Biochim. Biophys. Acta*, **137**, 498.
41. Nystrom, R. F. and Brown, W. G. (1947). *J. Am. Chem. Soc.*, **69**, 2548.
42. Nevenzel, J. C. and Howton, D. R. (1957). *J. Org. Chem.*, **22**, 319.
43. Kritchevsky, D., McCandless, F. J., Knoll, J. E., and Eidinoff, M. L. (1955). *J. Am. Chem. Soc.*, **77**, 6655.
44. Kates, M. and Sastry, P. S. (1969). *In Methods in enzymology*, Vol. 14 (ed. K. Mosbach and J. Loewenstein), p. 197. Academic Press, New York.
45. Warner, T. G. and Benson, A. A. (1977). *J. Lipid Res.*, **18**, 548.
46. Pugh, E. L. and Kates, M. (1977). *J. Biol. Chem.*, **252**, 68.
47. Stoffel, W. (1965). *J. Am. Oil Chemists' Soc.*, **42**, 583.
48. Dauben, W. G. (1948). *J. Chem. Soc.*, **70**, 1976.
49. Anker, H. S. (1952). *J. Biol. Chem.*, **194**, 177.
50. Anker, H. S. (1957). In *Methods in enzymology*, Vol. 4 (ed. S. P. Colowick and N. O. Koplan), p. 779. Academic Press, New York.
51. Bergstrom, S., Borgstrom, B., and Rottenberg, M. (1952). *Acta. Physiol. Scand.*, **25**, 120.
52. Sgoutas, D. S. and Kummerow, F. A. (1964). *Biochemistry*, **3**, 406.

7

High-resolution ^{1}H and ^{13}C NMR

FRANK D. GUNSTONE

1. Introduction

This chapter is intended to show how structural information can be obtained from high-resolution ^{1}H and ^{13}C NMR spectra of fatty compounds. There is no description of NMR theory or of the instruments required to produce spectra. It is assumed that the researcher has obtained a spectrum and wants to know what information can be derived from it. The following discussion is based on the results obtained with a 200–300 mHz field for ^{1}H spectra and the ability to enhance ^{13}C spectra when required.

^{1}H spectra contain less information than ^{13}C spectra but the former provide better quantitative data. The quantitative aspects of a ^{13}C spectrum can be improved by adjusting delay times and/or by the addition of chemicals which promote relaxation.

In his review on 'Nuclear magnetic resonance spectroscopy (high resolution)', Pollard (1) discussed the general methods of assigning NMR signals to specific structural features (hydrogen atoms or carbon atoms as appropriate) under four general categories:

- use of model compounds and existing tabulations of data,
- labelling with ^{2}H or ^{13}C,
- use of chemical shift reagents, and
- special NMR techniques such as ^{1}H decoupling, off-resonance experiments, spectral editing, and 2-D NMR

some of which require special equipment. These are not detailed here and reference must be made to textbooks devoted to NMR spectroscopy.

There is now enough information to determine the structure of most single compounds whether these be acids, esters, acylglycerols, or phospholipids. The greater challenge is to discover what can be learned from ^{1}H and ^{13}C NMR spectra of the mixtures of compounds which occur naturally or of fractions (simpler, but still mixtures) which can be obtained by a variety of separation procedures. For example, the ^{13}C NMR spectrum of a natural oil or fat shows 50–100 signals. This represents a wealth of information not all of which can yet be derived from the spectrum. This chapter will show how to decode the information provided by ^1H and ^{13}C NMR spectra.

2. ^1H NMR spectroscopy

2.1 Deshielding effects

The ^1H NMR spectrum of the straight-chain alkane [**1**] contains only two signals: signal *a* for the two end-groups representing 6H and signal *b* for the methylene groups representing 2*n*H.

$$\underset{a}{CH_3}\underset{b}{(CH_2)_n}\underset{a}{CH_3} \qquad [\mathbf{1}]$$

These signals have chemical shifts of 0.89 and 1.23 p.p.m. respectively. The integral which also appears in the spectrum indicates the area under the peak and is proportional to the number of hydrogen atoms. For alkane [**1**] these integrals will be in the proportion 6:2*n*, and it will be possible to determine the value of *n* if this is not too large. The spectrum becomes more complex (and more informative) when functional groups are incorporated into the molecule. Additional signals arise from the CH or CH_2 unit to which the group is attached, and also from nearby CH_2 and CH_3 groups because the functional group exerts its influence along a chain of a few atomic centres.

The effect of many functional groups has been determined by examination of a large number of examples and some of these are collected together in *Table 1*. Since it is difficult to detect changes below 0.1 p.p.m. it is apparent that the influence of a functional group rarely goes beyond the β-position. With the exception of the methyl group however, it is always significant in the α-position.

Small effects, not apparent on their own, may become so when a CH_2 or CH group is under the combined influence of two or three nearby functional groups and examples of this will be pointed out. Though the effects of more than one functional group are generally additive they may not be so in a strictly numerical sense.

Notice also from the figures in *Table 1* that there are small but not very useful differences between *cis* (*Z*) and *trans* (*E*) double bonds and between monoenes and polyene systems. The figures in this table can be used to calculate chemical shifts for a wide range of acids and ester. The calculated values for methyl oleate would be:

$$\underset{a}{CH_3}\underset{b}{(CH_2)_6}\underset{d}{CH_2}CH{=}CH\underset{d}{CH_2}\underset{b}{(CH_2)_4}\underset{c}{CH_2}\underset{e}{CH_2}COOCH_3$$

a 0.89 (3H), *b* 1.23–1.30 (18H), *c* 1.55 (2H), *d* 1.96 (4H), *e* 2.19 (2H)

Table 1. Deshielding effects of functional groups (p.p.m. in CCl_4 solution) along an alkyl chain: basic chemical shifts 0.885 (CH_3) and 1.225 (CH_2)

Functional group	α	β	γ	δ
COOH	1.04	0.36	0.10	0.06
$COOCH_3$	0.96	0.32	0.06	0.04
$COOC_2H_5$	0.94	0.32	0.06	0.03
CH_3	0.03	0.00	0.00	0.00
CH=CH (*c*)	0.73[a]	0.07	0.02	0.02
CH=CH (*t*)	0.69[b]	0.05	0.05	0.00
$CH{=}CHCH_2CH{=}CH$	0.77	0.08	0.03	0.03
–CH–CH– (epoxide, O bridging) *c*	0.17	0.19	0.10	0.06
–CH–CH– (epoxide, O bridging) *t*	0.17	0.15	0.08	0.04

[a] 0.69 for a CH_3 group
[b] 0.74 for a CH_3 group

2.2 Methyl alkanoates

Instead of the two signals shown by the alkane [**1**] the methyl alkanoate [**2**] shows five signals. The methyl ester groups has its own 3 proton signal and its

$$\underset{a}{CH_3}\underset{b}{(CH_2)_n}\underset{c}{CH_2}\underset{d}{CH_2}CO\underset{e}{OCH_3} \qquad [2]$$

a 0.90 (3H), *b* 1.31 (2*n*H), *c* 1.58 (2H), *d* 1.95 (2H), and *e* 3.65 (3H)

influence in the alkane chain is apparent at the α(C-2) and β(C-3) positions. The observed figures cited above are in accordance with values calculated from data in *Table 1*. The integrals further show the proportion of hydrogen atoms in each signal (namely, 3:2*n*:2:2:3) and the splitting pattern is as expected. The OCH_3 signal is a singlet, the CH_3 and α-CH_2 signals are triplets, and the β-CH_2 a quintet. The signal for the remaining methylene groups is complex and broad because it consists of a large number of overlapping signals.

2.3 Methyl alkenoates and the polyene esters

In addition to the five signals in the ^1H NMR spectrum of methyl stearate the spectrum of methyl oleate shows additional signals for the two olefinic (—*CH*=*CH*—) and the four allylic (—$\mathit{CH_2}$CH=CH$\mathit{CH_2}$—) hydrogen atoms at 5.35 and 2.05 p.p.m. respectively (*Figure 1*). The common methylene-interrupted polyene esters also show a signal at 2.77 for CH_2 groups adjacent to two *cis* double bonds as in the spectrum of methyl linoleate (*Figure 2*):

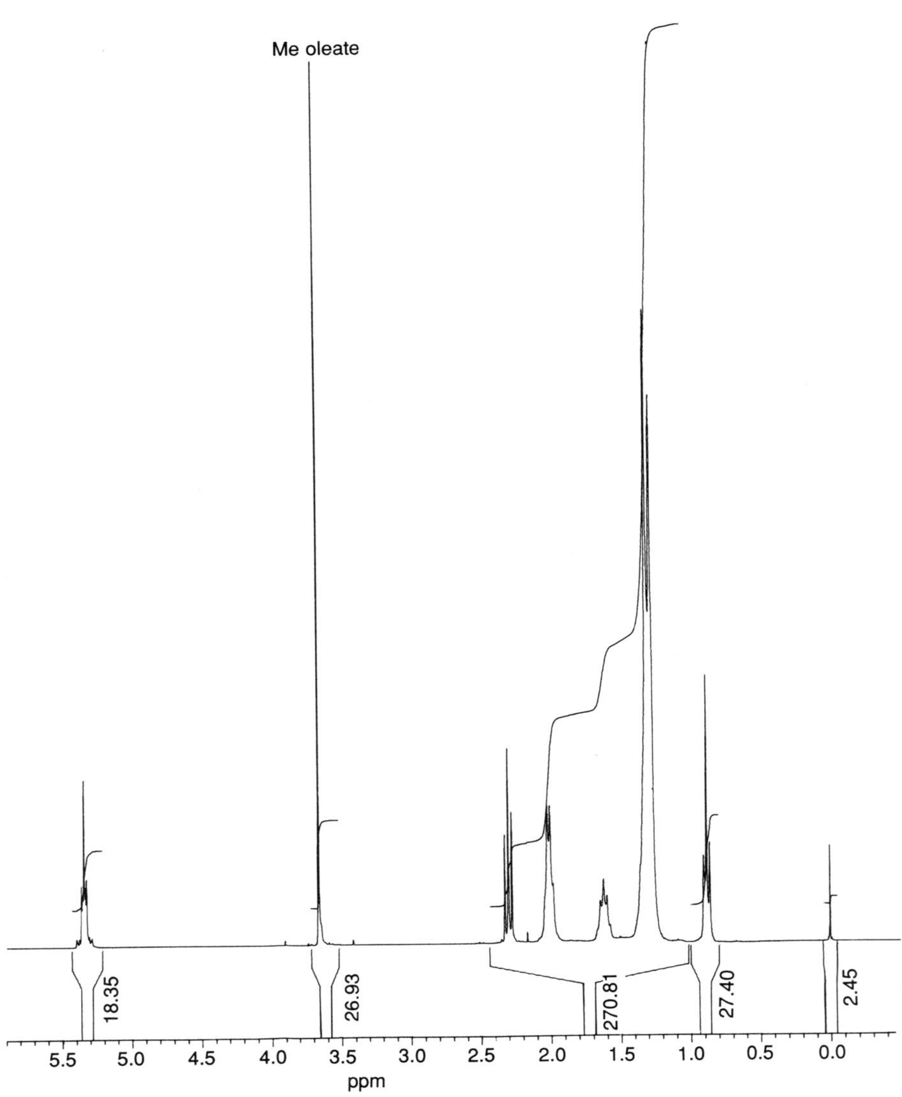

Figure 1. ^{1}H NMR spectrum of methyl oleate.

$CH_3(CH_2)_3CH_2CH{=}CHCH_2CH{=}$
a b d h f h
$CHCH_2(CH_2)_4(CH_2)_4CH_2CH_2COOCH_3$
h d b c e g

a 0.89 (3H), *b* 1.31 (14H), *c* 1.62 (2H), *d* 2.05 (4H),
e 2.30 (2H), *f* 2.77 (2H), *g* 3.67 (3H), *h* 5.35 (4H)

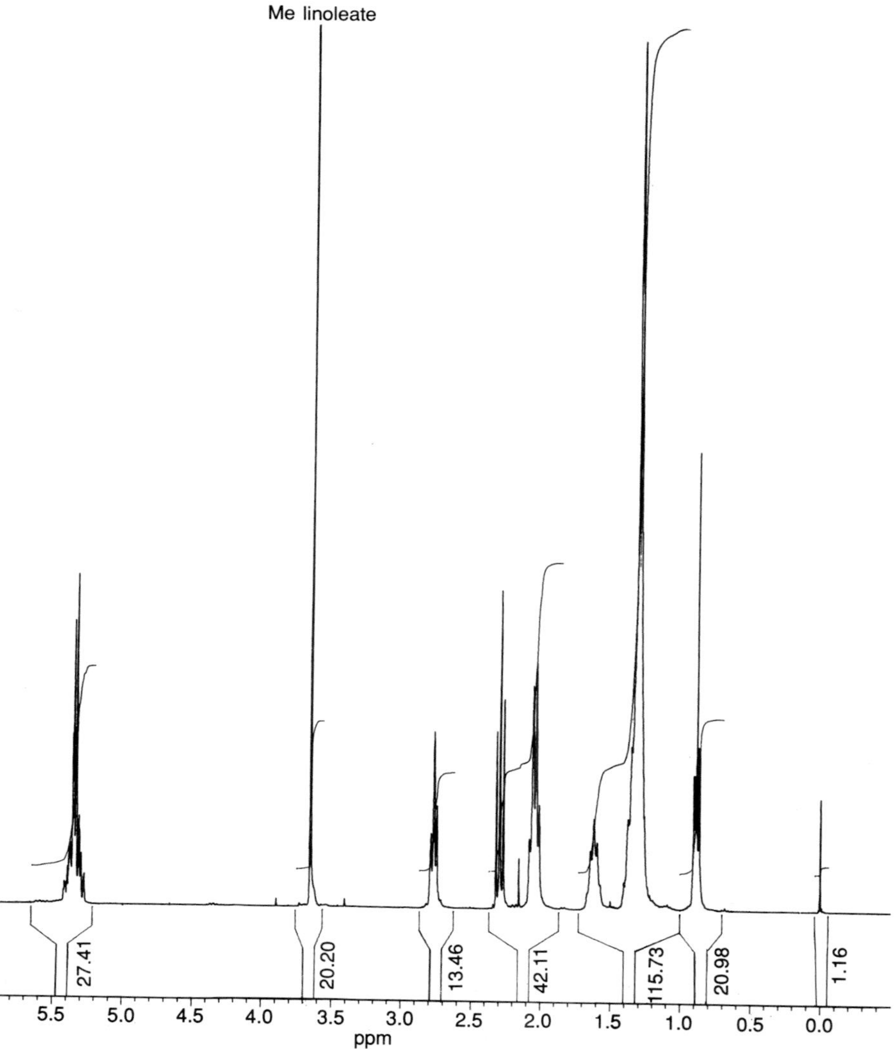

Figure 2. ^{1}H NMR spectrum of methyl linoleate.

Table 2. ^{1}H NMR signals calculated for monoene esters

	Methylene group						
Double bond	**2**	**3**	**4**	**5**	**6**	**7**	**8**
Δ^9	2.19	1.55	1.29	1.29	1.25	1.30	1.96
Δ^8	2.19	1.55	1.31	1.29	1.30	1.96	–
Δ^7	2.19	1.57	1.31	1.30	1.96	–	
Δ^6	2.21	1.57	1.38	1.96	–		
Δ^5	2.21	1.62	2.02	–			
Δ^4	2.26	2.18	–				
Δ^3	2.92	–					

From the integrals of the spectrum of a methylene-interrupted polyene ester it is easy to determine the number of double bonds but it is only possible to fix the position of the polyene unit in the chain if it gets close to the $COOCH_3$ or the CH_3 group. This happens with all *n*-3 polyene acids where the final double bond is close enough to the methyl group to shift the signal from the usual 0.89 to the higher value of 0.98 p.p.m. The NMR spectrum of a natural lipid containing *n*-3 acids along with the unusual range of *n*-6, *n*-9 and saturated acids will show two distinct signals for the CH_3 group, and the integrals will show the ratio of *n*-3 acyl chains to all the rest. This can be seen in the spectrum for cod liver oil (*Figure 3*).

When the double-bond system gets close to the COOH group the overlapping influences give rise to distinct signals. Some calculated values are shown in *Table 2*. These suggest that it should be easy to distinguish unsaturation at Δ^3 andΔ^4 from that at Δ^8 and Δ^9 and that careful inspection of good spectra will show differences for compounds with unsaturation at Δ^5 or Δ^6 or Δ^7. These expectations have been confirmed (2). Monoene acids with unsaturation at these positions are rare [apart from 18.1 (6*c*)] but some important polyene acids with unsaturation starting close to the carboxyl group have the structures shown below and evidence for their presence is to be found in the signals for the hydrogen atoms on C-2, C-3, C-4, and the allylic carbon atom (see *Figure 3*).

18:1	6*c*	$HO_2C(CH_2)_4CH{=}CH(CH_2)_{10}CH_3$	petroselinic
18:3	6*c*9*c*12*c*	$HO_2C(CH_2)_4(CH{=}CHCH_2)_3$ $(CH_2)_3CH_3$	GLA
20:5	5*c*8*c*11*c*14*c*17*c*	$HO_2C(CH_2)_3(CH{=}CHCH_2)_5CH_3$	EPA
22:6	4*c*7*c*10*c*13*c*16*c*19*c*	$HO_2C(CH_2)_2(CH{=}CHCH_2)_6CH_3$	DHA

where GLA refers to gamma linolenic acid, EPA to eicosapentaenoic acid and DHA to docosahexaenoic acid.

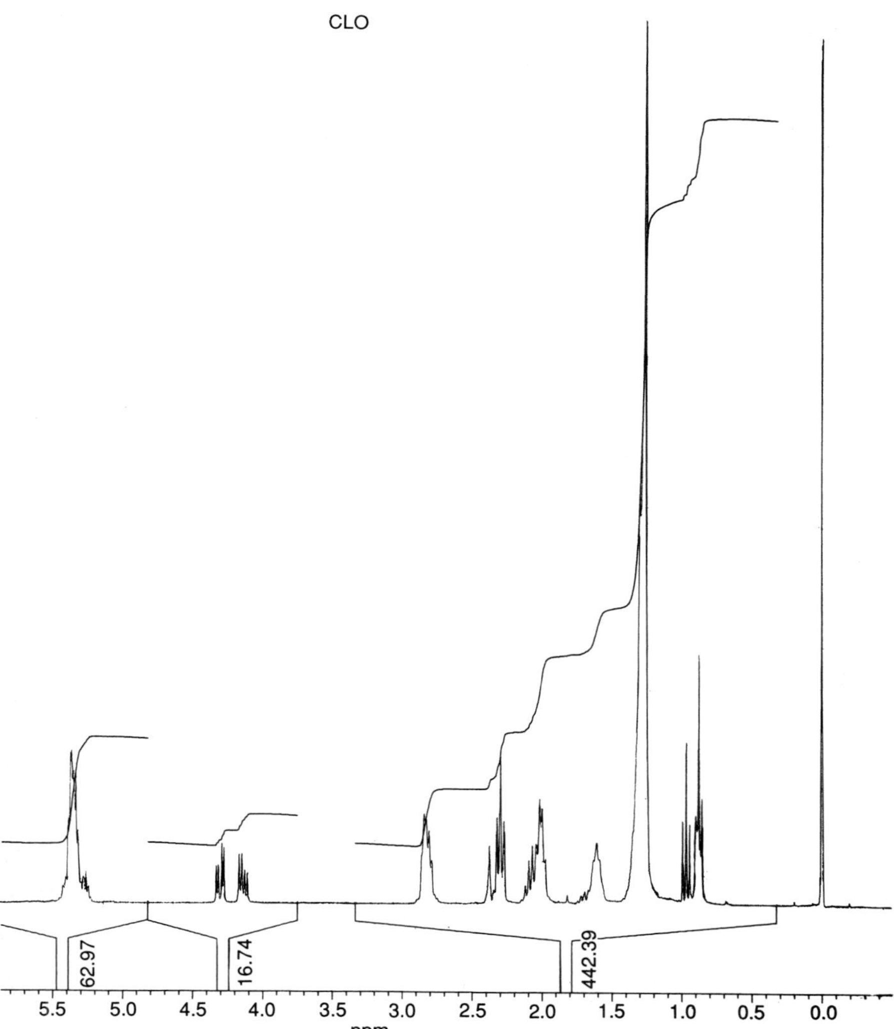

Figure 3. ^{1}H NMR spectrum of cod liver oil.

2.4 Esters containing cyclopropane, cyclopropene, or oxygenated functions

Long-chain esters with cyclopropane and cyclopropene systems can be recognized by characteristic NMR signals. Cyclopropene compounds show a sharp two proton singlet at 0.8 but the cyclopropanes give more complex spectra. The *cis* isomer [**3**] has a multiplet at −0.3 (1H, H*b*) and a broad band at 0.6 (3H, H*a*, *c*, *d*). The *trans* isomer shows a 4-proton band between 0.6 and −0.1.

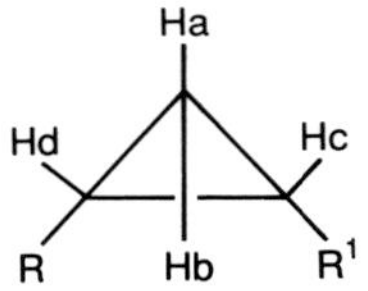

[3]

Epoxy esters have characteristic signals at 2.70 (*cis*) and 2.45 (*trans*) p.p.m. and the effect of the epoxide function on neighbouring CH_2 groups is indicated in *Table 1*. Hydroxy esters have a one proton signal at 3.4 for the C*H*OH proton in carbon tetrachloride solution. In basic solvents this appears at 3.76 (pyridine) and 4.03 (quinoline), and there are additional signals at 1.58 (in pyridine) and 1.77 (in quinoline) for the CH_2 groups α and *β* to the CHOH group. These latter signals are not present when carbon tetrachloride is used as solvent.

2.5 Acyl glycerols and phospholipids

The discussion so far has been in terms of acids and simple esters. Consideration must now be given to the additional signals present in acyl glycerols [**4**] and phospholipids [**5**].

$$\begin{array}{c} \quad\quad\quad\quad\quad\quad\; a \quad\; b \quad\; a \\ RCOOCH_2CHCH_2OCOR \\ | \\ OCOR \end{array} \quad [\mathbf{4}]$$

$$\begin{array}{c} O \\ a \quad b \quad c \quad \| \quad d \quad e \quad f \\ RCOOCH_2CHCH_2OPOCH_2CH_2NR^1{}_3 \\ \quad | \quad\quad\quad\quad | \\ RCOO \quad\quad O^- \end{array} \quad [\mathbf{5}]$$

In acyl glycerols a 1-proton signal at 5.25 p.p.m. results from proton *b* and overlaps with the signal for olefinic =CH groups. The four protons on glycerol carbons 1 and 3 (a) give two peaks at 4.12 and 4.28 pm. These complex signals have been discussed in more detail in Pollard's review (1). They can be seen in the spectrum of cod liver oil (*Figure 3*). The signals *a–f* in phospholipids differ for phosphatidylethanolamines (R^1=H) and phosphatidylcholines ($R^1=CH_3$) and are summarized in *Table 3*. The 9-proton singlet at 3.35 provides clear evidence of the $N^+(CH_3)_3$ group in phosphatidylcholines.

Table 3. ^{1}H NMR signals for phospholipids (3)

	a	*b*	*c*	*d*	*e*	*f*
PE R^1=H	4.0–4.2	5.2	3.9	4.0–4.2	3.1	7–8.5
PC $R^1=CH_3$	4.13 4.40	5.20	3.94	4.31	3.79	3.35

```
                  O
    a  b  c      ||     d    e + f
RCOCH2CHCH2OPOCH2CH2NR'3
           |          |
        OCOR    O−
```

2.6 Shift reagents

Some metal complexes based on europium or praseodymium—known generally as lanthanide-induced shift reagents (LIS) or chemically induced shift reagents (CSR)—interact with oxygen-containing groups to produce a marked effect on nearby CH_2 signals so that the neighbouring five or six CH_2 groups show separate signals. The shift from the normal position depends on the closeness of the CH_2 group to the oxygen-containing function and on the concentration of the shift reagent. Chiral shift reagents have been used to determine absolute configuration and enantiomeric purity and this is important in the synthesis of acylglycerols and phospholipids.

3. ^{13}C NMR spectroscopy

3.1 Introduction

A ^{13}C NMR spectrum contains more signals than a ^{1}H spectrum because the effects of each functional group are larger and are effective through more atomic centres. Some typical and useful values are set out in *Table 4* and comparison of these values with those listed for ^{1}H spectra in *Table 1* confirms this claim.

Since these effects are approximately additive, quite specific signals are observed when an unsaturated system gets close to the ester group or the methyl group. It is possible to distinguish acyl groups with unsaturation at Δ^4, Δ^5, and Δ^6 and at the *n*-3 and *n*-6 positions though specific identifications are not confined to these cases. A typical spectrum is shown in *Figure 4*. (The list of numbers which came with the spectrum are more useful than the line drawing so other spectra will not be presented in this form.)

3.2 Methyl esters

Table 5 shows typical values for ^{13}C NMR signals for methyl stearate, oleate, linoleate, and α- and γ-linolenates. Different investigators give slightly

Table 4. Influence of functional groups (p.p.m. in $CDCl_3$) on methylene (29.7) and methyl (14.1) signals (ref. 4)

Functional group	α	ß	γ	δ	ε	ξ	Functional group
CH_3	−7.0	+2.3	−0.3				14.1
$COOCH_3$	+4.3	−4.7	−0.5	−0.4	−0.2	−0.1	174.1
CH=CH *c*	−2.5	0	−0.4	−0.2			–
CH=CH *t*	+2.8	−0.1	−0.5	−0.2			–
$(CH{=}CH)_3$*c*[a]	−1.8	0	−0.4	−0.1			–
$(CH{=}CH)_3$*t*[a]	+3.1	0	−0.5	−0.2			–
CH—CH *c* (epoxide, O bridging)	−1.7	−2.9	−0.4				
CH—CH *t* (epoxide, O bridging)	+2.6	−3.5	−0.2				
CH_2OH	+3.1	−3.9					62.9
CH_2OAc	−3.8	−0.5					64.7 (171.1, 20.9)
CH_2OOH	−1.9	−3.6					
CHOH	+7.8	−4.0	+0.1	−0.1	−0.1	−0.1	42.2
CHOAc	+4.4	−4.4	−0.2	−0.2	−0.1	−0.1	44.7
C=O	+13.1	−5.8	−0.4	−0.3	−0.2	−0.1	181.1
CHO	+13.4	−8.4					201.7
C≡C	−11.4	−1.2	−1.0	−0.6	−0.2	−0.1	
CH_2Cl	+3.0	−0.9					45.1
CH_2Br	+3.1	−1.0					
CH_2I	+3.9	+0.8					
$CH_2OCH_2CH_3$	−4.7	−0.5					70.8 (66.0, 15.3)

[a] These figures depend on the configuration of the nearest double bond in the conjugated triene system

different figures depending on the solvent and concentration of the solution examined but differences between signals which are fairly close together are fairly reproducible. An appreciation of the values for these common methyl esters assists in the identification of signals in glycerol esters (see Section 3.3). Attention is drawn to the following points.

(a) Signals for the C-1, C-2, C-3, and ω3, ω2, and ω1 carbon atoms are readily distinguished from the remainder (see *Figure 4*). This is true for most acyl chains and changes from the normal values can usually be interpreted in structural terms. An obvious example is the signal for C-17 (ω2) in *n*-3 esters like α-linolenate. This carbon is now allylic and has a much lower resonance than usual. A smaller change is also apparent at the end (ω1) carbon atom for *n*-3 compounds.

(b) Olefinic systems produce characteristic signals for all or most of the olefinic carbon atoms, for allylic carbon atoms (with marked differences between *cis* and *trans* compounds), and for allylic groups lying between

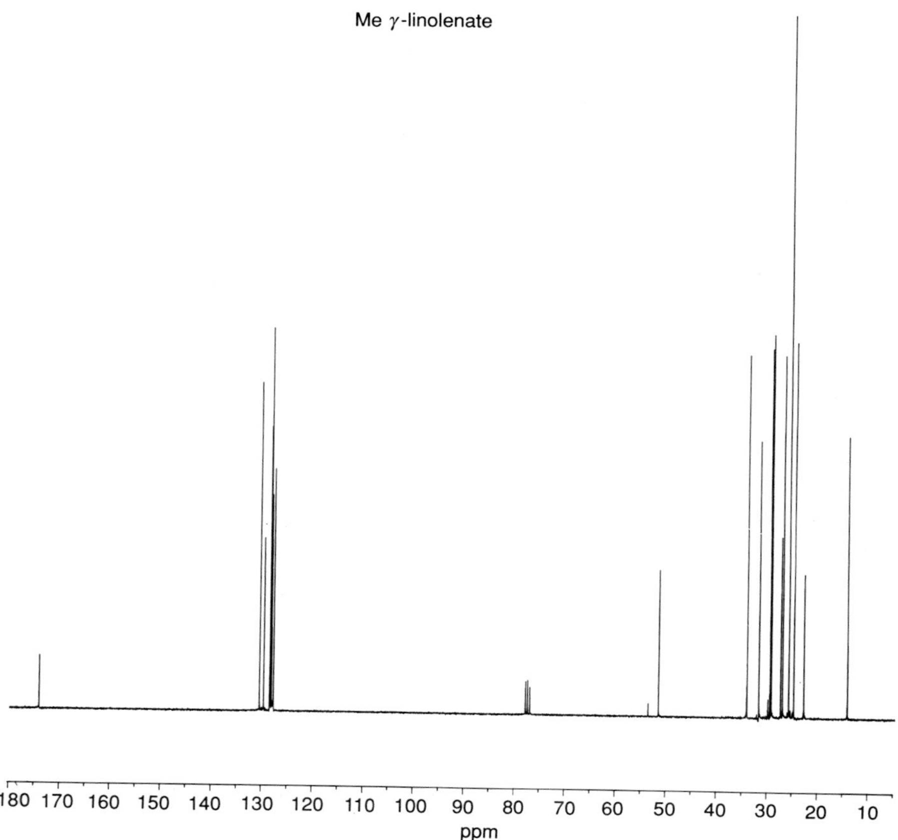

Figure 4. ^{13}C NMR spectrum of methyl γ-linolenate.

two double bonds. These signals lie in the boxes in *Table 5* and represent four, seven, and ten carbon atoms in the monoene, diene and triene esters respectively. Further comments on the olefinic signals are made later.

(c) Signals between 29.0 and 29.8 (sometimes referred to as the methylene envelope) are not always fully resolved and little attention is usually paid to them. They represent 12, 8, 5, 4, and 2 of the carbon atoms in the five esters in *Table 5*. More thorough study will sometimes produce useful information about these signals. The highest value is usually associated with carbon atoms ß to a double-bond and the lowest (in the Δ^9 esters) with C-4, 5, and 6.

(d) For *cis* monoenes the difference between the two olefinic signals diminishes as the double bond moves away from the ester group and is

Table 5. NMR spectra of methyl stearate, oleate, linoleate, and α- and γ-linolenate: assignment of signals

	carbon atom 2	3	4	5	6	7	8	9	10	11	12	13	14	15	16	17	18
18:0	34.08	24.97	29.16	29.26	29.45	29.59	29.64	29.67	29.68	29.71	29.71	29.71	29.66	29.35	31.92	22.69	14.08
18:1	34.11	24.98	29.21	29.18	29.13	29.72	27.19	129.73	129.99	27.25	29.81	29.37	29.58	29.37	31.96	22.73	14.13
18:2	34.10	25.01	29.24	29.24	29.19	29.66	27.27	130.00	128.09	25.69	127.96	130.14	27.27	29.44	31.61	22.66	14.10
18:3 (α)	34.08	24.99	29.23	29.18	29.15	29.63	27.25	130.19	127.78	25.67[a]	128.26	128.26	25.57[a]	127.17	131.87	20.61	14.31
18:3 (γ)	33.96	24.63	29.16	26.92	129.57	128.28	25.68	128.07	128.39	25.68	127.65	130.39	27.27	29.41	31.59	22.65	14.10

[a] These two signals could be interchanged

not apparent from about Δ^9 onwards. As the values in *Table 6* show they can be used to determine double-bond position in monoene esters.

(e) For monoenes and methylene-interrupted polyenes such as **[6]** the

$$\underset{a}{CH_3(CH_2)_nCH}=\underset{b}{CHCH_2}(CH=CHCH_2)_pCH=\underset{c}{CH}(CH_2)_mCOOCH_3 \quad \textbf{[6]}$$

olefinic carbon atom *a* has the highest chemical shift followed by *c* whilst *b* has the lowest value. This is true for the important C_{18} esters listed in *Table 5* but it is not known how widely this generalization holds.

^{13}C NMR spectra have been useful in the study of seed oils containing conjugated trienoic acids. These are mainly 18:3 $\Delta^{9,11,13}$ or $\Delta^{8,10,12}$ acids in various stereoisomeric forms. Complete assignments have been made for the six carbon olefinic unit [7] and the effects of this unit on methylene groups on either side have been assessed (see *Table 4*). Original papers must be consulted for the details (example, ref. 5) but the most obvious difference is between CH_2 groups on the *cis* or *trans* end of triene unit thus:

```
   −0.4      −1.8        c                      t              −0.4      0
—CH₂CH₂CH₂CH₂CH=CHCH=CHCH=CHCH₂CH₂CH₂CH₂—   [7]
−0.1            0                                         +3.1     −0.5
```

Results from the study of hydroxy, acetoxy, and oxo acids are incorporated into *Table 4* (6).

Table 6. Chemical shifts induced in *cis* olefinic carbon atoms (129.90 p.p.m. when undisturbed) in monoenoic acids and esters by COOH, $COOCH_3$, and CH_3 groups

	acids[a]			esters[a]		
			diff.			diff.
Δ^2	–	–	–	−10.49	+20.79	31.3
3	−9.85	+4.27	14.1	−8.90	+3.50	12.4
4	−2.84	+2.05	4.9	−2.37	+1.64	4.0
5	−1.70	+1.53	3.2	−1.46	+1.31	2.8
6	−0.88	+0.72	1.6	−0.72	+0.58	1.3
7	−0.44	+0.41	0.9	−0.41	+0.32	0.7
8	−0.25	+0.25	0.5	−0.22	+0.23	0.5
9	−0.11	+0.15	0.3	−0.12	+0.12	0.2
10	−0.07	+0.10	0.2	–	–	–
11	−0.01	+0.06	0.1	0	0	0
ω^4	+0.23	−0.25	0.5	–	–	–
ω^3	−0.53	+1.63	2.2	–	–	–
ω^{2b}	+1.01	−6.33	7.3	–	–	–

[a] Shift refers to carbon atoms *n* and *n* + 1 for the Δ^n series and *m* + 1 and *m* for the ω*m* series (see example below)

[b] ω1 values (acid) are 139.12 and 114.15 for ω1 and ω2 carbon atoms respectively.

Example: in 18:1 (5) the olefinic carbon atoms in the acid will show signals at 129.90–1.70 (128.20) and at 129.90 + 1.53 (131.43) for C-5 and C-6 respectively.

3.3 Acylglycerols

The ^{13}C NMR spectra of acylglycerols, whether individual (synthetic) compounds, natural mixtures (oils, fats), or fractions derived from these, contain signals arising from the acyl chains along with additional signals from the glycerol carbon atoms. The signals coming from the acyl chains may have slightly different chemical shifts from those of methyl esters and they may also show small differences depending on whether the acyl chains are attached to glycerol carbons 1 and 3 or to glycerol carbon 2. These differences are so small (up to 0.05 p.p.m.) that they cannot be recognized from single values obtained with pure compounds. The differences are less than the range of variation of reported values. But when both signals are present then under appropriate NMR conditions they may be resolved though not always fully. Two examples are discussed briefly refering to acyl carbon atoms in the first example and to olefinic carbon atoms in the second. In the latter example the resolved signals are at least nine carbon atoms from the ester function. It is expected that the differences would be even more apparent with double bonds closer to the ester function.

(a) Mixtures of symmetrical and unsymmetrical glycerol oleate dipalmitate (POP and PPO) and of symmetrical and unsymmetrical glycerol dioleate palmitate (OPO and OOP) show two clear signals for C-1 at about 173.2 and 172.8. Since the signal at 173.2 is about twice the size of the other signal they must be associated with the 1/3 and 2 positions of glycerol respectively. When the spectra are expanded both signals split into two depending on whether the acyl group is palmitic or oleic. From the relative sizes of these signals in the two mixtures it is possible to assign one value to palmitic acid and the other to oleic acid. The results are summarized in *Table 7*, from which it is apparent that the separation is about 0.03 p.p.m. even though the actual resonances differ somewhat between various spectra. Similar results were reported by Ng (7) in his studies of palm oil. The difference between two saturated chains is likely to be smaller (except perhaps if there is a considerable difference in chain-length) but separations may also be observable with other unsaturated acids.

Table 7. C-1 resonances for some triacylglycerols containing saturated and unsaturated acyl groups

Glycerol position	Acyl group	POP+ PPO	OPO+ OOP	Ng (ref.7)
1/3	16:0	173.24	173.22	173.16
	18:1	173.21	173.19	173.13
2	16:0	172.83	172.81	172.76
	18:1	172.80	172.78	172.72

(b) Awl *et al.* (8) have reported the ^{13}C NMR spectra of four synthetic glycerides containing both linoleic (L) and α-linolenic (Ln) acyl chains. The two symmetrical glycerol esters (LLnL and LnLLn) show signals for 10 olefinic carbons but there is no splitting of any of these. Splitting is observed in some of the olefinic signals in the unsymmetrical esters (LLLn and LnLnL). In LLLn the linoleic olefinic carbon atoms 9, 10, and 12 (but not 13) show two signals, and in LnLnL the linolenic olefinic carbon atoms 10, 12, and 13 (but not 9, 15, and 16) are split. Since these differences arise from the position of the acyl chains on glycerol it is not surprising that the olefinic carbon atoms furthest away are those in which it is hardest to detect splitting.

The ^{13}C NMR spectrum of an oil or fat contains a very large number of signals but these can usually be divided into 11 groups using the figures given in *Table 8*. There may be several signals within each group because (a) acyl groups are attached to glycerol 1/3 or glycerol 2 and (b) there are differences in the acyl chains themselves of which the most important features seem to be the extent of unsaturation and its relation to the ester group or the end methyl group. Some comments which may be of value in interpreting a spectrum are now given. Where figures are quoted it should be remembered that small but reproducible differences may be more significant than the actual figures themselves.

(a) C-1 signals (usually small in size because they are quaternary carbon atoms) are normally at 173.2 (glycerol 1/3) and 172.8 (glycerol 2). The possibility of these being split further for different acyl chains has already been discussed (see *Table 7*). These values refer to saturated chains and chains with Δ^9 unsaturation. More significant shifts are observed for Δ^6 acids (173.0 and 172.6) and for Δ^5 and Δ^4 acids (172.7–172.9 and 172.1–172.4).

Table 8. Major grouping of signals in the ^{13}C NMR spectrum of a natural oil or fat

Approximate value (p.p.m.)	**Source of signal**	**Discussed in Section 3.3, para:**
173	C-1	(a)
127–132	olefinic carbon atoms	(b)
[77	solvent ($CDCl_3$)]	
69 and 62	glycerol carbons	(c)
33.3–34.3	C-2	(d)
31.6–32.0	ω3	(e)
28–30	methylene envelope (see text)	(f)
26.6–27.3	allylic	(g)
25.6–25.7	doubly allylic	(h)
24.5–25.0	C-3	(i)
22.6–22.8	ω2	(j)
14.1–14.3	ω1	(k)

Table 9. Olefinic signals in methyl oleate (O), linoleate (L), and α-linolenate (Ln) in numerical order

p.p.m.	Carbon atom
131.87	Ln16
130.19	Ln9]
130.14	L13]
130.00	L9]
129.99	O10]
129.73	O9
128.26	Ln12 + 13
128.09	L10
127.96	L12
127.78	Ln10
127.17	Ln15

(b) Most olefinic carbon atoms have their own signals and the values for methyl oleate, linoleate, and α-linolenate (the most common C_{18} unsaturated acids) are listed in numerical order in *Table 9*. Because these values are subject to small variations depending on experimental conditions three pairs (Ln9 and L13), (L9 and O10), and (Ln12 and 13) may overlap or may even appear in reverse order. Most seed oils contain oleic and linoleic acids and it is usually easy to distinguish the four linoleic signals which, apart from overlap, should be approximately the same size. The presence of α-linolenic acid is easily recognized by the signals at 131.9 (Ln16) and 127.2 (Ln15) which are well-separated from the other signals in this list. Other *n*-3 polyenes may, however, show similar signals.

(c) The glycerol carbon atoms show signals at 69.0 and 62.1 p.p.m. Since the latter signal is the larger it must be associated with glycerol 1/3 and the other peak with glycerol 2.

(d) For C-2, saturated and Δ^9 acyl chains have signals at 34.19–34.24 (glycerol-2) and 34.02–34.08 (glycerol-1/3). In Δ^6 acids the signals are shifted to 33.91–33.95 (splitting not yet reported) whilst Δ^4 and Δ^5 acids give values at 33.58–33.62 and 33.34–33.46.

(e) The signal for the ω3 carbon atom at 31.96–32.00 for saturated and *n*-9 acyl chains appears at 31.56–31.57 for *n*-6 acyl chains, and as an olefinic signal in *n*-3 acids.

(f) Many of the CH_2 groups give very similar signals in the range 29.0–29.9. Some of these are sometimes resolved but mostly they are not and the mixed signals are described as the methylene envelope. This group would probably repay further study with expanded spectra.

(g) Allylic signals are usually in the range 26.6–27.3. These relate only to the allylic groups at each end of a monoene or polyene system. Doubly allylic carbons are discussed in the next section. Frequently, there are two allylic signals resulting from the influence of the carboxyl group on one allyl group and the end methyl on the other allyl group. This is most obvious in *n*-3 unsaturated acids when one allylic group is also ω2 and is likely to appear at about 20.58–20.68 p.p.m. Acyl chains with Δ^5 and Δ^6 unsaturation show allyl resonances at 26.56–26.57 (C-4 in Δ^5 acids) and 26.87–26.89 (C-5 in Δ^6 acids). Allylic signals are very different if the closest double bond has *trans* configuration. This is illustrated in some values for methyl oleate and elaidate (see also *Table 4*).

11 *cis* 8	11 *trans* 8
$—CH_2CH{=}CHCH_2—$	$—CH_2CH{=}CHCH_2—$
27.19 27.25	32.66 32.64

(h) Doubly allylic carbon atoms ($—CH{=}CHCH_2CH{=}CH—$), present in methylene interrupted polyenes, are usually in the range 25.67–25.70. A lower value of 25.60–25.61 is also present in the spectra of fish oils, and probably results when such doubly allylic groups are part of Δ^4 polyenes (22:6, possibly C-6) and/or Δ^5 polyenes (20:5, possibly C-7).

(i) C-3 Carbon atoms give signals between 24.8 and 25.0. Split signals have been observed (24.90–24.95 and 24.88–24.90) but these have not been fully assigned.

(j) ω2 Carbon atoms have signals at 22.70–22.80 (saturated and *n*-9 acyl chains), 22.60–22.68 (*n*-6 compounds), and 20.58–20.68 (*n*-3 compounds). This last signal is now also allylic [see para. (g) above].

(k) Finally, the signal for the end methyl group ranges from 14.10–14.14 (saturated and *n*-9), 14.06–14.09 (*n*-6), and 14.26–14.27 (*n*-3).

As an example of this analysis the ^{13}C NMR spectrum for olive oil (more than 60 signals) is set out and analysed in *Table 10*). This oil is known to be rich in oleic acid (56–80%) and to contain also linoleic (3–20%), palmitic (7–20%), hexadecenoic (0–4%), and stearic (1–4%) acids. Some small peaks (and the methylene envelope) have not been assigned. The small peaks may result from minor acyl components or from unsaponifiable material present in olive oil (1.5%). Some interesting splitting is already apparent (for example, the signals for L9, O9, and L10) but more would probably be observed with expanded spectra.

3.4 Phospholipids

The ^{13}C NMR spectra of some phospholipids have been reported. These are mainly obtained from synthetic compounds (single compounds which may be enantiomeric or racemic) and include both phosphatidylcholines and phosphatidylethanolamines. Some typical results are given in *Table 11*. Signals for the acyl groups do not differ greatly from those in simple esters or

Table 10. ^{13}C NMR spectrum of olive oil

p.p.m.	Intensity		Assignment
173.24	1.4	16:0	C-1 glycerol 1/3
173.21	4.0	18:0	
172.80	2.6		C-1 glycerol 2
130.20	1.8		L13
130.01	10.5		O10
129.97	1.9		L9
129.92	0.7		
129.82	0.6		–
129.71	9.4		O9
129.68	6.0		
128.10	1.2		L10
128.08	1.2		
127.91	1.2		L12
68.91	5.0		glycerol 2
65.05	0.5		–
62.11	7.5		glycerol 1/3
34.21	5.2		C-2 glycerol 2
34.06	4.2		C-2 glycerol 1/3
34.04	9.2		
31.95	4.7	sat + *n*-9	ω3
31.93	15.2		
31.90	0.8	–	
31.81	0.5	–	
31.55	2.1	*n*-6	
29.85 to 29.01	25 peaks		–
27.29	0.6		–
27.28	1.1		–
27.24	15.6		O11, L14
27.22	6.0		
27.19	16.1		O8, L8
27.14	0.7		–
25.65	2.2		C-11 (linoleic)
24.90	7.4		C-3
24.89	6.4		
24.86	10.0		
22.70	14.5	*n*-9, sat	ω2
22.60	1.8	*n*-6	
14.12	13.9	*n*-9, sat	ω1
14.08	2.1	*n*-6	

triacylglycerols. There are, however, small differences between the acyl chains in position 1 and those in position 2. Additional signals in the range 40–70 p.p.m. are derived from the carbon atoms in glycerol and the choline or ethanolamine unit. There are very clear differences between the two kinds of phospholipids.

Table 11. ^{13}C NMR spectra of some phospholipids

	sn-SLPC[b]	*sn*-POPC[b]	*sn*-SAPC[b]	PC[ab]	PE[ab]
$N^+(CH_3)_3$	54.40	54.46	54.44	54.27	–
$CH_2\underline{C}H_2N$	66.48	66.49	66.79	66.27 66.20	41.0
$\underline{C}H_2CH_2N$	59.40	59.51	59.39	59.37 59.31	62.1
CH_2OP	63.34	63.63	63.30	63.33 63.27	64.1
CH_2OCOR	63.07	63.13	62.98	63.01	63.0
CHOCOR	70.60	70.65	71.03	70.65 70.54	70.9
C-1	173.53 (S) 173.17 (L)	173.57 173.23	173.50 (S) 172.92 (A)	173.51 173.15	173.34 173.07
C-2	34.18 (S) 34.34 (L)	34.41 34.26	34.13 (S) 34.76 (A)	34.34 34.14	34.37 34.18
C-3	24.97	25.03	24.90	24.98 24.89	25.09 24.97
allylic	27.23	27.31	26.55 (A4) 27.24 (A16)	–	–
doubly allylic	25.65	–	25.64	–	–
ω3	31.96 (S) 31.55 (L)	32.03	31.92 (S) 31.42 (A)	31.88	31.98
ω2	22.71 (S) 22.57 (L)	22.74	22.67 (S) 22.58 (A)	22.67	22.72
ω1	[14.08][c]	[14.14][c]	[14.08][c]	14.10	14.09
olefinic	129.98 (L9) 128.10 (L10) 127.92 (L12) 130.23 (L13)	129.75 (O9) 130.09 (O10)	128.87 (A5) 128.87 (A6) 128.33 (A8) 128.11 (A9) 127.84 (A11) 128.65 (A12) 127.56 (A14) 130.49 (A15)		

[a] Saturated cmpds (unpublished observations by the author)
[b] $CDCl_3$ as solvent, S = stearic, L = linoleic, P = palmitic, O = oleic, A = arachidonic
[c] In the original papers shifts are given relative to ω1 = 0.0. The values given here have been taken from our earlier tables and the other values have been adjusted to make comparisons easier.

References

1. Pollard, M. (1986). *Analysis of oils and fats* (ed. R. J. Hamilton and J. B. Rossell), p. 401. Elsevier Applied Sciences, London and New York.
2. Frost, D. J. and Gunstone, F. D. (1975). *Chem. Phys. Lipids*, **15**, 53.

3. Birdsall, N. J. M., Feeney, J., Lee, A. G., Levine, Y. K., and Metcalfe, J. C. (1972). *J. Chem. Soc., Perkin II*, 1441.
4. Gunstone, F. D., Pollard, M. R., Scrimgeour, C. M., Gilman, N. W., and Holland, B. C. (1976). *Chem. Phys. Lipids*, **17**, 1; Bus, J., Sies, I., and Lie Ken Jie, M. S. F. (1976). *Chem. Phys. Lipids*, **17**, 501; Gunstone, F. D., Pollard, M. R., Scrimgeour, C. M., and Vedanayagam, H. S. (1977). *Chem. Phys. Lipids*, **18**, 115; Bus, J., Sies, I., and Lie Ken Jie, M. S. F. (1977). *Chem. Phys. Lipids*, **18**, 130.
5. Bergter, L. and Seidl, P. R. (1984). *Lipids*, **19**, 44.
6. Tulloch, A. P. and Mazurek, M. (1976). *Lipids*, **11**, 228.
7. Soon, Ng. (1983). *J. Chem. Soc. Chem. Commun.*, 179; (1984) *Lipids*, **19**, 56; (1985) *Lipids*, **20**, 778.
8. Awl, R. A., Frankel, E. N., and Weisleder, D. (1989). *Lipids*, **24**, 866.
9. Santaren, J. F., Rico, M., Guilleme, R., and Ribera, A. (1982) *Org. Mag. Res.*, **18**, 98; Santaren, J. F., Rico, M., and Ribera, A. (1983) *Org. Mag. Res.*, **19**, 238.

This chapter describes a new and rapidly developing field. Since it was prepared (1990) the following significant publications have appeared.

Gunstone, F. D. (1990). ^{13}C-NMR spectra of some synthetic glycerol esters alone and as mixtures. *Chem. Phys. Lipids*, **56**, 195.

Gunstone, F. D. (1990). The ^{13}C-NMR spectra of oils containing γ-linolenic acid. *Chem. Phys. Lipids*, **56**, 201.

Gunstone, F. D. (1990). ^{1}H and ^{13}C-NMR spectra of six *n*-3 polyene esters. *Chem. Phys. Lipids*, **56**, 227.

Gunstone, F. D. (1991). The ^{13}C-NMR spectra of six oils containing petroselinic acid and of aquilegia oil and meadowfoam oil which contain Δ^5 acids. *Chem. Phys. Lipids*, **58**, 159.

Gunstone, F. D. (1991). ^{13}C-NMR studies of mono-, di-, and tri-acylglycerols leading to quantitative and semiquantitative information about mixtures of the glycerol esters. *Chem. Phys. Lipids*, **58**, 219.

Gunstone, F. D. (1991). High resolution NMR studies of fish oils. *Chem. Phys. Lipids*, **59**, 83.

Gunstone, F. D. (1991). The composition of hydrogenated fats determined by high resolution ^{13}C-NMR spectroscopy. *Chem. & Ind. (London)*, p. 802.

Lie Ken Jie, M. S. F., Lam, C. C., and Yan, B. F. Y (1992). ^{13}C-NMR studies of some synthetic saturated glycerol triesters. *J. Chem. Res.*, (*S*) **12**, (*M*) 250.

Wollenberg, K. F. (1990). Quantitative high resolution ^{13}C nuclear magnetic resonance of the olefinic and carboxyl compounds of edible vegetable oils. *J. Am. Oil. Chem. Soc.*, **67**, 487.

8

Mass spectrometry of lipids

RICHARD P. EVERSHED

1. Introduction

Mass spectrometry is used for the structure elucidation of lipids, and their detection and quantification at trace levels. Through recent advances in sample introduction and ionization techniques probably all lipid classes can now be investigated using mass spectrometry. The coupling of chromatographic techniques (such as GC, HPLC, and SFC), to mass spectrometers allows separation of the individual molecular species with on-line mass spectrometric analysis of the eluting components. Accounts of gas chromatography (GC), and high-performance liquid chromatography (HPLC) of lipids appear in Chapters 4, and 5 so will not be discussed at length here. The reader is directed to refs 1–7 for accounts of the theory of mass spectrometry and ancillary techniques and reviews of the mass spectrometric analysis of lipids.

2. Basic principles

Considered in its simplest form a mass spectrometer (MS) is a device for producing and mass measuring ions. In the case of organic molecules the masses and relative abundances of molecular or pseudomolecular ions and fragment ions are a direct reflection of molecular structure. Hence, mass spectrometry is unique in its ability to yield often complete structure assignments without needing to use other physico-chemical techniques. Although isomeric structures may be distinguished on the basis of their mass spectra alone, such distinctions are often best made by considering elution orders in combined chromatographic/mass spectrometric analyses (for discussions of the GC and HPLC separation of isomeric lipid species, see Chapters 4 and 5).

The complexity of the mass spectrometric instrumentation has meant that analyses have traditionally been conducted by highly-trained operators. However, the increasing availability of less sophisticated computer-controlled bench-top GC-MS instruments, means that the lipid analyst is now able to perform many mass spectrometric analyses with only the minimum of expert assistance. A detailed knowledge of the workings of mass spectrometers is probably unnecessary for most analysts. However, an appreciation of the basics of operation of the vacuum system, ion source, mass analyser, and ion detection is undoubtedly of value; ensuring safe and efficient handling of the valuable instrumentation. Chapman's monograph *Practical organic mass spectrometry*[8] covers all mass spectrometric techniques commonly used in the analysis of lipids.

An overview of sample introduction and ionization methods commonly used in the mass spectrometric analysis of lipids is presented below (Section 3), as the choice of these can greatly affect the outcome of an analysis. Complementary to this is a knowledge of possible sample preparation strategies, including chemical transformations or derivatizations that can be used to enhance the structural information content of mass spectral data or improve detection limits in trace analyses. Additionally, the analyst needs to be aware of the capabilities of the different operational modes of the mass spectrometer that exist. For instance, the use of full mass range-scanning to provide mass spectra for structure investigations versus the use of selected ion monitoring (SIM) to enhance sensitivity of detection in trace analyses. Due consideration is given to these latter points in Section 4, which is devoted to applications. The extraction of lipids from biological materials is discussed in Chapter 2. The majority of the procedures described for the GC (Chapter 4) and HPLC (Chapter 5) of lipids also apply to combined GC-MS and LC-MS.

3. Basic instrumentation

The following section describes the components of a mass spectrometer giving due consideration of the specific requirements of lipid analysis. Viewed in its simplest form a mass spectrometer consists of: (*i*) sample inlet system; (*ii*) source of ions; (*iii*) mass measuring system; and (*iv*) means of ion detection, amplification and recording (*Figure 1*). Mass spectrometers operate at reduced pressure (typically $< 10^{-6}$ torr) to enable detection of the small amounts (sub-nanogram to microgram) of material that are analysed.

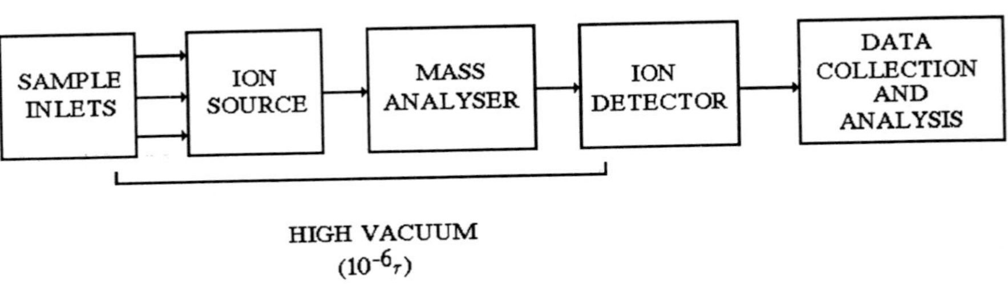

Figure 1. Components of a mass spectrometer.

3.1 Sample inlet systems

3.1.1 Direct insertion probe

The direct insertion probe provides the simplest means of sample introduction and is widely used in the analysis of lipids. The sample is added, usually as a dilute solution in an organic solvent, to a small glass crucible mounted in the tip of a metal probe. After evaporation of the solvent the probe is introduced into the ion source of the mass spectrometer through a vacuum lock. Most low-molecular-weight apolar lipids volatilize very readily at the low pressures encountered in the mass spectrometer under the influence of the heat of the ion source block(*c.* 200 °C). Less volatile substances may be volatilized by applying additional heat, using an integral heater in the probe tip. Some of the dedicated bench-top GC-MS instruments do not possess the facility for sample introduction by direct insertion probe.

The production of gas phase ions from an analyte is an essential part of any mass spectrometric analysis. Frequently, the heating of polar or thermally-unstable materials using a direct insertion probe causes surface catalysed thermal decomposition on the glass crucible. Amongst the methods that have been used to circumvent this problem are 'in-beam' techniques, where the sample is coated on the surface of an inert probe tip which is presented close to the electron beam (see refs 8 and 9 and references therein). Evaporation and ionization may then occur with minimal thermal decomposition. Another variant of this is to use an extended probe tip, which allows the sample to be introduced directly into a chemical ionization (CI) plasma (10) (see Section 3.2.2 for further discussions of the CI technique).

Desorption chemical ionization (DCI), developed from the direct insertion probe techniques of sample introduction, is effective in dealing with thermally-labile samples. In DCI the sample is coated directly onto an electrically heated coiled filament constructed from tungsten, rhenium, platinum, or platinum-iridium (8). The sample-coated filament is introduced into a conventional CI source and resistively heated at high rates to temperatures of 500–1000 °C. At these high temperatures and rapid heating rates evaporation is favoured over decomposition. The use of a CI source with reagent gases such as methane or ammonia, results in the production of quasi-molecular ions (see below Section 3.2.2, for further discussions of CI).

3.1.2 Fast atom bombardment (FAB)

FAB is the most commonly used mass spectrometric technique for producing gas phase ions from thermally-unstable biological substances, such as polar lipids (for example, phospholipids), and has largely superseded other techniques (for example, field desorption; ref. 11), that were used for the analysis of these 'difficult' materials. FAB mass spectrometry uses a beam of high-energy atoms, commonly xenon, produced by accelerating xenon ions through an electric field, to bombard a solution of the analyte dissolved in a relatively involatile liquid matrix (for example, glycerol; ref. 12). Dissipation of the kinetic energy of the bombarding atom beam (primary beam) induces volatilization and ionisation of the analyte and matrix molecules. Bombardment with a primary beam of Cs^+ produces similar spectra to those obtained using xenon (13). Gains in absolute sensitivity (i.e. enhanced secondary ion yields) have been observed for high-molecular-weight substances when employing Cs^+. However, this is inconsequential, as the molecular weights of the vast majority of lipids lie in the low to medium molecular weight range (200 to 1000 amu). The selectivity of response to different substances, which vary little in structure, is a disadvantage of FAB; necessitating the isolation of substances (for example, by reversed-phase HPLC), in a relatively pure state. The presence of chemical impurities, such as inorganic cations, can greatly reduce secondary yields, i.e. produce a loss of sensitivity.

3.1.3 Gas chromatography–mass spectrometry (GC-MS)

GC-MS is the most widely used sample introduction technique for the mass spectrometric analysis of lipids. GC and MS are especially complementary as because both are gas phase techniques, displaying optimum performance with sample sizes in the nanogram range. The GC analysis of lipids has been discussed at length in Chapter 4. Flexible fused silica capillary columns are now preferred for GC-MS work as they can be fed directly from the GC oven into the ion source of the mass spectrometer. This is the optimum arrangement as it ensures efficient transfer of analytes from the GC to the mass spectrometer. The temperature of the transfer line between the GC and MS must be maintained at, or slightly above, the maximum temperature used in the GC analysis, in order to avoid sample components condensing in this region. The carrier gas flows employed in capillary GC, *c*.2 ml min^{-1}, are readily accommodated by the vacuum pumping systems of modern mass spectrometers.

Packed columns are rarely used now in GC-MS work, as the higher carrier gas flow rates, *c*.50 ml min^{-1}, are incompatible with the mass spectrometer vacuum system. A separator must be used to remove the bulk of the carrier gas from the column effluent prior to its introduction into the mass spectrometer. These separation devices are notoriously troublesome as they provide sites for adsorption and decomposition of higher molecular weight and thermally unstable lipids. As with packed GC columns, the high flow-rates employed with the increasingly popular wide bore (0.53 mm i.d.) capillary columns are incompatible with high vacuum pumping systems of mass spectrometers.

3.1.4 Liquid chromatography-mass spectrometry (LC-MS)

The usefulness of LC-MS lies in its potential for the analysis of mixtures of lipids that are intractable by GC-MS without prior chemical or enzymatic degradation; for example, phospholipids (see Chapter 2). Combining high-performance liquid chromatography (HPLC) with mass spectrometry is much more difficult than with GC. The fundamental problem that exists in any LC-MS system is the enormous volume of solvent vapour that is generated from the evaporation of the liquid eluent in the MS. The vacuum pumping systems of conventional MS instruments are simply unable to deal with the volume of vapour produced. The proliferation of LC-MS interfaces that are now available commercially are a reflection of the problems that exist in combining HLPC and MS effectively. It is inappropriate to discuss all the variants of LC-MS that exist, as many will never find routine application to lipid analysis.

The moving belt (14), direct liquid introduction (DLI) (15), thermospray (16) and molecular beam (17) interfaces have all been used for the LC-MS of lipids. Of these the DLI interface has been most widely applied to the analysis of lipids (18, 19). With the DLI interface the HPLC effluent is introduced directly into the ion source. The problem of introducing large volumes of liquid is overcome by splitting the effluent such that only *c.*1% of the total effluent flow (1.5–2.0 ml min^{-1}) enters the ion source, via a small orifice (2–5 μm in diameter). The splitting of the effluent stream inevitably reduces the sensitivity of analyses. Another disadvantage of this system is that ionization is restricted to solvent-mediated chemical ionization (CI). However, coincidentally, some of the best HPLC eluents for lipids include solvents that are excellent CI reagents; for example, acetonitrile and propionitrile (see below, Section 3.2.2, for a more detailed description of CI). The 'soft' ionizing thermospray source also generates simple CI-type spectra. Operation of the thermospray technique will be discussed further in Section 4.1.13. Conventional electron impact and CI spectra are attainable with moving belt and molecular beam type interfaces, and both these interface systems are amenable to the LC-MS analysis of lipids.

3.2 Ion sources

Electron ionization (EI) and chemical ionization (CI) are the two most widely used ionization techniques in the mass spectrometric analysis of lipids. The following section describes the mechanism of operation of these two ionization techniques. The FAB process has already been discussed above in relation to sample inlet systems (Section 3.1.2). Those readers requiring more detailed discussion of ionization techniques are directed to the specialist mass spectrometry literature; for example, ref. 8.

3.2.1 Electron ionization (EI)

Electron ionization (EI) is produced by accelerating electrons from a hot filament through a potential difference, usually of 70 eV. Organic molecules present in this electron beam (for example, as a result of evaporation from a direct insertion probe, or elution from a GC column), will be ionized and fragmented. Extensive fragmentation occurs in the case of alkyl compounds, such as lipids. The initial product of EI in the ion source is a radical cation, resulting from the removal of an electron from the analyte molecule. If this singly-charged species is stable it will be the highest mass ion appearing in the mass spectrum for that substance. It is referred to as the molecular ion, denoted $M^{+\cdot}$, and it provides the molecular weight (measured in atomic mass units, amu) of that substance. The $M^{+\cdot}$ ion is present at *m/z* 270 in the case of methyl palmitate (*Figure 2a*). The electron beam usually possesses sufficient energy to induce fragmentation in organic compounds, by loss of radicals or neutral molecules (see *Scheme 1*).

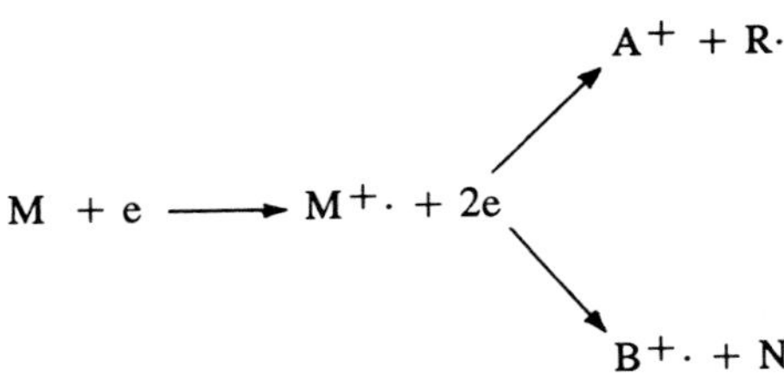

Scheme 1. Electron ionization (EI) induced by collision of an electron with a molecule to producing a molecular ion ($M^{+\cdot}$) which can give the fragment ions, A^+ (cation) or $B^{+\cdot}$ (radical cation), by a loss of a radical $R^{\cdot}$) or a neutral molecule (N), respectively.

Although EI produces extensive fragmentation which is of use in structural investigations of lipids, the $M^{+\cdot}$ ion is frequently weak or absent from spectra. The relative abundance of $M^{+\cdot}$ may be enhanced by decreasing the ion-source block temperature, or electron beam potential (say to 20 eV). When this latter approach is unsuccessful, chemical ionization (CI) is used to derive molecular weight information.

3.2.2 Chemical ionization (CI)

Chemical ionization (CI) is used to generate mass spectra by means of ionic reactions rather than electron bombardment. The technique is termed a 'soft' ionization technique, as the spectra usually contain essentially only molecular weight information, fragmentation being either absent, or much reduced compared to EI. Many variations of CI exist, but typically a reagent gas (for example, methane, ammonia, or isobutane), is introduced into the ion source at a pressure of around 1 torr. Bombardment of the gas with electrons yields a population of ions and neutral molecules. The reagent gas ions will take part in ion-molecule reactions, with sample molecules vapourizing into the source (*Scheme 2*). The most abundant reagent gas ions formed by the electron bombardment of methane are CH_5^+ and $C_2H_5^+$.

As can be seen from the mass spectrum of methyl palmitate, shown in *Figure 2b*, the most abundant ion arises by proton transfer (see *Scheme 2*) and

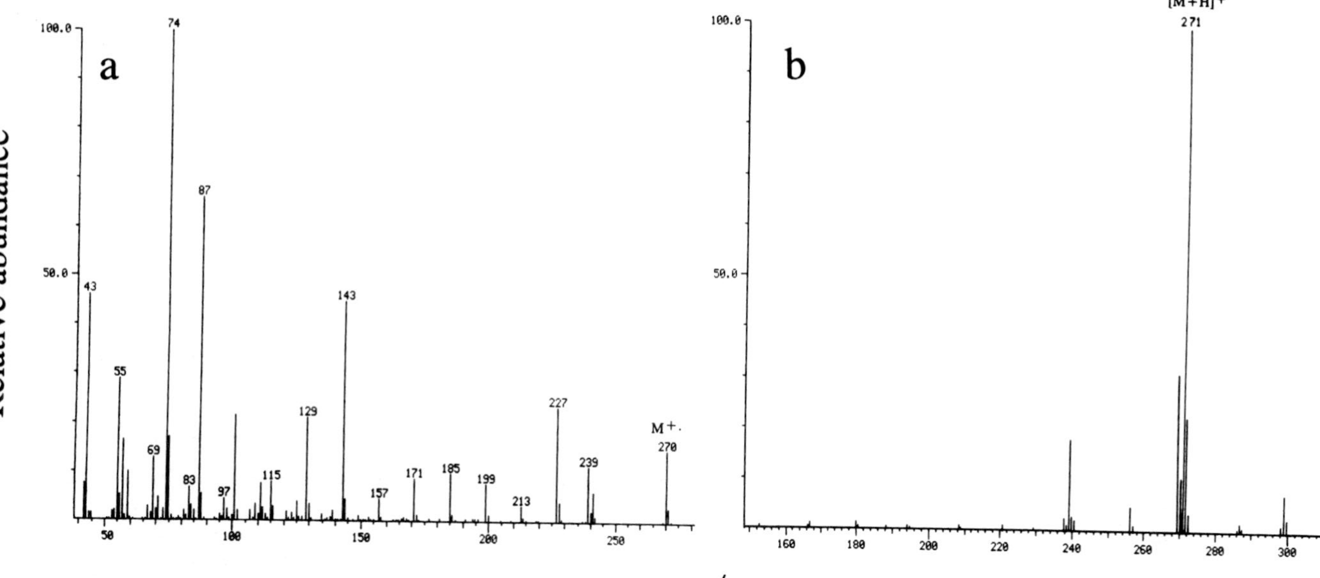

Figure 2. Electron ionization (a) (70 eV) and methane chemical ionization (b) mass spectra of methyl palmitate.

$$M + C_2H_5^+ \longrightarrow MH^+ + C_2H_4$$

Scheme 2. Reagent gas ions, produced by electron bombardment of methane, react with analyte molecules to yield positively charged quasi-molecular adduct ions.

is termed the quasi-molecular ion, $[M + H]^+$. The higher mass ions that are present in the spectrum correspond to molecular adduct ions, $[M + CH_5)^+$, $[M + C_2H_5]^+$, $[M + C_3H_5]^+$, etc., which arise by electrophilic addition of reagent gas ions to the sample molecules. The quasi-molecular ion and adduct ions are useful in infering the molecular weights of sample molecules. The lack of fragment ions in CI spectra results from the low energy transfer involved in the CI process compared to EI. Furthermore, even-electron $[M + H]^+$ ions are energetically more stable than the radical $M^{+\cdot}$ ions produced by EI. Those readers requiring more detailed discussions of CI should refer to refs 8 and 9 and references therein.

3.3 Mass analysers

A number of types of analyser are now available that will measure the mass of ions according to their mass-to-charge ratio, m/z. The instruments most commonly employed in the investigation of lipids use quadrupole mass filters and magnetic sector mass analysers. The ion trap analysers, which are now commercially available as components of certain dedicated bench-top GC-MS systems, although of limited mass range capability, fulfil the requirements of many lipid analyses. The following sections discuss the basic principles and advantages and disadvantages of mass analysis in magnetic sector and quadrupole analysers for the study of lipids.

3.3.1 Magnetic sector analysers

The double-focusing analyser is the most common magnetic sector instrument configuration currently in use for the analysis of organic compounds. The usual arrangement is for an electrostatic sector to precede a magnetic sector (conventional Nier–Johnson geometry). The electrostatic sector is an energy focusing device. Hence, ions produced and accelerated by the ion source are focused by the electrostatic sector according to their translational energies, irrespective of their mass-to-charge ratio (m/z). The magnet on the other hand is a momentum analyser, where the ions are separated according to their m/z ratio. The combination of an electrostatic sector with a magnetic analyser is highly versatile, providing the capability for operation at high resolution. The latter property enables accurate mass measurements to be performed, from which the elemental compositions of ions can be determined. Although double focusing magnetic sector instruments are well-suited to GC-MS work, they are more difficult to use with other sample introduction and ionization techniques since the ion source operates at a high voltage (several kilovolts). The direct insertion probe, DCI and FAB techniques discussed above are all readily used in conjunction with magnetic sector instruments. A range of alternative arrangements of electrostatic and magnetic sectors are available in commercially produced instruments, introducing the possibility for a wide range of scan modes (8); further discussion of these is inappropriate here.

3.3.2 Quadrupole analysers

Quadrupole mass spectrometers are constructed and function in entirely different ways to magnetic sector instruments. The quadrupole analyser comprises four parallel rods of hyperbolic or circular cross-section. A radiofrequency (RF) potential and DC voltage is connected between opposite pairs of rods. Ions are injected into the oscillating electric fields by a small accelerating voltage (10–20 V) and, under the influence of the fields begin to undergo complex oscillations. Mass separation is achieved by varying (scanning) the voltages on the quadrupole rods, while keeping the RF/DC ratio constant. At any one point in the scan only one mass can pass through the system.

Quadrupole instruments are very popular, owing to their relatively simpler operation compared to magnetic sector instruments. They are also less expensive than sector instruments and compatible with a wider range of inlet systems, including LC-MS. Although the mass range of quadrupole instruments is substantially less than that of many magnetic sector instruments, it is sufficient for the vast majority of lipid analyses. A number of the properties of magnetic sector and quadropole instruments are compared in *Table 1*.

3.4 Ion detection

The small current (typically in the range 10^{-9} to 10^{-17} A) generated by ions passing through the mass analyser must be amplified and converted into a

Table 1. Advantages/disadvantages of mass spectrometers using magnetic sector or quadrupole mass analysers

Magnetic sector	Quadrupole
Advantages	*Advantages*
Greater versatility	Compact
Accurate mass measurement	Easy to operate
More specific forms of selected ion monitoring	Source at earth potential makes interfacing less problematical
Metastable ion analysis	Easy to look at positive and negative ions
High mass capabilities > 4000 daltons	Much less expensive than sector instruments
Disadvantages	*Disadvantages*
Highly sophisticated	Limited mass range
Relatively more expensive	Only low resolution capabilities
High source potentials present interfacing problems	Limited data types compared to magnetic sector instruments

voltage that can be digitized or displayed on some form of recording device. Electron multipliers, are widely used for this purpose. These devices comprise copper–beryllium dynodes which emit electrons when impacted by highly energetic ions, focused from the mass analyser. Cascade emission of electrons through a series of dynodes produces gains of the order of 10^6. The final dynode of the electron multiplier is connected to a pre-amplifier, to convert the output current to a voltage suitable for recording.

3.5 Data collection and interpretation

Computers are essential to the efficient operation of modern mass spectrometers. The type of computers employed range from large, powerful computers to desk-top personal computers. The computer is usually fully integrated into modern instruments, controlling the various mass spectrometer, inlet, ionization, and scan functions. In addition, the computer provides an essential data collection and storage facility. The large volume of mass spectral data produced in GC-MS and LC-MS analyses of complex biological mixtures requires substantial post-run data processing.

The interpretation of this large volume of GC-MS data is aided by automated searching of mass spectral databases held in the permanent memory of the mass spectrometer computer. A number of database libraries of EI mass spectra are available from proprietary sources. Most of these libraries contain substantial numbers of mass spectra of lipids and their commonly used derivatives. *Figure 3* shows part of the output from a commercially available GC-MS data processing package. In addition to showing the closest matching library mass spectrum the library search also shows a range of statistical evaluations of the search. Although potentially very useful the results of library searches are in no way definitive. The search results must be carefully assessed on the basis of the statistical evaluation of the match and consideration of GC elution orders, where mass spectra have been obtained by combined GC-MS analyses. The possibility exists for installing mass spectral data processing software and libraries on personal computers for off-line data processing.

4. Applications

The following section (4.1) describes the mass spectral behaviour of a range of commonly encountered lipids and their derivatives based on the ions observed in full-scan mass spectra. SIM mass spectrometry is an alternative mode of analysis that is used to enhance sensitivity and selectivity in trace analyses. Section 4.2 describes the use of SIM for the detection and quantification of lipids present as minor constituents of complex biological mixtures.

25409 SPECTRA IN LIBRARYNB SEARCHED FOR MAXIMUM PURITY
128 MATCHED AT LEAST 6 OF THE 16 LARGEST PEAKS IN THE UNKNOWN

RANK	IN	NAME
1	1380	HEXADECANOICACID, METHYLESTER
2	8272	PENTADECANOICACID, 14-METHYL-, METHYLESTER
3	1588	TETRADECANOICACID, METHYLESTER
4	8829	HENEICOSANOICACID, METHYLESTER
5	5497	UNDECANOICACID, METHYLESTER

RANK	FORMULA	M. WT	B. PK	PURITY	FIT	RFIT
1	C17. H34. O2	270	74	839	991	839
2	C17. H34. O2	270	74	816	947	854
3	C15. H30. O2	242	74	713	905	766
4	C22. H44. O2	340	74	701	942	737
5	C12. H24. O2	200	74	670	897	724

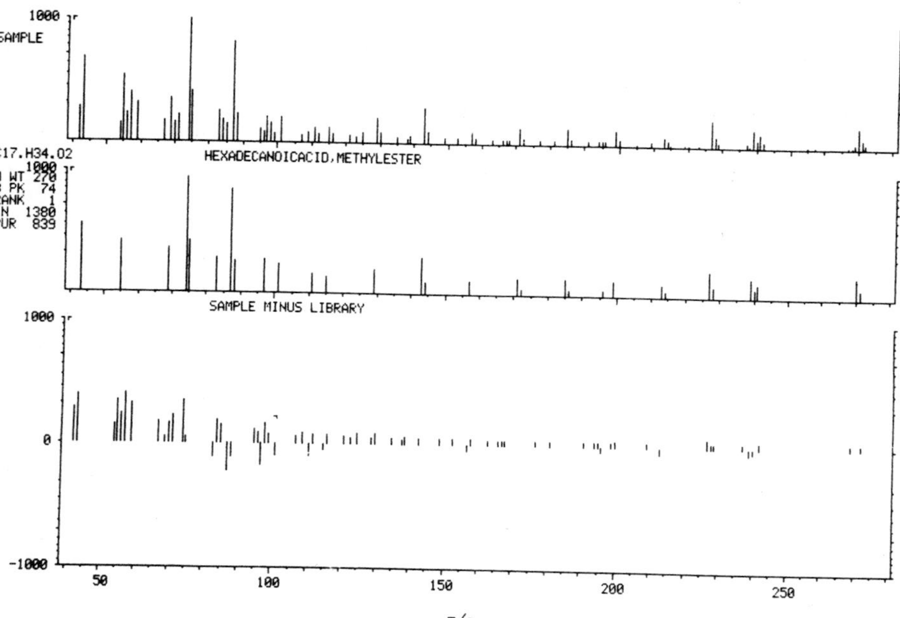

Figure 3. Part of the output from a commercially available library search package showing the best match of the analyte spectrum with the library database spectrum for methyl palmitate. (Reproduced with permission of Finnigan MAT Ltd.)

4.1 Structure elucidation

Electron ionization (EI) is the most widely used ionization technique in the structure investigation of lipids by mass spectrometry. The following section discusses the trends in the EI mass spectra of some of the most commonly encountered classes of lipid. Descriptions will be given for the preparation of special derivatives of certain lipids that can be employed to provide more detailed structural information. Some examples are given of the use of CI mass spectrometry for the structure investigation of lipids. Examples are also presented for the use of FAB and LC-MS in the analysis of some polar and thermally-unstable lipids.

4.1.1 Fatty acids

i. Methyl esters

Fatty acids are most commonly analysed by GC-MS as their methyl ester derivatives (20) (FAMEs; see Chapter 2 for methods of preparing fatty acid methyl esters). *Figure 2a* shows the 70 eV EI spectrum of methyl palmitate. The spectrum is typical of that which would be obtained for the analysis of any fully saturated straight-chain FAME. The molecular ion ($M^{+\cdot}$) at *m/z* 270 in the case of methyl palmitate, is generally present in FAME spectra. The mass of the $M^{+\cdot}$ ion confirms the carbon number and degree of unsaturation of the fatty acid in question. The molecular weights of other fatty acid methyl esters can be calculated by adding or subtracting multiples of 14 amu (the mass of one methylene group, CH_2) in order to establish the molecular weight of higher and lower homologues of palmitic acid (16:0). The presence of double bonds (or cyclopropane rings) are indicated by a reduction in molecular weight of 2 amu. Hence, the $M^{+\cdot}$ ions of methyl stearate (18:0), methyl oleate (18:1), methyl linoleate (18:2), and methyl linolenate (18:3) appear at *m/z* 298, 296, 294, and 292 respectively. Other high mass ions, occurring at M-31 and M-43 in the spectra of saturated fatty acid methyl esters, correspond to the loss of the methoxy (CH_3O—) and propyl ($CH_3(CH_2)_2$—) groups respectively. The loss of the propyl group involves a complex intra-molecular rearrangement. The base peak, *m/z* 74, in the spectrum of saturated FAME's (for example, methyl palmitate; *Figure 2a*) arises through the six-centre McLafferty re-arrangement, which results in proton transfer and cleavage of the bond linking the C-2 and C-3 carbons (see *Scheme 3*). The relative abundance of the *m/z* 74 ion decreases with increasing unsaturation of the alkyl chain as a result of competitive fragmentations. The mass spectra of unsaturated FAMEs differ from their saturated counterparts by exhibiting an M-32 (loss of methanol) ion rather than M-31 (loss of a methoxy radical).

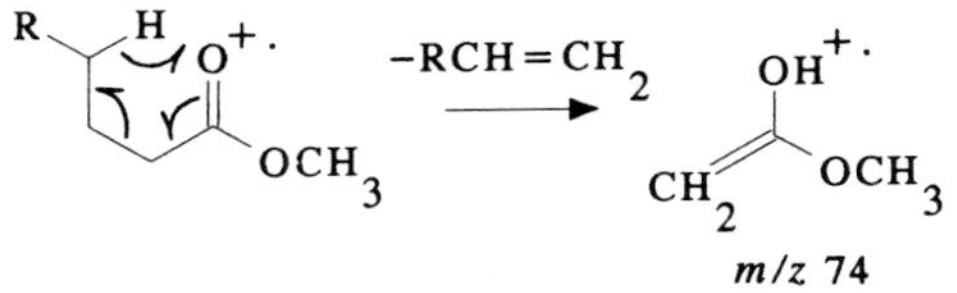

R H $O^{+\cdot}$ OCH_3 $\xrightarrow{-RCH=CH_2}$ $OH^{+\cdot}$ CH_2 OCH_3

m/z 74

Scheme 3. The six-centre hydrogen transfer reaction commonly referred to as the McLafferty rearrangement.

ii. Trimethylsilyl esters

Trimethylsilyl (TMS) esters of fatty acids are also commonly prepared for GC-MS analysis (see Chapter 2 for trimethylsilylation procedures). The TMS derivative of palmitic acid is shown in *Figure 4*. While an $M^{+\cdot}$ ion (*m/z* 328) is present, much more prominent is the $[M—CH_3]^+$ ion (*m/z* 313). This latter fragment ion is present in the mass spectra of the TMS derivatives of all

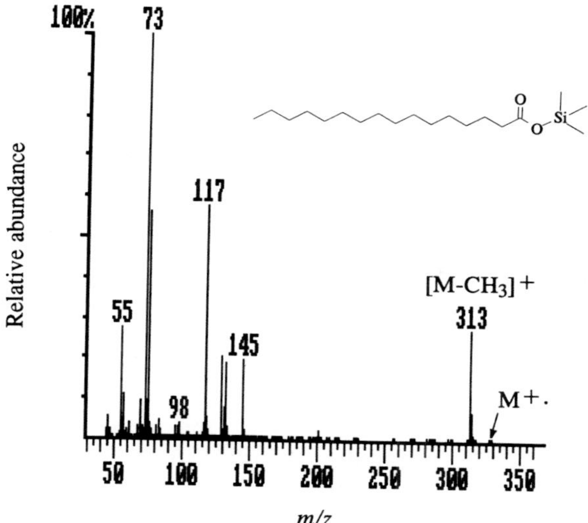

Figure 4. Electron ionization (70 eV) mass spectrum of the trimethylsilyl (TMS) ester of palmitic acid.

saturated and unsaturated fatty acids (although being of somewhat reduced relative abundance in the latter) and is used to assign carbon number and the degree of unsaturation in the case of unknowns. The ions at *m/z* 73 and 75 corresponding to $[(CH_3)_3Si]^+$ and $[(CH_3)_2SiOH]^+$ respectively, are very common in the mass spectra of TMS derivatives. The fragment ions at *m/z* 117, 129, 132, and 145 are also prominent in the spectra of saturated and monounsaturated fatty acid TMS esters, and help to distinguish them from the TMS ether derivatives of fatty alcohols (see later). The *m/z* 132 and 145 ions are related to the *m/z* 74 and 87 McLafferty rearrangement ions seen in FAME spectra (see above). Also in common with methyl esters, the fragmentation of the TMS derivatives becomes less distinctive with increasing unsaturation of the alkyl chain.

iii. Tert.-butyldimethylsilyl esters

Unlike methyl esters, TMS esters are susceptible to rapid hydrolysis. The *tert.*-butyldimethylsilyl (TBDMS) esters however, are sufficiently resistant to hydrolysis (10^4 times more stable than the corresponding TMS derivatives) that they will withstand rigorous purification conditions if necessary. Like the TMS esters the EI mass spectra of fatty acid TBDMS esters do not generally exhibit a molecular ion; the highest mass ion usually being $[M-15]^+$. In common with the TBDMS derivatives of many other classes of compound, the most abundant high mass ion corresponds to loss of the *tert.*-butyl moiety from the *tert.*-butyldimethylsilyl group.

Abundant low mass ions are seen at *m/z* 73 and 75; the structures of these ions have been discussed above. Unlike the methyl and TMS esters, and mass spectra of the TBDMS derivatives contain little evidence for fragment ions arising by McLafferty rearrangement. Presumably the six-centre rearrangement is restricted by the bulk of the TBDMS group. However, in common with the other derivatives of fatty acids, increasing unsaturation increases the abundance of ions in the spectrum resulting from alkyl chain fragmentations (21).

4.1.2 Derivatizations of the carboxyl group for the determination of double-bond positions

The EI mass spectra of the methyl esters, TMS, and TBDMS esters of unsaturated fatty acids can not be used to deduce the positions of unsaturation in fatty acids, owing to migration of the double bond(s) during the ionization process. Two approaches are commonly used to overcome this problem and locate double-bond positions. One approach is to derivatize the carboxyl moiety in order to stabilize the positive charge and prevent double-bond migration. The alternative strategy is to form derivatives of the double bond(s).

i. Pyrrolidides (22)

Pyrrolidide derivatives of fatty acids were the first heterocylic nitrogen derivatives of fatty acids prepared for mass spectrometric analysis. The procedure in *Protocol 1* is used to prepare pyrrolidide derivatives of free fatty carboxylic acids or fatty methyl esters (23).

Protocol 1. Preparation of pyrrolidide derivatives

1. Dissolve the fatty acid or methyl ester mixture (10 mg) in pyrrolidine (1 ml) in a screw-capped vial, add glacial acetic acid (0.1 ml), and heat for 1 h at 100 °C.
2. After cooling, dissolve the reaction mixture in dichloromethane (8 ml) and wash with 2 M hydrochloric acid (3 × 4 ml) and water (2 × 4 ml).
3. Dry the dichloromethane solution of pyrrolidides over anhydrous sodium sulfate, and use either directly for GC-MS, or after evaporation of the dichloromethane.

Pyrrollidide derivatives can also be prepared directly from complex acyl lipids; however, longer reaction times (up to 4 h) and an increased temperature (120 °C) are required to achieve maximum yields. The EI mass spectrum of the pyrrolidide derivative of oleic acid (*N*-octadec-9-enoylpyrrolidine) is shown in *Figure 5*. The base peak in the spectrum, *m/z* 113, is the McLafferty rearrangement ion. A molecular ion, *m/z* 335 is present in reasonable abundance, the odd mass being due to the presence of a nitrogen atom in the molecule. The $M^{+\cdot}$ ion of saturated fatty acids is always

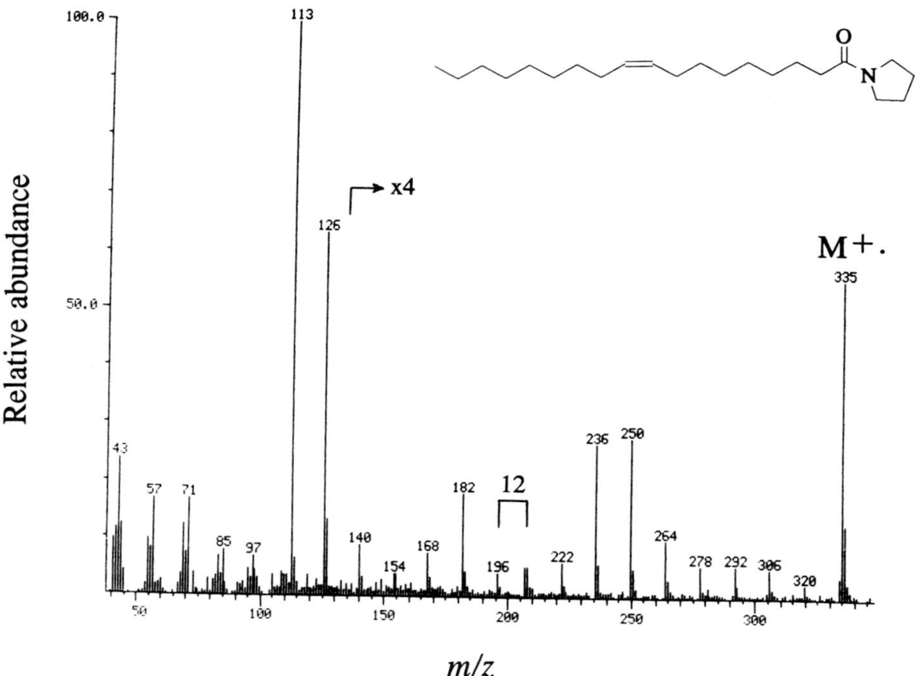

Figure 5. Electron ionization (20 eV) mass spectrum of *N*-octadec-9-enoylpyrrolidine.

accompanied by an $[M\text{-}1]^+$ of *c*.40% relative abundance, due to the loss of a single proton. The pyrrolidide spectra of straight-chain saturated fatty acids are further typified by the presence of a regular series ions separated by 14 daltons, which decreases in abundance from *m*/*z* 126 to 322, with the exception of *m*/*z* 154, 194, and 308. In the spectrum of the mono-unsaturated derivative (*Figure 5*) this regular pattern of fragmentation is disturbed in such a way as to allow the position of the double bond to be deduced. The main peak in each cluster of the series of ions is separated by 14 daltons, except in the vicinity of the double bond, then the separation is 12 daltons (see *Figure 5*).

A number of rules relating to the fragmentation of fatty acid pyrrolidides have been formulated which can be used to deduce the positions of unsaturations and other structural features. A complete explanation of these rules appears in ref. 24. It must be emphasized that the identification of unknowns is no straightforward matter. Most reliable assignments are made by comparing the spectra of unknowns with those of authentic compounds obtained under similar analytical conditions. The identification of fatty acid structures from pyrrolidide mass spectra must taken into account GC-MS elution orders. Care must also be taken when using the relative intensities of ions to draw conclusions concerning molecular structure, that these intensities have not been affected by the co-elution of compounds of similar structure, or by the background subtraction routines commonly employed in the processing of GC-MS data.

ii. Picolinyl (3-hydroxymethyl-pyridine) esters (25)

Picolinyl ester derivatives have been compared to the pyrrolidides for the structure investigation of fatty acids in natural samples (26). Although the pyrrolides were deemed to have marginally better chromatographic properties, the picolinyl esters were preferred for their mass spectral properties, particularly in terms of the markedly higher abundance of the diagnostic ions used to provide evidence for the number and position of double bonds. Picolinyl esters are prepared as set out in *Protocol 2* (6).

Protocol 2. Preparation of picolinyl esters

1. Dissolve free fatty acids (5 mg) in trifluoroacetic anhydride (0.5 ml) then heat at 50 °C for 30 min in a screw-capped vial.
2. After cooling, and evaporation of the excess reagent under a stream of nitrogen immediately add 3-hydroxymethylpyridine (20 mg) and 4-dimethylaminopyridine (4 mg) as a solution in dichloromethane (0.2 ml), allow the mixture to stand (3 h at room temperature).
3. Evaporate solvent, dissolve the residue in hexane (8 ml), wash with water (3 × 4 ml), then evaporate the hexane to dryness.
4. To remove residual free fatty acids, dissolve reaction products in diethyl ether (1 ml) add a few milligrams of amino bonded silica (Bond Elut™ or Sep-Pak™) and shake the mixture. After 10 min centrifuge the mixture, decant the solvent and analyse the picolinyl esters by GC-MS.

Purification of the picolinyl esters can also be carried out using a small column of Forisil™ eluting with a mixture of hexane-diethyl ether mixture (1:4 v/v) (6).

The mass spectral behaviour of the picolinyl esters is similar in many respects to that of the pyrrolidide derivatives. The major advantage of the picolinyl esters derives from the greater abundance of the diagnostically important high mass fragment ions. The EI mass spectrum of the picolinyl ester derivative of oleic acid is shown in *Figure 6a*. A molecular ion is present and can be used to assign the carbon number and degree of unsaturation of unknown fatty acids. The ions at *m/z* 92, 93, and 108 results from fissions of the picolinyl ester moiety as shown, while the fragment ion at *m/z* 151 results from McLafferty rearrangement with cleavage of the C-2/C-3 bond. Other

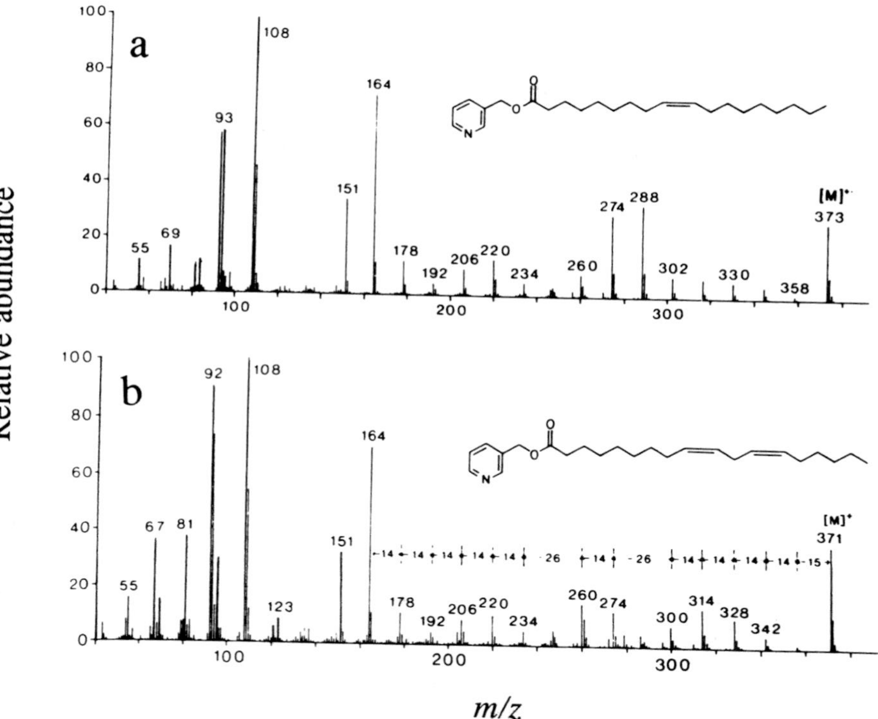

Figure 6. Electron ionization (25 eV) mass spectra of the picolinyl derivative *Z*-9-octadecenoic (a) and *Z*-9,12-octadecadienioc (b) acids. (Reproduced with permission, from ref. 28.)

ions are produced by cleavages of carbon–carbon bonds, with the position of the double bond being indicated by the 13 amu separation between the fragment ions at *m/z* 234 and 247, and between *m/z* 247 and 260.

The mass spectra of the picolinyl derivatives of polyunsaturated fatty acids show the same general features. The mass spectrum of *Z,Z*-9, 12-octadecadienoic acid is shown in *Figure 6b*. Although the spectrum is more complex, the positions of the double bonds are determined by the 26 dalton separation, between ions representing cleavage adjacent to the double bond, i.e. *m/z* 234 and 260, and between *m/z* 274 and 300. Increasing the unsaturation in the fatty acid increases the complexity of the mass spectra obtained, however, the spectra can still be used to determine the position of double bonds. For more detailed discussions of the spectra of the picolinyl derivatives of more complex fatty acids, containing methyl branches, cyclopropane rings, triple bonds, etc., the reader is directed to refs 25 and 27–35.

4.1.3 Derivatization of the double bond to determine position

An alternative approach to the location of unsaturation in fatty acids involves chemical derivatization at the double bond. The aim is to produce EI mass spectra from GC-MS analyses that contain fragment ions diagnostic of the double bond position. Two of the most commonly used techniques are described below. Those readers requiring information regarding other possible derivatives of the double bond should consult refs 6, 36, 37, and 38 and references cited therein.

i. Oxidation to vicinal diols (39, 40)

Although one of the earliest methods developed for determining the position of double bonds in unsaturated fatty acids by GC-MS this method continues to be used. The diols and multiple diols of mono- and polyunsaturated fatty acids or their methyl esters are prepared as shown in *Protocol 3*.

Protocol 3. Preparation of diols

1. Dissolve 1 mg of fatty acid methyl ester in 0.2 ml of a mixture of pyridine and dioxane (1:8 v/v).
2. Add 6 ml of a suspension of OsO_4 (2% w/v solution in dioxane, freshly prepared) and allow to stand for 2 h at room temperature.
3. Add 6 ml of a suspension of Na_2SO_3 (prepared by adding 8.5 ml of 16% w/v Na_2SO_3 water to 2.5 ml of methanol) and allow to stand for 1.5 h, with occasional shaking.
4. Centrifuge, remove supernatant, and dry over anhydrous Na_2SO_4.
5. Decant, evaporate the solvent, and trimethylsilylate the residue according to the procedure given in Chapter 2.

The TMS ethers of the diols of fatty acid methyl esters possess excellent GC properties. The EI mass spectrum of the bis-(TMS) ether derivative of the vinical diol formed from methyl oleate is shown in *Figure 7*. Although no molecular ion is seen in the spectrum the weak $[M-15]^+$ ion at m/z 459 may be used to confirm the presence of the bis-TMS derivative of the C_{18} fatty acid methyl ester. The abundant fragment ions in the spectrum at m/z 215 and 259 characterize the position of the double bond in the original compound unambiguously, as shown. The mass spectra of the TMS ethers of the multiple diols of polyunsaturated fatty acid methyl esters are much more complex; however, they can be used to deduce the structures of unknowns (37). The simplicity of the mass spectra means this method is very conveniently applied to the determination of double bond positions in mixtures of mono-unsaturated fatty acids.

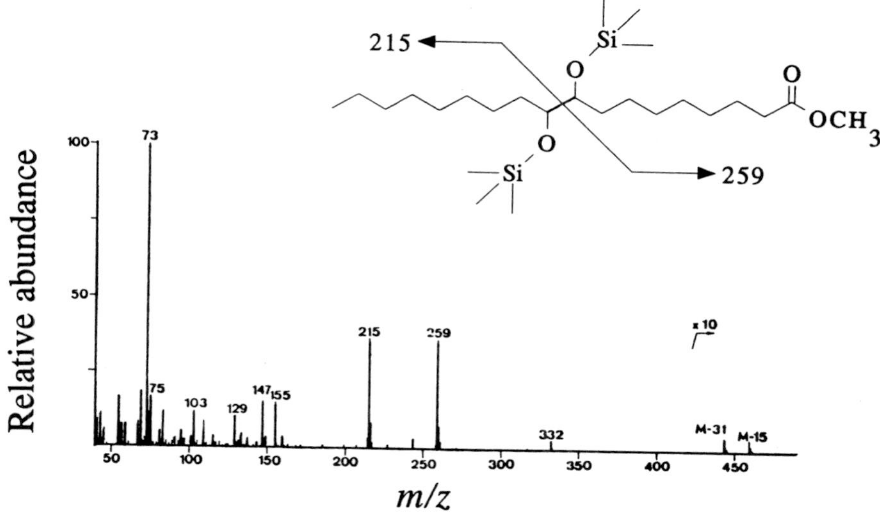

Figure 7. Electron ionization (70 eV) mass spectrum of methyl 9,10-*bis*(trimethylsiloxy) octadecanoate. (Reproduced, with permission, from ref. 39.)

ii. Dimethyl disulfide adducts

Dimethyl disulfide (DMDS) adducts were first used analytically for the location of double bonds in linear alkenes (41). However, the technique is equally well applied to fatty acid methyl esters (42). The derivatization is a one-step reaction, performed as set out in *Protocol 4*.

Protocol 4. Preparation of dimethyl disulfide adducts

1. Dissolve 100 μg of fatty acid methyl ester in hexane (100 μl) then add 100 μl of dimethyl disulfide and 2 drops of an iodine solution (6% w/v in diethyl ether), and allow to stand overnight (15 h) at room temperature.
2. Add 0.5 ml of hexane and 5% w/v sodium thiosulphate (0.5 ml) and shake to destroy the iodine.
3. Remove the organic layer and repeat the extraction of the aqueous layer with two further 0.5-ml portions of hexane.
4. Combine the hexane washings, dry over anhydrous Na_2SO_4, decant and submit to GC-MS, either directly, or after concentrating the hexane extract.

The EI mass spectrum of the DMDS adduct of methyl oleate is shown in *Figure 8*. The molecular ion, which is visible at *m/z* 390, establishes the

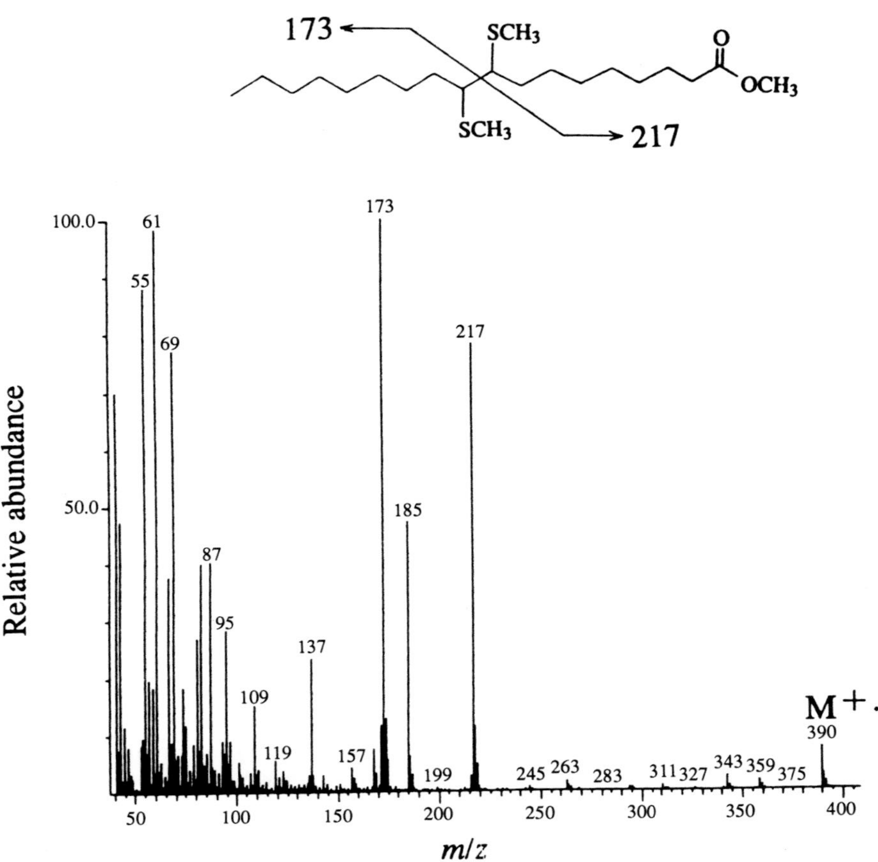

Figure 8. Electron ionization mass spectrum of the dimethyldisulfide adduct of methyl oleate (*Z*-9-octadecenoate).

carbon number of the original fatty acid. The abundant ions at *m/z* 159 and 231 locate the position of the double bond unambiguously. As with the bis-TMS derivatives of the diols, the DMDS derivatives can be applied very reliably to the determination of the positions of double bonds in mixtures of positional isomers of monounsaturated fatty acids. Moreover, as the addition of the derivatizing agent to the double bond is stereospecific *E*- and *Z*-isomers of fatty acids can be discriminated in GC-MS analyses (*Scheme 4*). Although the mass spectra of the DMDS derivatives are congruent, the erythro- isomer (produced from the *E*-fatty acid) elutes after the threo- isomer (produced from the *Z*-fatty acid) on an apolar dimethyl polysiloxane (OV-1 type) coated capillary column (43). This technique is less easily applied to the study of polyunsaturated compounds. For example, diunsaturated compounds yield

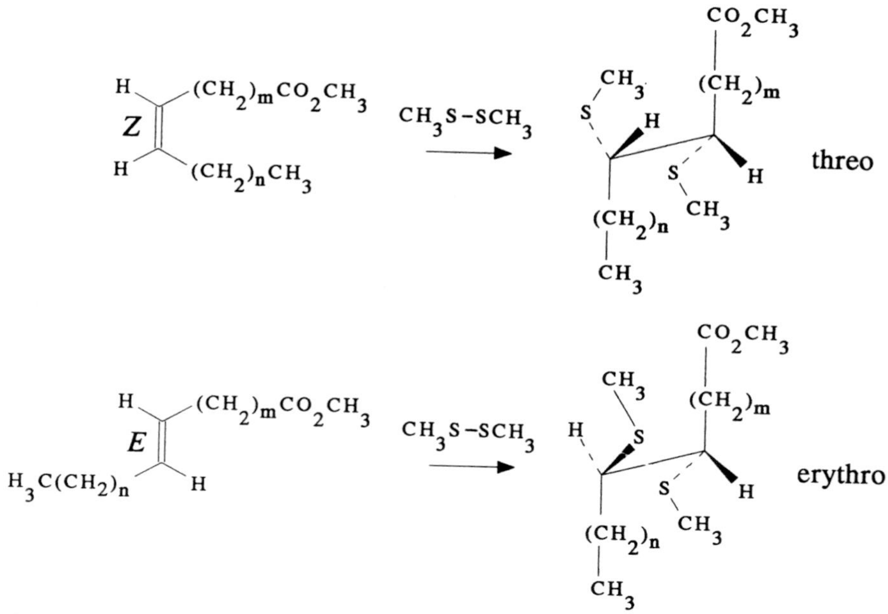

Scheme 4. The stereospecific addition of dimethyldisulfide to geometric isomers of monounsaturated fatty acid methyl esters.

linear or cyclic polyethers, depending on the number of methylene groups separating the two double bonds (44).

4.1.4 Triacylglycerols

Most of the early analyses of triacylglycerols by mass spectrometry used the direct insertion probe technique for sample introduction. There now exists the option for carrying out on-line high-temperature capillary GC-MS (see Chapter 4 for a description of high-temperature GC of lipids), allowing the separation and characterization of complex mixtures of triacylglycerols. When using high-temperature GC-MS it is essential to ensure that the GC-MS transfer line temperature is set at a temperature equivalent to, or slightly higher than, the maximum GC oven temperature used in the temperature programming (usually around 350 °C). The same problems, i.e. of poor recoveries of triacylglycerols bearing polyunsaturated fatty acyl moieties, exist in high-temperature GC-MS work as in high-temperature GC (see Chapter 4).

Figure 9 shows the total ion chromatogram for the GC-MS analysis of a mixture of authentic triacylglycerols. The experimental conditions and the peak identities are given in the figure caption. The partial mass spectrum for tristearin is shown in *Figure 10*. A full mass spectrum will generally contain an $M^{+\cdot}$ ion together with an $[M\text{-}18]^{+}$ ion due to the loss of water, but as these ions

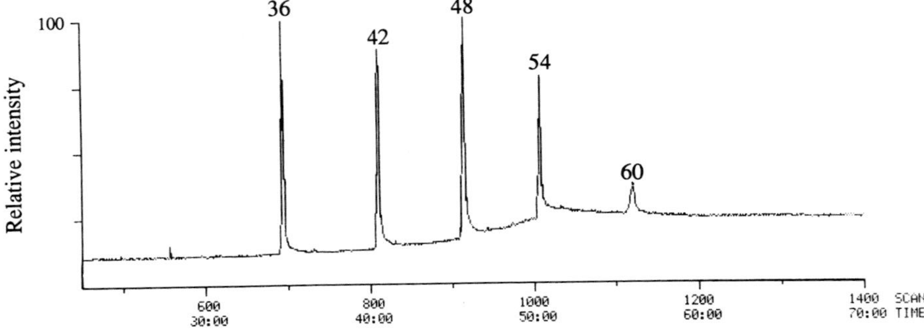

Figure 9. Total ion current (TIC) chromatogram obtained for the high temperature GC-MS analysis of a mixture of authentic triacylglycerols. The numbers on the peaks refer to the total number of acyl carbon atoms in each compound: 36 = trilaurin; 42 = trimyristin; 48 = tripalmitin; 54 = tristearin, and 60 = triarachidin. The compounds were dissolved in hexane to give a concentration of 100 ng μl^{-1}. One microlitre was introduced using on-column injection into a 25 m × 0.25 mm i.d. aluminium-clad capillary column coated with methyl 65% phenyl polysiloxane (Alltech Assoc. Inc.; 0.1 μm film thickness). The GC oven temperature was held for 2 min at 50 °C before programming to 350 °C at 10 °C min^{-1}. The GC-MS transfer-line and ion source block were maintained at 360 at 300 °C respectively. Electron ionization (70 eV) afforded the mass spectrum shown in *Figure 10*. NB: The distal 5 cm of the aluminium cladding of the column must be removed (by dissolution with aqueous alkali) when such columns are used with magnetic sector mass spectrometers to prevent electrical discharging (77).

are of very low abundance they are often difficult to detect in GC-MS analyses. The lower mass ions are of greater value in analytical work. An abundant acylium ion, $[RCO]^+$, is present at *m/z* 267 ($[CH_3(CH_2)_{16}CO]^+$) in the mass spectrum of tristearin (*Figure 10*). Where other acyl moieties are present the masses of their acylium ions can be used to deduce their carbon number and degree of unsaturation. It is generally accepted that the presence of unsaturation in the acyl moiety gives fragment ions of *m/z* [RCO-1], namely a ketene structure, $R{-}CH{=}C{=}O^+$, of greater abundance than the acylium ions described above. However, this is not always the case, since the mass spectra of triolein and trilinolein recorded by high temperature GC-MS, under the conditions used for the analysis shown in *Figure 9*, yielded spectra containing prominent acylium ions at *m/z* 265 and 263 respectively. These latter ions were of substantially greater abundance than the corresponding ketene-type ions discussed above.

Abundant fragment ions are seen at higher mass corresponding to the loss of an acyloxy group from the molecular ion, $[M{-}RCO_2]^+$. Thus, the fragment ion seen at *m/z* 607, $[M\text{-}283]^+$, in the EI spectrum of tristearin is valuable in confirming the deductions made concerning the nature of the acyl moieties, based on the lower mass acylium ion appearing at *m/z* 267. A range of other fragment ions are seen in the mass spectra of triacylglycerols, which

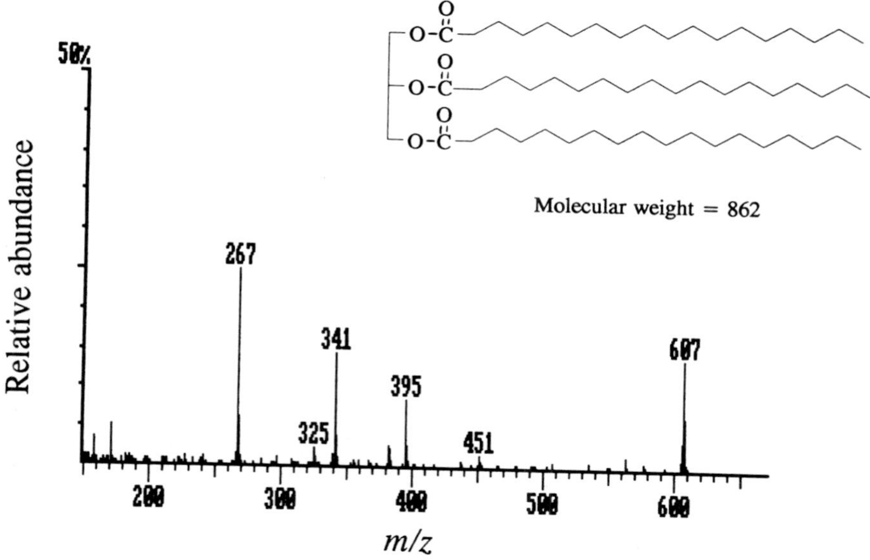

Figure 10. Partial electron ionization (70 eV) mass spectrum of tristearin (the peak labelled '54' in the TIC shown in *Figure 9*.

can be used to deduce the structure of unknowns present in complex biological mixtures. The fragmentation pattern observed in the EI mass spectra of triacylglycerols is summarized schematically in *Figure 11* (45). The EI mass spectra of triglycerides cannot be used to determine the positions of substitution of the acyl moieties on the glycerol backbone.

Triacylglyceride mixtures from biological materials are extremely complex, since the number of possible molecular species is equal to the number of fatty acids cubed. Hence, any triacylglyceride analysis will only provide partial compositional information. The cross-linked dimethyl polysiloxane coated capillary GC columns are only capable of carbon number resolution. More polar stationary phases are available that can resolve triglyceride species according to their differing degrees of unsaturation (see Chapter 4).

Where structure interpretations by EI are uncertain, positive-ion and negative-ion CI mass spectra provide useful alternatives. Positive-ion CI mass spectra, recorded using ammonia as the reagent gas at an ion source block temperature of 230 °C, yield quasi-molecular $[M + NH_4]^+$ ions for triacylglycerols that are *c*.20-fold more abundant than the $M^{+\cdot}$ ions obtained when using EI (46). The only fragment ions observed correspond to $[MH - RCO_2H]^+$ (base peak). Hence, these spectra are useful for determining the carbon number and degree of unsaturation of the triacylglycerol and individual acyl moieties. In contrast, the spectra obtained in the negative-ion scanning mode, using ammonia as reagent gas at an ion source block

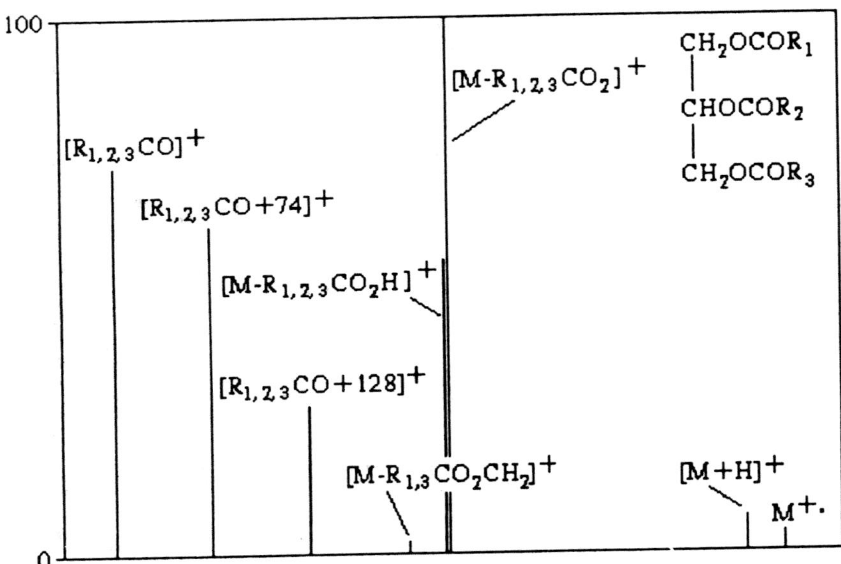

Figure 11. Generalized pattern of EI fragmentation of triacylglycerols. (Reproduced, with permission, from ref. 45.)

temperature of 300 °C, exhibit only abundant $[RCO_2]^-$, $[RCO_2 - H_2O]^-$ and $[RCO_2 - H_2O - H]^-$ ions (47). Although negative ion chemical ionization (NICI) spectra lack information concerning the carbon number of the intact triacylglycerol, they can be used very conveniently to determine the nature of the acyl groups, where a complex mixture of molecular species is present. Moreover, as the spectra are obtained at an ion-source block temperature of 300 °C the technique is compatible with high temperature GC–MS (47) (see also Chapter 4). When using NICI, the assignment of carbon number is made by comparing the retention times of peaks in GC-MS analyses to those of authentic compounds of known carbon number (see *Figure 9*).

As alluded to above, combined LC-MS is a potentially attractive method for the analysis of high-molecular-weight or thermally-unstable lipids, such as the high-molecular-weight or polyunsaturated triacylglycerols. However, LC-MS instrumentation is not widely available. Reversed-phase HPLC provides a very effective means of separating mixtures of triacylglycerols (48). The LC-MS analysis of the triacylglycerols from corn oil using a DLI interface (see Section 3.1.4 above) is shown in *Figure 12* together with the mass spectrum of one of the major peaks. The mass spectra obtained exhibit quasi-molecular $[M + H]^+$ ions, and $[MH - RCO_2H]^+$ fragment ions corresponding to *sn*-1,2-, *sn*-2,3-, and *sn*-1,3-diacylglycerol moieties. Hence, the carbon number and degree of unsaturation of the molecular species and their fatty acyl moieties contained in the HPLC peaks can be determined.

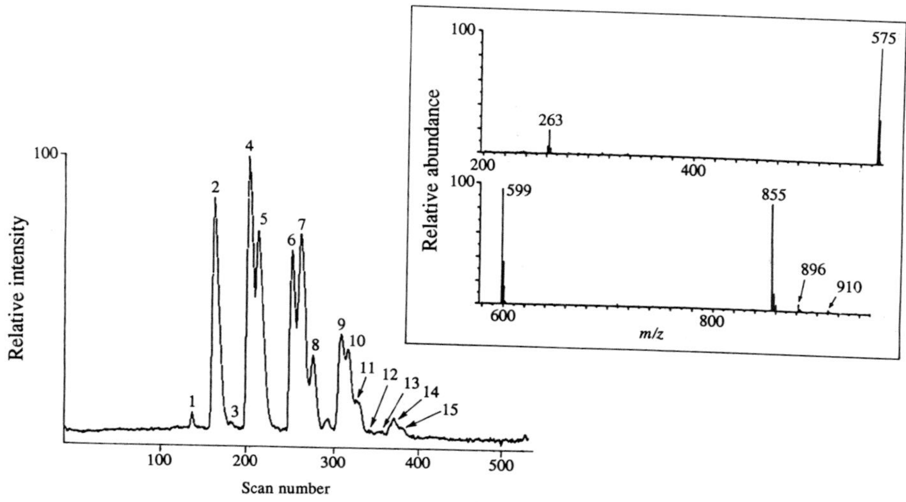

Figure 12. TIC for the reversed-phase LC-MS analysis of the triacylglycerols of corn oil using a DLI interface (Hewlett Packard). Elution was achieved using a 30–90% v/v gradient of propionitrile in acetonitrile at a flow rate of 1.5 ml min^{-1}. The column temperature was maintained at 30 °C. Mass spectra were recorded every 7 s with scanning from *m/z* 200. The inset shows the solvent mediated CI mass spectrum for peak 5 in the TIC. (Reproduced with permission from ref. 48.) Peak identities are: 1 = 18:2, 18:2, 18:3; 2 = 18:2, 18:2, 18:2; 3 = 16:0, 18:2, 18:3; 4 = 18:1, 18:2, 18:2; 5 = 16:0, 18:2, 18:2; 6 = 18:1, 18:1, 18:2; 7 = 16:0, 18:1, 18:2; 8 = 16:0, 16:0, 18:2; 9 = 18:1, 18:1, 18:1; 10 = 16:0, 18:1, 18:1; 11 = 16:0, 16:0, 18:1; 12 = 16:0, 16:0, 16:0; 14 = 18:0, 18:1, 18:1; 15 = 16:0, 18:0, 18:1.

4.1.5 Diacylglycerols

The enzymatic treatment of phospholipids with phospholipase C yields mixtures of diacylglycerols which are readily analysed by GC and GC-MS as acetate, TMS, or TBDMS derivatives. The EI behaviour of the acetate derivatives (49, 50) follow similar general rules as for the triacylglycerols (see above). The TMS and TBDMS derivatives are more popular derivatives for the GC-MS analysis of diacylglycerols.

The EI mass spectra of the TMS derivatives (51) of 1,2-diacylglycerols do not generally contain $M^{+\cdot}$ ions. However, high mass ions corresponding to the loss of a methyl group, $[M-15]^+$, and trimethylsiloxy group, $[M-90]^+$ are useful in assigning the carbon number and degree of unsaturation of the intact diacylglycerols. As these ions are often rather weak in GC-MS analysis, carbon number is more reliably determined by retention time comparisons with authentic compounds.

The nature of the individual acyl moieties can be determined by considering the ions resulting from the loss of an acyloxy moiety, $[M - RCO_2]^+$, or $[M - RCO_2H]^+$, when the acyl moiety is unsaturated. As with triacylglycerols, the more abundant acylium ion, $[RCO]^+$ is the most useful fragment ion for determining the nature of the individual acyl moieties. Assignment of the nature of the acyl moieties can be corroborated using the $[RCO + 74]^+$ and $[RCO + 90]^+$ ions. Low mass ions are observed at *m/z* 129, $[CH_2CHCHOTMS]^+$, and 145, $[CH(O)CHCH_2OTMS]^+$, which contain the TMS group and part of the glycerol backbone. The EI mass spectra of the TMS ethers of 1,3-diacylglycerols contain a major $[M\text{–}RCOOCH_2]^+$ ion which does not appear in the spectra of 1,2-diacylglycerols (52, 53).

TBDMS derivatives are more chemically stable than the TMS derivatives and are finding increasing application to the analysis of diacylglycerols. The TBDMS derivatives yield EI spectra that are similar in many respects to the TMS derivatives (54, 55). The TBDMS derivative yields an abundant $[M\text{-}57]^+$, ion corresponding to the loss of the *test.*-butyl radical from the intact derivative. The carbon number and degree of unsaturation of the diacylglycerol is readily deduced from the *m/z* of the $[M\text{-}57]^+$ ion. Although the relative abundance of the $[M\text{-}57]^+$ ion diminishes with increasing unsaturation, it still has diagnostic value in diacylglycerols containing docosahexaenoate ($C_{22:6}$) moieties. The carbon number and degree of unsaturation of the individual acyl groups are readily determined from the ion corresponding to loss of RCO_2 and RCO_2H radicals. If the acyl moiety is unsaturated then the $[M - RCO_2]^+$ ion is significantly more abundant than $[M - RCO_2H]^+$. The abundance of the $[M - RCO_2]^+$ ion containing the acyl group at position 1 will be more abundant than the corresponding fragment bearing the acyl moiety from position 2.

4.1.6 Monoacylglycerols

Both acetate and TMS ether derivatives can be used in the GC-MS analysis of monoacylglycerols (56). The spectra of the TMS ethers are more useful, often displaying weak $M^{+\cdot}$ and $[M\text{—}CH_3]^+$ ions, from which the carbon number and degree of unsaturation can be deduced. An $M^{+\cdot}$ ion is usually observed when the acyl moiety is unsaturated. Where these are not observed then this information can be gleaned from the abundant $[M\text{-}90]^+$ (loss of HOTMS) ion at lower mass. The appearance of characteristic fragment ions at *m/z* 218 in the EI mass spectra of the bis-TMS ethers of 2-monoacylglycerols, and $[M\text{-}103]^+$ in the spectra of 1-monoacylglycerols allows isomeric species to be reliably distinguished. The mass spectra of the TBDMS derivatives of monoacylglycerols have not been so extensively investigated. The trends in fragmentation will be largely analogous to those of the TMS derivatives with an increase in the abundance of the diagnostic high mass ions; for example, $[M\text{-}57]^+$.

4.1.7 Wax esters

The EI mass spectrum of the major wax ester of beeswax (see Chapter 4) is shown in *Figure 13*. Although a molecular ion is seen (*m/z* 592) in this

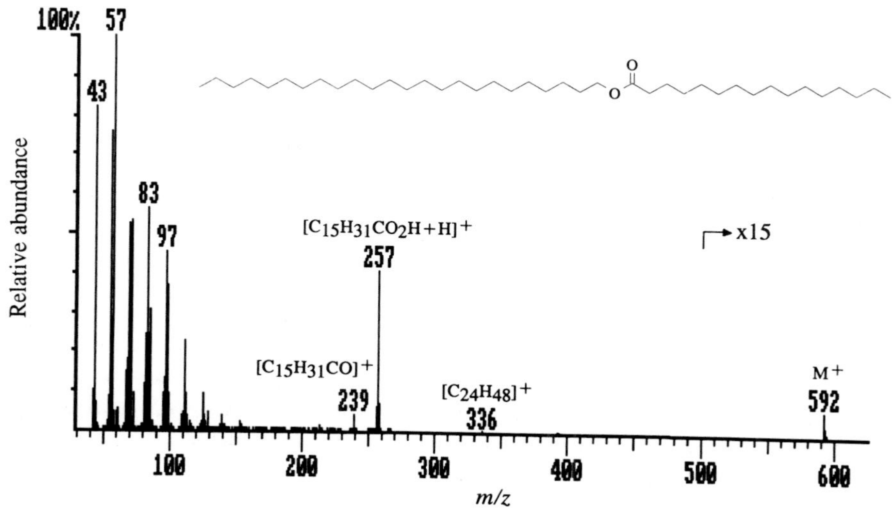

Figure 13. Electron ionization (70 eV) mass spectrum of tetracosanyl palmitate.

example, they are not always detectable in the mass spectra of wax esters obtained by GC-MS. Where an $M^{+\cdot}$ ion is seen it can be used to determine the carbon number and degree of unsaturation of the wax esters. The molecular weight of the wax ester whose mass spectrum is shown in *Figure 13* is 592 mass units, hence giving a molecular formula of $C_{40}H_{80}O_2$. The peak at *m/z* 257 corresponds to the $[C_{15}H_{31}CO_2H + H]^+$ ion, produced by cleavage of the alkyl oxygen bond, and provides information concerning the carbon number and degree of unsaturation of the fatty acyl moiety. This intense fragment ion is accompanied by an acylium ion at *m/z* 239 of empirical formula $[C_nH_{2n+1}C{=}O]^+$, which further corroborates the nature of the fatty acyl moiety. The only fragment ion derived directly from the fatty alcohol moiety is a very weak, even mass alkyl fragment ion, $[C_nH_{2n}]^+$, at *m/z* 336; which arises by fission of the alkyl oxygen bond with concomitant hydrogen transfer. Further indication of the nature of the fatty alcohol moiety can be deduced from the mass difference between the $M^{+\cdot}$ and $[RCO_2H + H]^+$ ions. In practice, however, the low abundance of the $M^{+\cdot}$ ion limits the reliability of this approach.

When diagnostic ions concerning the nature of the alcohol moiety are very weak or undetectable then the carbon number can be determined by comparing the GC retention times of wax ester peaks in a total ion chromatogram with those of compounds of known structure. The carbon numbers of the homologous wax esters found in beeswax are well-established and can be used for this purpose (see Chapter 4). Alternatively, CI can be used to enhance the molecular weight information content of mass spectra through the production of diagnostic quasi-molecular ions (57). The methane

CI spectra of wax esters contain a variety of quasi-molecular ion species, i.e. $[M + H]^+$, $[M + C_2H_5]^+$, $[M + C_3H_5]^+$ and $[M - H]^+$, the latter arising by hydride abstraction. The $[M - 1]^+$ ion predominates in saturated wax esters, whereas monounsaturated esters display $[M - 1]^+$ and $[M + 1]^+$ ions in approximately equal abundance. CI spectra are affected by variations in a wide range of source parameters. Hence, as a general rule interpretations are best made by comparing the mass spectra of authentic and unknown compounds recorded under the same experimental conditions.

4.1.8 Long-chain alcohols

The major high mass ion in the EI spectra of aliphatic long-chain primary alcohols arises through the loss of water, $[M\text{-}18]^+$, in the ion source. Thus, as no $M^{+\cdot}$ ion is observed, the spectra are easily confused with those of mono-alkenes, although the confusion is readily resolved in GC-MS analyses by comparing the retention times of the alcohol and alkene.

A more desirable approach to the mass spectrometric analysis of long-chain alcohols involves derivatization of the hydroxyl moiety. Acetates are conveniently prepared as described in Chapter 2 and display excellent GC properties. The mass spectra of acetates possess a relatively weak $M^{+\cdot}$ ion and ions of lower mass appear due to the loss of acetyl moiety, $[M\text{-}60]^+$, and ethylene group, $[M\text{-}60\text{–}28]^+$. The low mass, *m/z* 60 ion, $[CH_3CO_2H]^+$, is also very distinctive for acetate derivatives.

TMS ethers are also readily formed, possess excellent GC properties, and produce useful EI mass spectra. The mass spectrum of the TMS derivative of *n*-hexacosanol is shown in *Figure 14*. Although the $M^{+\cdot}$ ion is weak the very

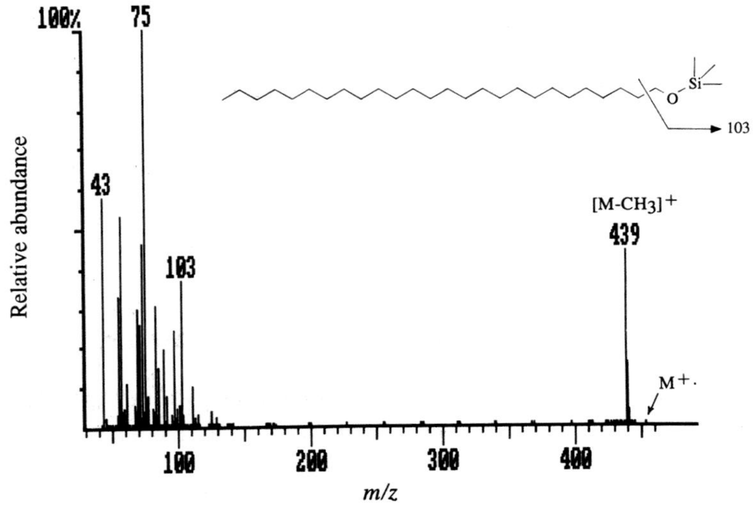

Figure 14. Electron ionization (70 eV) mass spectrum of the TMS ether derivative of hexacosanol.

abundant $[M-15]^+$ ion can be used to assign the carbon number and degree of unsaturation of the alcohol. The ions at lower mass (for example, *m/z* 103), are useful in distinguishing between the TMS ethers of long-chain alcohols and TMS esters of fatty acids. NB: The $[M-15]^+$ ion of the TMS ester of a fatty acid (for example, palmitic acid) has the same nominal mass (isobaric) with the TMS ether of the next highest carbon number fatty alcohol (*n*-heptadecanol). The mass spectra of long-chain alcohols give little other additional structural information. The TMS ethers of secondary alcohols display prominent fragment ions due to α-cleavages that usually allow precise assignment of the position of hydroxyl groups in the alkyl chain (see *Figure 7*).

Nicotinate esters of long-chain alcohols are prepared to obtain structure information concerning the position of other functional groups. The nicotinate derivatives are prepared as shown in *Protocol 5* (58, 59).

Protocol 5. Preparation of nicotinate derivatives (60)

1. Add 200 μl of saturated (*c.* 5%) nicotinyl chloride hydrochloride in dry acetonitrile to a long-chain primary or secondary alcohol (100 μg) in a screw-capped vial.
2. Seal and heat for 10 min at 100 °C.
3. Allow to cool and add 1 ml each of water and hexane.
4. Decant the hexane layer and use directly for GC-MS analysis.

The spectra given by the nicotinate derivatives resemble those of the picolinyl esters (see *Figure 6* above). Spectra have been recorded for saturated and unsaturated, branched and cyclopropane alcohols (58, 59, 61).

4.1.9 Long-chain ketones

Long-chain ketones occur widely as constituents of plant epicuticular waxes. The mass spectrum of nonacosa-15-one, derived from a *Brassica* leaf wax, is shown in *Figure 15*. The mass spectrum is typical of long-chain ketones, and displays a weak molecular ion, *m/z* 422, which provides an indication of carbon number. The more abundant ions, appearing at *m/z* 225 and 241, arise by cleavage of carbon–carbon bonds α- and *β*- to the ketone group, and serve

to identify unambiguously the position of the carbonyl group (62). The ion at *m/z* 240 arises via a six-centre McLafferty rearrangement, as shown in *Scheme 3*.

4.1.10 Long-chain aldehydes

Long-chain aldehydes are found as constituents of some plant epicuticular

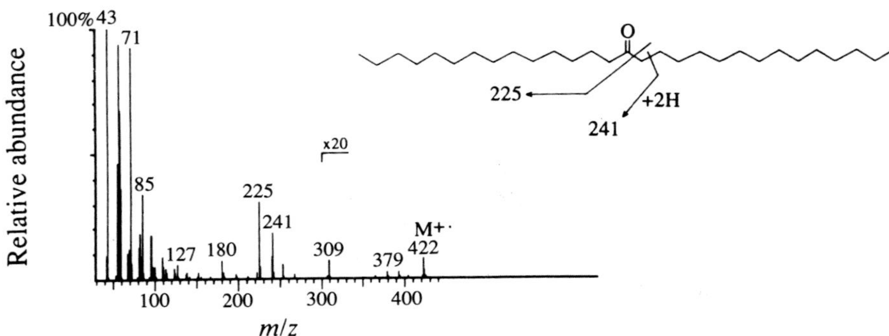

Figure 15. Electron ionization (70 eV) mass spectrum of nonacosa-15-one.

leaf waxes and in the acid hydrolysates of plasmalogens (63). While short-chain aldehydes exhibit a prominent molecular ion, a gradual decrease is seen in the relative abundance of $M^{+\cdot}$ for the longer-chain compounds.

Characteristic high mass fragment ions are seen which correspond to the loss of water ($[M-18]^+$), ethylene ($[M-28]^+$) and an enol ion radical ($[M-44]^+$). The latter ion arises via a six-centre McLafferty rearrangement, analogous to that described for fatty acid derivatives (see *Scheme 3*). Although mass spectrometry can be used to characterize aldehydes, an alternative approach is to reduce them to alcohols, then analyse them as acetate, TMS ether or other derivatives (see above Section 4.1.8 and ref. 63).

N,N-dimethylhydrazones can be prepared for long-chain chain aldehydes

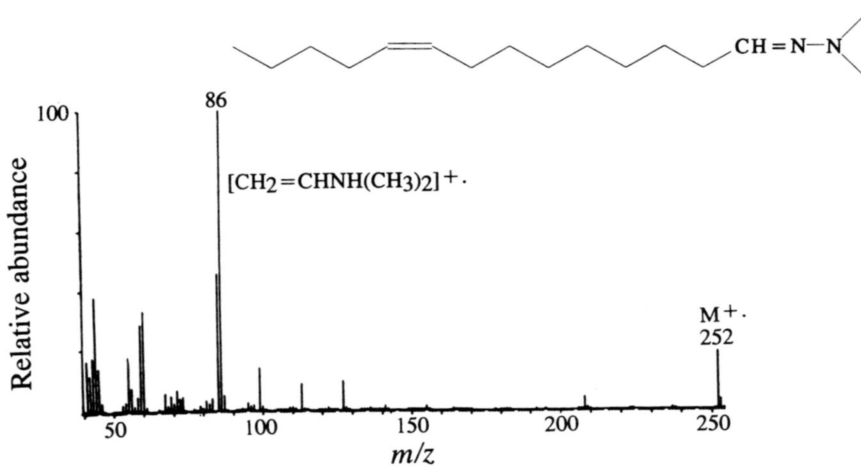

Figure 16. Electron ionization (70 eV) mass spectrum of the *N,N*-dimethylhydrazone derivative of *Z*-9-hexadecenal. (Reproduced, with permission, from ref. 64.)

as an aid to their mass spectrometric investigation (64). The *N,N*-dimethylhydrazone of *Z*-9-tetradecenal shown in *Figure 16* displays a prominent $M^{+\cdot}$ ion, and a base peak at *m/z* 86 which arises by a McLafferty rearrangement.

4.1.11 Hydrocarbons

The EI spectrum of *n*-nonacosane (C_{29} *n*-alkane) is shown in *Figure 17*. The molecular weight is established from the molecular ion, $M^{+\cdot}$, at *m/z* 408. The relative abundance of the $M^{+\cdot}$ in decreases with increasing chain branching. The remainder of the *n*-alkane spectrum is characterized by a series of ions of empirical formula C_nH_{2n+1}. These are odd mass ions, separated by 14 mass unit increments which decrease in relative abundance with increasing mass. A 29 dalton increment is always seen between the $M^{+\cdot}$ and the first C_nH_{2n+1} alkyl fragment.

Deducing the position of structural features in hydrocarbons from EI mass spectra is difficult. The positions of double bonds are most easily deduced by reactions at the sites of unsaturation, as in the case of fatty acids (Section 4.1.3). The occurrence of chain branching, such as *iso-* or *anteiso-*, is best deduced from elution orders in GC-MS analyses (see Chapter 4). Prominent peaks at even mass in EI spectrum provide further indications of chain branching.

4.1.12 Sterols

Combined GC-MS is widely used to identify the components of mixtures of sterols commonly found in lipid extracts of biological materials. EI is the most widely used ionization technique, with sterols being most frequently analysed as the free compounds, acetate, or TMS ether derivative. The free 3β-hydroxy compounds can yield abundant molecular ions, $M^{+\cdot}$, from which the carbon number and degree of unsaturation of the sterol nucleus can be deduced. However, excessive dehydration in the ion source can be problematical. Moreover, the GC behaviour of the free compounds is more sensitive

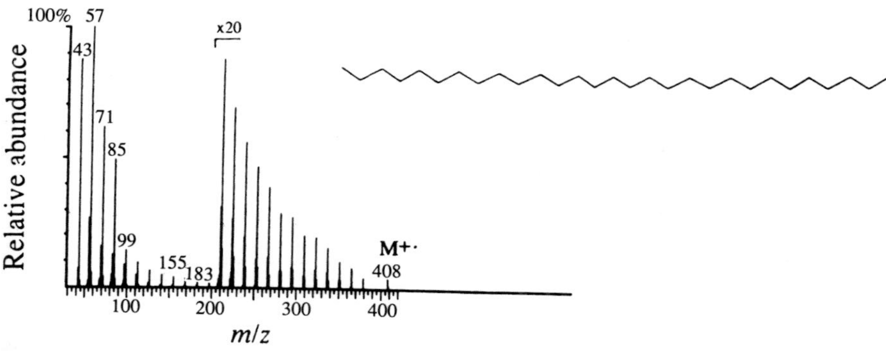

Figure 17. Electron ionization (70 eV) mass spectrum of *n*-nonacosane.

to the presence of active sites in the column and injection system, than that of their derivatives. The mass spectrometry of sterols and their derivatives has been extensively reviewed (65–67). The fragmentations of sterols are conveniently considered in terms of those arising from the ring system (including derivatizing group), and side-chain.

Acetic acid is lost very readily by 1,2-elimination from acetate derivatives of 3β-hydroxy sterols. Hence, the highest mass ion observed is frequently $[M\text{-}60]^{+\cdot}$, particularly in the case of Δ^5 sterols. The $[M\text{-}60]^{+\cdot}$ ion is isobaric with the $[M\text{-}18]^{+\cdot}$ ion, produced by the dehydration of free sterols. The TMS ethers exhibit abundant $M^{+\cdot}$ ions, and other high mass ions corresponding to loss of angular methyl groups, $[M\text{-}15]^+$, and 1,2-elimination of the trimethylsilanol group, $[M\text{-}90]^+$. These latter ions are clearly visible in the mass spectrum of the cholesterol, shown in *Figure 18*. Also seen in this spectrum are ions at *m/z* 129 and 329 which correspond to the loss of the TMS group, together with a three carbon fragment of ring A containing the C-1, C-2, and C-3 carbons, with charge retention on either fragment ion; this latter fragmentation is especially characteristic of Δ^5 sterols. The TMS ethers of $\Delta^{5,7}$ sterols exhibit an analogous fragment ion $[M\text{-}131]^+$, involving the loss of two additional hydrogen atoms. The loss of a methyl group, largely from C-19, gives rise to the ions at *m/z* 353 ($[M{-}CH_3{-}TMSOH]^+$) and 443 ($[M{-}CH_3]^+$) in the mass spectrum of cholesterol shown in *Figure 18*. The relative abundances of ions resulting from the loss of the methyl group and hydroxyl moiety (including derivatizing group) are dependent on the presence and position of other structural features, such as nuclear double-

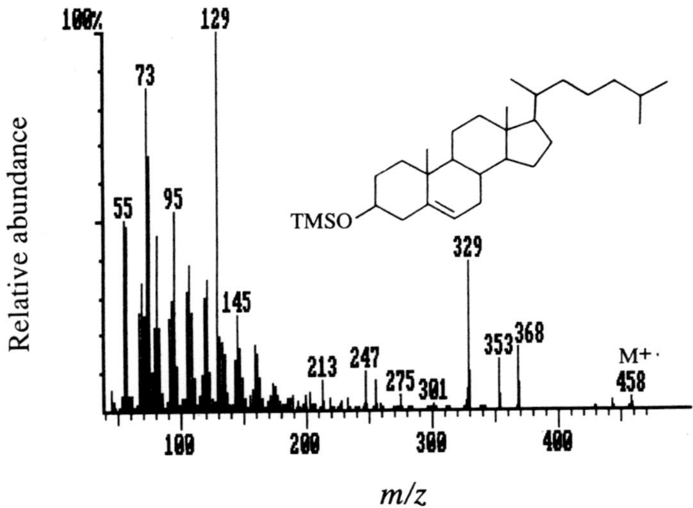

Figure 18. Electron ionization (70 eV) mass spectrum of the TMS ether derivative of cholesterol.

bonds. For example, sterols possessing Δ^8 or $\Delta^{9(11)}$ double bonds and a 14α-methyl group exhibit strong $[M–CH_3]^+$ and $[M–CH_3–ROH]^+$ (where R = $CH_3C{=}O$ or $–Si(CH_3)_3$, in acetate or TMS ether derivatives respectively).

The fragment ion at *m/z* 213 in the cholesterol spectrum (*Figure 18*) arises through loss of the side-chain together with carbons 15–17 from ring D. This fragmentation occurs commonly in sterols and is especially useful for establishing the number of double bonds in rings B and C of the sterol nucleus. Hence, stanols display an abundant ion at *m/z* 215, Δ^5, Δ^7, and Δ^8 sterols an ion at *m/z* 213, $\Delta^{5,7}$ and $\Delta^{5,8}$ sterols an ion at m/z 211, and the $\Delta^{5,7,9(11)}$ sterols an ion at *m/z* 209. The reader is directed to ref. 67, and references cited therein, for more detailed discussions of the characteristic fragmentations of other sterol nuclei that can be encountered in biological materials.

Fragmentations involving the sterol side-chain can provide information diagnostic of the side-chain itself and of the sterol nucleus. For example, the ion at *m/z* 255 in the mass spectrum of cholesterol (*Figure 18*) arises by loss of the side-chain and silanol group, $[M—C_8H_{17}—TMSOH]^+$. Hence, cleavage of the side-chain provides evidence for the carbon number and degree of unsaturation of the side-chain and the sterol nucleus. Moreover, the appearance of characteristic fragment ions will be strongly dependent upon the position of double bonds and alkyl substituents in the side-chain. For example, sterols with a $\Delta^{24,28}$-bond produce strong ions resulting from fragmentation by McLafferty-type rearrangement of the side-chain. The reader is again directed to ref. 67, for more detailed discussions of the specific fragmentations of the wide range of other side-chain structures that occur in sterols.

4.1.13 Phospholipids

Phospholipids are the most commonly encountered class of polar lipids found in the total lipid extracts of biological tissues. Until recently, the common approach to their analysis was to perform a class separation by TLC, followed by a chemical degradation (for example, transmethylation), prior to analysis of the resulting fatty acid methyl esters by GC and/or GC-MS (see above, Section 4.1.1). An alternative approach has been to release diacylglycerols from phospholipids by treatment with phospholipase C, followed by their GC and/or GC-MS analysis, after appropriate derivatization (see Section 4.1.5 and Chapter 2). The preferred approach is to analyse phospholipids as their intact molecular species. Both FAB-MS and reversed-phase LC-MS can be used to analyse intact phospholipids.

i. Fast atom bombardment-mass spectrometry (FAB-MS)

Both positive-ion (68, 69) and negative-ion (70, 71) FAB-MS can yield structural information from intact phospholipids. The positive-ion FAB mass

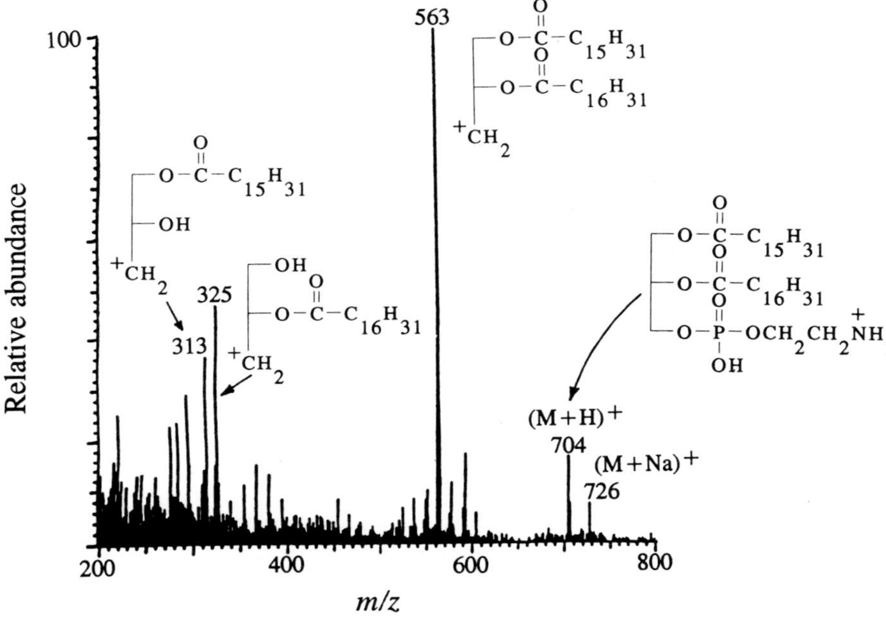

Figure 19. Positive-ion FAB mass spectrum of phosphatidyl ethanolamine ($C_{16:0}/C_{17:1}$) recorded using glycerol/thioglycerol as the liquid matrix. (Reproduced, with permission, from ref. 69).

spectrum of phosphatidylethanolamine shown in *Figure 19* was recorded using a mixture of glycerol/thioglycerol. Four groups of diagnostic ions can be discerned:

(a) Quasi-molecular ion region, containing protonated molecular ions, $[M + H]^+$, or alkali metal cation adducts; for example, $[M + Na]^+$.

(b) Diacylglycerol fragment ions resulting from the loss of the phosphate ester moiety.

(c) The monoacylglycerol fragments arising by loss of the phosphate esters plus a fatty acyl moiety.

(d) Low mass ions corresponding to the phosphate ester moiety (not shown in *Figure 19*).

By careful consideration of these fragment ions considerable progress can be made in deriving detailed compositional information on complex phospholipid mixtures (69). Partial purification of crude phospholipid mixtures by thin-layer chromatography can be used to simplify the interpretation of FAB spectra.

A somewhat different picture, although no less useful analytically, is obtained by use of negative-ion FAB-MS (70). Again, four diagnostically useful groups of ions can be discerned:

(a) The deprotonated molecular anion region, $[M - H]^-$.
(b) An intermediate mass region containing, amongst others, ions due to the loss of a single fatty acyl moiety.
(c) Fatty acid region containing prominent carboxylate anions, $[RCO_2]^-$, usually as the base peak.
(d) Low mass ions, characteristic of the phosphate ester moiety.

It should be noted that as the highest mass ion observed in the negative-ion spectra of phosphatidylcholines is a quasi-molecular anion, resulting from the elimination of $[CH_3]^+$, which is isobaric (of the same mass) with the highest mass ion, $[M - H]^-$, of a phosphatidylethanolamine, bearing the same fatty acyl moieties. In order to resolve this ambiguity the positive-ion FAB spectra must be recorded, as both the cholines and the ethanolamines display $[M + H]^+$ ions; i.e. the $-N(CH_3)_3$ moiety is retained intact in the quasi-molecular ion of cholines.

ii. Liquid chromatography/mass spectrometry (LC-MS)

Thermospray LC-MS can be used for molecular species analysis of phospholipids (72), with the advantage over FAB-MS of on-line HPLC separation, combined with MS detection and characterization. The thermospray mass spectra of 16:0, 18:1-phosphatidylcholine and 16:0, 18:1-phosphatidylethanolamine are shown in *Figure 20*. The fragmentation patterns are simple, and similar to those obtained by positive-ion FAB-MS (see above *Figure 19*).

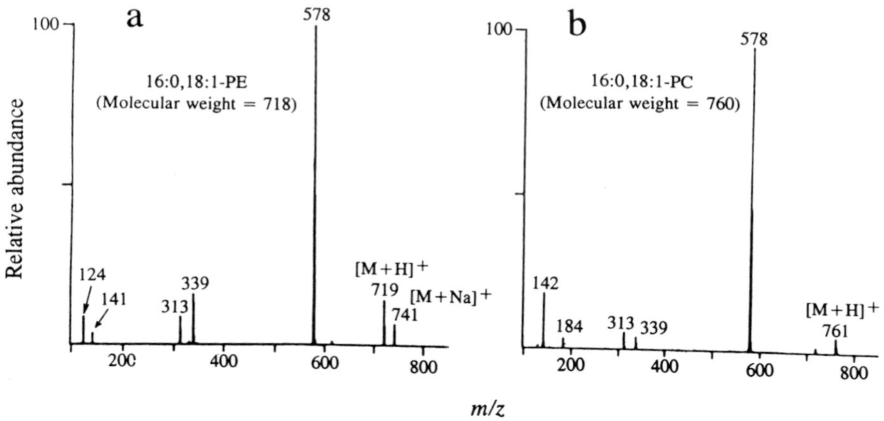

Figure 20. Positive ion-thermospray (electron emitting) filament turned on (0.2 mA)) mass spectra of authentic 16:0, 18:1-phosphatidylcholine (b) and 16:0, 18:1-phosphatidylethanolamine (a) using 95:5 methanol/0.1 M ammonium acetate as solvent. (Reproduced with permission from ref.72.)

Information is readily obtained concerning the molecular weight, fatty acyl content, and nature of the phosphate ester head group. Both compounds display quasi-molecular ions ($[M + H]^+$ and $[M + Na]^+$) at high mass, from which the molecular weight can be assigned. The base peak, *m/z* 578, in each of the spectra corresponds to the diglyceride fragment ion. The difference in mass between this ion and the glyceride ion occurring at lower mass, *m/z* 313 and 339, allow the nature of the fatty acyl moieties to be reliably assigned. The ions at low mass, *m/z* 142 and 184, in *Figure 20b*, derive from the phosphocholine head group. In contrast the spectrum of the phosphatidylethanolamine shown in *Figure 20a* shows low mass ions at *m/z* 124 and 141 which allow unambiguous identification of the head group. The structures of these latterly mentioned fragment ions are shown in *Scheme 5*.

The chromatogram shown in *Figure 21* shows the total ion chromatogram obtained for the LC-MS analysis of a mixture of synthetic phosphatidylcholines. The full experimental conditions are given in the figure caption. The structures of the individual molecular species were assigned by analysing the spectrum of each of the peaks in the chromatogram. The individual molecular species are well-separated according to their carbon number and degree of unsaturation. Although the natural mixtures may be less well resolved owing

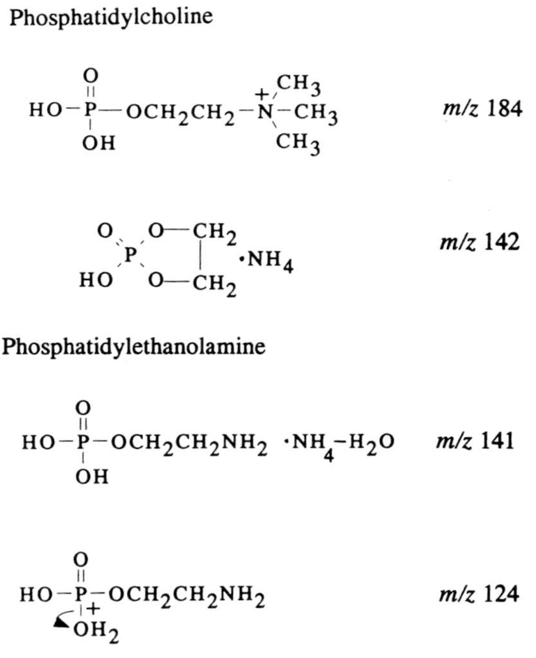

Phosphatidylcholine

$HO-P(=O)(OH)-OCH_2CH_2-N^+(CH_3)_3$ — *m/z* 184

$O{=}P(OH)(OCH_2CH_2O)\cdot NH_4$ — *m/z* 142

Phosphatidylethanolamine

$HO-P(=O)(OH)-OCH_2CH_2NH_2 \cdot NH_4-H_2O$ — *m/z* 141

$HO-P(=O)(\overset{+}{O}H_2)-OCH_2CH_2NH_2$ — *m/z* 124

Scheme 5. Proposed structures of low mass fragment ions in the positive-ion thermospray mass spectra of phospholipids.

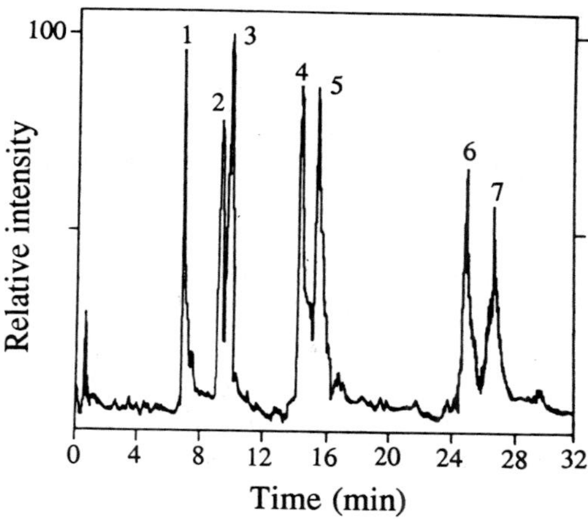

Figure 21. Reversed phase LC-MS analysis of a mixture of authentic phosphatidylcholines (5 μg of each component) using a 4.6 mm × 7.5 cm Ultrasphere-ODS (3 μm) packed column), eluting with 71:5:7 methanol/hexane/0.1 M ammonium acetate at a flow rate of 1 ml min^{-1}. The mass spectrometer was set to scan the range *m/z* 120 to 820. The peaks in the chromatogram correspond to phosphatidylcholines containing the following fatty acyl moieties: 1 = 14:0, 20:4; 2 = 14:0, 16:0; 3 = 14:0, 18:1; 4 = 14:0, 18:0; 5 = 16:0, 18:1; 6 = 16:0, 18:0; 7 = 18:1, 18:0. (Reproduced, with permission, from ref. 72.)

to their greater complexity, co-eluting or partially resolved components can be revealed by plotting ion chromatograms based on the prominent diglyceride fragment ions. Thermospray LC-MS can also be applied to phosphatidylinositols, phophatidylserines, sphingomyelins, triglycerides, and platelet-activating factor (73).

4.2 Quantification and trace analysis

Frequently in metabolic investigations the analyst is faced the task of detecting and quantifying lipid species, present as minor components of complex biological extracts. Mass spectrometry provides amongst the most sensitive and selective physico-chemical analytical technique available for the analysis of lipids.

4.2.1 Specificity and sensitivity

The major advantage of using mass spectrometry in trace analyses derives from the ability to monitor pre-selected ion(s) in the spectrum of a compound. This increases both sensitivity, as the instrument does not then spend time scanning redundant regions of the mass spectrum, and selectivity, through the monitoring of an ion(s) (molecular or fragment) that is characteristic of the analyte of interest. This technique is commonly referred to as selected ion monitoring (SIM). Greatest sensitivity is achieved in SIM analyses when a single intense ion is monitored. In such instances picogram

(10^{-12} g) detection limits are achieved routinely, with sub-picogram limits possible under favourable circumstances.

SIM is frequently used in conjunction with GC-MS or LC-MS, thereby adding extra selectivity to the analysis through the provision of a high-resolution chromatographic separation step. In a trace analysis using GC-MS with SIM, confirmation of the presence of the chosen analyte would rest on the detection of the characteristic ion, as a peak in an ion chromatogram, at the expected GC retention time. The expected retention time is determined by the prior analysis of an authentic sample of the analyte of interest. Still further enhancements in analytical selectivity can be achieved by monitoring more than one ion from the mass spectrum of an analyte. The possibility then exists for confirming the presence of the analyte, by comparing the ratio of the responses of these ions determined by SIM to their abundance ratios in the full-scan mass spectrum. *Figure 22* shows a number of possible hypothetical results that can be obtained from the search for a particular analyte using SIM for three ions.

4.2.2 Quantification

The effectiveness of mass spectrometry in quantitative analysis derives from the ability to make simultaneous measurements, with equivalent specificity, of the concentrations of the chosen analyte and added internal standard, or calibrant. Optimum precision is attained in quantitative mass spectrometry when the internal standard is a stable isotope labelled (most commonly ^{2}H, ^{13}C, or ^{18}O) analogue of the analyte. This method of quantitative mass spectrometry is known as isotope dilution mass spectrometry.

The high precision that can be attained in quantitative mass spectrometric analyses has lead to its adoption as a reference technique for use in assessing the performance of other less specific analytical techniques; for example,

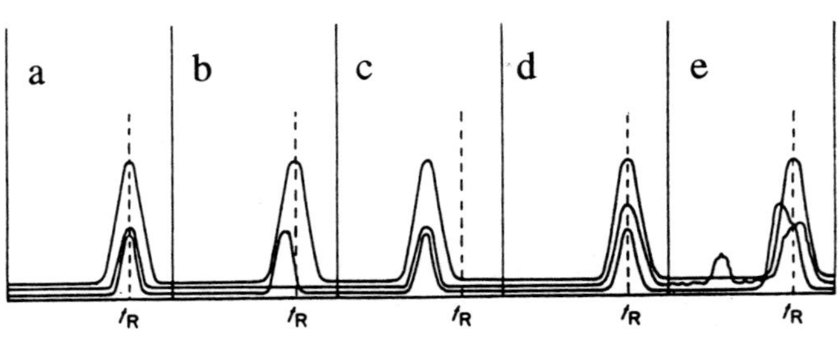

	a	b	c	d	e
Relative intensities:	Correct	Incorrect	Correct	Distorted	Distorted
Retention times:	Correct	Incorrect	Incorrect	Correct	Distorted
Compound present?	Yes	No	No	Probably	Possibly

Figure 22. Hypothetical results from the GC or LC-MS selected ion monitoring analysis for a targeted comopound. (Reproduced with permission, from ref. 8.)

radioimmunoassay. An overview of the methodological aspects of quantitative mass spectrometry can be found in ref. 74.

It is clearly impractical to attempt to present here specific protocols for the quantitative analysis of even a small fraction of the wide variety of lipid species that exist. Instead, the following section will discuss the general principles underlying the quantitative analysis of lipids by isotope dilution mass spectrometry. Two analytical protocols are then presented as examples of the use of stable isotope dilution mass spectrometry, one employing GC-MS and the other FAB-MS.

Figure 23 shows the analytical protocol recommended for use in quantitative analysis by mass spectrometry. The recovery of the internal standard during the isolation procedure should be identical to that of the analyte, and it should not be differentiated from that of the analyte until the final mass spectrometric detection stage. Stable isotopically-labelled analogues generally fulfil this latter requirement. However, the positioning of the isotopic label must be carefully considered in order to reduce the risk of isotopic exchange. Other factors that must be considered include the proportion of unlabelled material in the synthetic isotopically labelled internal standard; and the

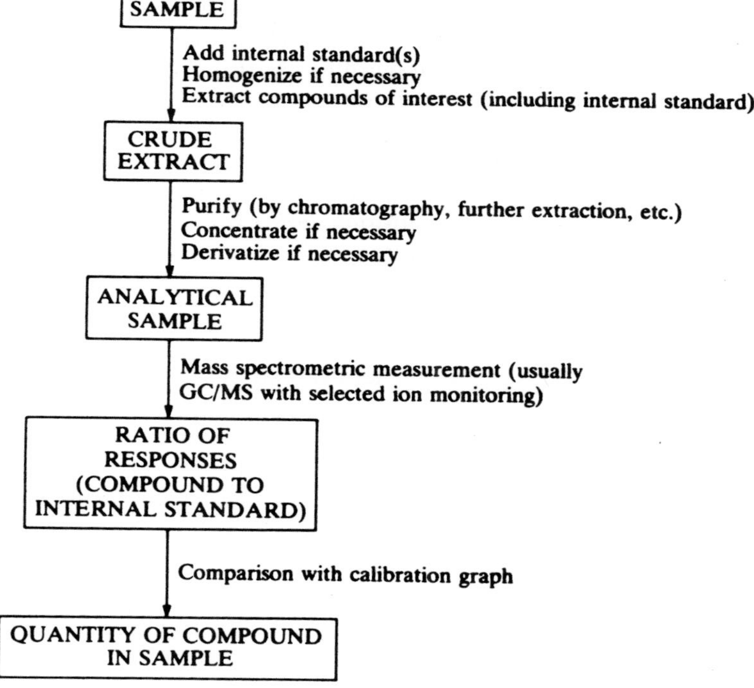

Figure 23. The analytical protocol used in quantitative mass spectrometry. (Reproduced, with permission, from ref. 9.)

possibility for chromatographic separation between the analyte and the isotopically labelled internal standard. Internal standards incorporating 3–4 isotopic labels are considered ideal, although 2 labels may be used in many instances.

The amount of internal standard added should be of a comparable concentration to the analyte to be analysed. The homogenization (equilibration) step is important to ensure that the physico-chemical state of the analyte and internal standard are identical. An acceptable equilibration method would be the addition of the calibration material in a small volume (< 0.02 volume fraction of the sample) of ethanol to a plasma sample followed by several hours equilibration at 4–20 °C (74). Provided the calibration material has been well-chosen then standard isolation and chemical derivatization procedures are employed, as in any analytical protocol, the prime concern is the maximization of the recovery of the internal standard and analyte.

GC-MS, employing isotope dilution, has been widely used in quantitative analyses due to the advantages of including a high-resolution chromatographic step on-line with the mass spectrometer; this combination reduces the possibility for interference from ions of the same *m*/*z* as the analyte and internal standard. Capillary columns are preferred over packed columns, as their higher efficiencies produce enhanced signal-to-noise ratios which greatly improve detection limits. Similar advantages are offered by LC-MS and this technique is finding increasing application as instruments becomes more widely available. Probe techniques are much less used than for sample introduction in isotope dilution analyses than GC-MS and LC-MS, owing to the greater chance of interference of ions from impurities of the same *m*/*z* as the internal standard and analyte.

Low-resolution selected ion monitoring (SIM) is the most commonly used technique for detecting the internal standard and analyte. Before SIM can be performed the mass spectrum of the compound(s) of interest must be carefully inspected; the ions chosen should of reasonable abundance and must not coincide with potential interferences, such as column bleed ions. High mass ions are preferred as this will generally limit the chances of interferences from background ions and co-eluting impurities. The fragment ion at *m*/*z* 368 in the EI spectrum of cholesterol (*Figure 18*) would be a good choice for use in a SIM study. The use of even mass ions is recommended as these occur relatively infrequently in the mass spectra of organic compounds. In quantitative investigations using isotope dilution mass spectrometry it is vital to choose ions which contain the stable isotope labels. The generation of high mass ions can be greatly influenced by the choice of derivatizing agent where compounds are functionalized. When EI does not produce abundant high mass ions CI can be used, with the added advantage that the number of potentially interfering ions will be markedly reduced.

The high scan rates used in SIM studies ensure the frequent sampling of individual capillary GC-MS peaks necessary to maintain quantitative precision.

Quadrupole instruments are particularly effective for use in SIM work, as rapid switching between the selected masses is easily performed. However, single-stage quadrupole instruments are somewhat limited in their performance, where greater selectivity is required to eliminate interfering ions mass spectrometrically. Enhanced selectivity can be achieved very effectively in double-focusing magnetic sector mass spectrometers, by the use of high-resolution SIM. In this mode of operation the mass spectrometer is adjusted to monitor ions of selected elemental compositions. Other methods of enhancing selectivity include metastable peak and selected reaction monitoring; however, these techniques require specialized instrumentation (refs 8 and 9, and references therein.)

Protocol 6. Quantitative analysis of testosterone in saliva by GC/MS–SIM (75)

1. Add [16,16,17–$^2H_3$]testosterone (250 pg for male saliva or 100 pg for female saliva and saliva from prostatic cancer patients) to 5 ml of saliva and allow to equilibrate for 2 h at 20 °C.
2. Add 0.2 ml of a 1:10 dilution in water of anti-testosterone serum coupled to activated cellulose, and mix for 30 min on a roller mixer.
3. Recover the solid phase by centrifugation, wash four times with distilled water, and recover the steroid with methanol.
4. After drying the extract under a stream of dry nitrogen prepare the steroid oximes by adding 10 μl of hydroxylamine hydrochloride in pyridine (10 mg/ml), and heat at 60 °C for 2 h.
5. Evaporate solvent, dissolve residue in methanol and apply to a column (2 × 0.6 cm) of triethylaminohydroxypropyl (TEAP)-Sephadex LH-20. Elute with 2 ml of methanol.
6. Apply this eluate directly on to a column (4 × 0.6 cm) of sulfoethyl (SE)-Sephadex LH-20, and elute with methanol (2 ml), methanol/water (95:5; 2 ml) then methanol/pyridine (95:5; 2 ml).
7. Collect the last eluting fraction from above, evaporate to dryness.
8. Convert to TBDMS derivatives by treatment with 20 μl *tert.*-butyldimethylchlorosilane/imidazole/dimethylformamide at room temperature overnight.
9. Chromatograph on a column (4 × 0.6 cm) of Sephadex LH-20, eluting with 1 ml of chloroform/hexane/methanol.
10. Evaporate the solvent and redissolve in ethyl acetate (20 μl) and take a portion (5 μl) for capillary GC-MS analysis on a 15 m × 0.25 mm i.d. DB-1 coated (0.25 μm film thickness; dimethyl polysiloxane equivalent) column, using a falling needle injector (NB: on-column or split/splitless

Protocol 6. *Continued*

injection systems may be used but dilutions and injection volumes will need to be adjusted accordingly); hold the GC oven temperature at 270 °C.

11. A double-focusing magnetic sector mass spectrometer, operating in the EI mode (70 eV) at an ion-source block temperature of 220 °C, and an MS resolution of 2500 (10% valley) is used to monitor ions of *m/z* 474.32 and 477.34 (the $[M–C_4H_9]^+$ ions of endogenous and trideuteriated internal standard testosterone respectively) with dwell times of 100 ms at each mass.

This method provides a rigorous quantification of testosterone in saliva. The SIM assay is enhanced by the inclusion of the immunoadsorption method of extracting the analyte. The final quantification uses a standard curve relating the response ratio (*m/z* 474/477) to establish the quantity of testosterone in each assay. Sensitivity of detection when the assay was applied to the analysis of saliva was *c.* 4 pmol/litre and the precision of quantification 3.4% CV for a value of 160 pmol/litre.

Protocol 7. Quantitative analysis of cholesterol sulfate in plasma by FAB-MS (76)

1. Add internal standard $[6,6,7–^2H_3]$cholesterol sulfate (2.5 μg dissolved in methanol) to plasma (2 ml), together with methanolic potassium hydroxide (2 M, 2 ml), then heat at 70 °C for 30 min to hydrolyse neutral lipids.
2. Allow to cool, then extract the neutral lipids and cholesterol sulfate by shaking with chloroform (1 × 4 ml) and 15% methanol in chloroform (2 × 4 ml). Disperse any emulsion that forms by centrifugation.
3. Combine extracts from above, evaporate to dryness, redissolve in chloroform/methanol (2 ml; 2:1 v/v), add 10 ml of cyclohexane, and apply to a silica packed Sep-Pak (Waters Assoc., UK) pre-equilibrated by washing with 10 ml each of methanol, chloroform, and cyclohexane).
4. Elute Sep-Pak with cyclohexane/chloroform (6 ml; 2:1 v/v), then chloroform/methanol (6 ml; 2:1 v/v). The cholesterol sulfate elutes in the latter eluate.
5. Reduce the latter eluate to *c.*2 ml, dissolve in cyclohexane and re-chromatograph on a second Sep-Pak using the same sequence of solvents as above.
6. Reduce the final eluate to a small volume, transfer to a small vial and reduce to dryness.

7. Dissolve the residue from above in methanol (10 µl) and mix with glycerol on the FAB probe target, together with 0.5 µl of 5 M NaOH.
8. Record low-resolution (*c.*1000) negative-ion FAB spectra, employing xenon as the primary bombarding beam and accelerating voltage switching to scan the range *m/z* 460–472. Average of 50 scans to obtain quantitative data.

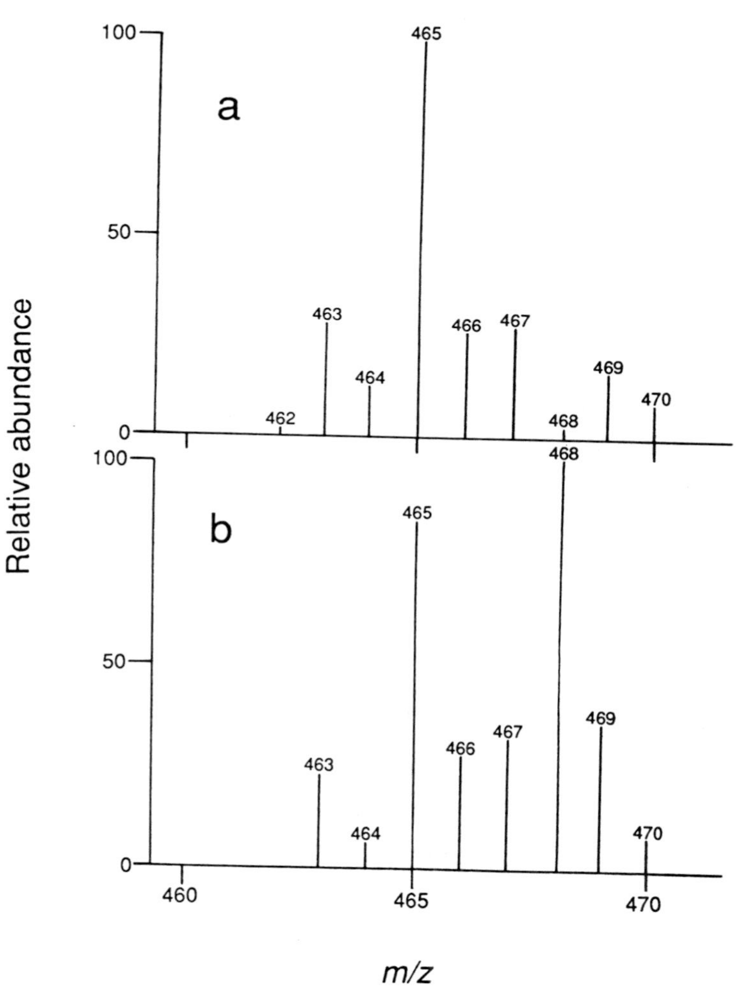

Figure 24. Negative-ion FAB mass spectra of (a) cholesterol sulfate isolated from human plasma and (b) cholesterol sulfate from human plasma containing 2.5 µg of trideuteriated cholesterol sulfate internal standard.

This protocol provides an example of an application of isotope dilution mass spectrometry using a probe technique for sample introduction. The selectivity of the negative-ion FAB ionization technique helps to simplify the sample work-up. *Figure 24* shows the results obtained from the analysis of plasma cholesterol sulfate with and without added deuterium labelled internal standard. The plasma concentration of cholesterol sulfate is calculated by comparing the relative abundances of the *m/z* 465 and 478 ions (after subtraction of a glycerol background) to a calibration curve prepared by serial dilution of pure cholesterol sulfate (5–50 μg) with trideuteriated cholesterol sulfate (2.5 μg) added to each dilution. A new calibration curve should be prepared for each batch of analyses. For healthy human subjects their mean cholesterol sulfate concentration was 1.34 ($\pm$ 0.29, $n = 9$) mg l^{-1}.

References

1. Waller, G. R. (ed.) (1972). *Biochemical applications of mass spectrometry*. John Wiley, New York.
2. Waller, G. R. and Dermer, O. C. (ed.) (1980). *Biochemical applications of mass spectrometry*, first supplementary volume. John Wiley, New York.
3. Lawson, A. M. (ed.) (1989). *Mass spectrometry*. Walter de Gruyter, Berlin.
4. Kuksis, A. (ed.) (1987). *Chromatrography of lipids in biomedical research and clinical diagnosis*. Elsevier, Amsterdam.
5. Christie, W. W. (1987). *HPLC and lipids: a practical guide*. Pergamon Press, Oxford.
6. Christie, W. W. (1989). *Gas chromatography and lipids: a practical guide*. The Oily Press, Ayr.
7. McEwan, C. N. and Larsen, B. S. (ed.) (1990). *Mass spectrometry of biological materials*. Marcel Dekker, New York.
8. Chapman, J. R. (1985). *Practical organic mass spectrometry*. John Wiley, Chichester.
9. Rose, M. E. and Johnstone, R. A. W. (1982). *Mass spectrometry for chemists and biochemists*. Cambridge University Press, Cambridge.
10. Baldwin, M. A. and McLafferty, F. W. (1973). *Org. Mass Spectrom.*, **7**, 1353.
11. Beckey, H. D. (1977). *Principles of field ionisation and field desorption mass spectrometry*. Pergamon Press, Oxford.
12. Barber, M., Bordoli, R. S., Sedgwick, R. D., and Tyler, A. N. (1981). *Nature*, **293**, 270.
13. Aberth, W., Straub, K. M., and Burlingame, A. L. (1982). *Anal. Chem.*, **54**, 2029.
14. McFadden, W. H., Schwartz, H. L., and Evans, S. (1976). *J. Chromat.*, **122**, 389.
15. Arpino, P. J., Guiochon, G., Krien, P., and Devant, G. (1979). *J. Chromat.*, **185**, 529.
16. Blakley, C. R., Carmody, J. J., and Vestal, M. L. (1980). *J. Am. Chem. Soc.*, **102**, 5931.
17. Winkler, P. C., Perkins, D. D., Williams, W. K. and Browner, R. F. (1988). *Anal. Chem.*, **60**, 489–92.

18. Kuksis, A., Marai, L., Myher, J. J., and Pino, S. (1987). In *Chromatography of lipids in biomedical research and clinical diagnosis* (ed. A. Kuksis), pp. 1–47. Elsevier, Amsterdam.
19. Kuksis, A. and Myher, J. J. (1989). In *Mass spectrometry* (ed. A. M. Lawson), pp. 265–351. Walter de Gruyter, Berlin.
20. Odham, G. (1980). In *Biochemical applications of mass spectrometry* (ed. G. R. Waller and O. C. Dermer), p. 153. Wiley–Interscience, New York.
21. Woolard, P. M. (1983). *Biomed. Mass Spectrom.*, **10**, 143.
22. Vetter, W., Walther, W., and Vecchi, M. (1971). *Helvetica Chim. Acta*, **54**, 1500.
23. Anderson, B. A. and Holman, R. T. (1974). *Lipids*, **9**, 185.
24. Anderson, B. A. (1978). *Prog. Chem. Fats other Lipids*, **16**, 279.
25. Harvey, D. J. (1982). *Biomed. Mass Spectrom.*, **9**, 33.
26. Christie, W. W., Brechany, E. Y., Johnson, S. B., and Holman, R. T. (1986). *Lipids*, **21**, 657.
27. Harvey, D. J. (1984). *Biomed. Mass Spectrom.*, **11**, 187.
28. Harvey, D. J. (1984). *Biomed. Mass Spectrom.*, **11**, 340.
29. Harvey, D. J. and Tiffany, J. M. (1984). *J. Chromat.*, **301**, 173.
30. Harvey, D. J., Tiffany, J. M., Duerden, J. M., Pander, K. S., and Menger, L. S. (1987). *J. Chromat.*, **414**, 253.
31. Harvey, D. J. (1989). *Adv. Mass Spectrom.*, **11**, 1472.
32. Christie, W. W., Brechany, E. Y., and Holman, R. T. (1987). *Lipids*, **22**, 224.
33. Christie, W. W., Brechany, E. Y., Gunstone, F. D., Lie Ken Jie, M. S. F., and Holman, R. T. (1988). *Lipids*, **22**, 664.
34. Christie, W. W., Brechany, E. Y., and Lie Ken Jie, M. S. F. (1988). *Chem. Phys. Lipids*, **46**, 225.
35. Christie, W. W., Brechany, E. Y., and Stefanov, K. (1988). *Chem. Phys. Lipids*, **46**, 127.
36. Jensen, N. J. and Gross, M. L. (1987). *Mass Spectrom. Rev.*, **6**, 497.
37. Dommes, V., Wirtz-Peitz, F., and Kunau, W.-H. (1976). *J. Chromat. Sci.*, **14**, 360.
38. Minnikin, D. E. (1978). *Chem. Phys. Lipids,*, **21**, 313.
39. Capella, P. and Zorzut, C. M. (1968). *Anal. Chem.*, **40**, 1458.
40. Niehaus, W. G. and Ryhage, R. (1968). *Anal. Chem.*, **40**, 1840.
41. Francis, G. W. and Veland, K. (1981). *J. Chromat.*, **219**, 3799.
42. Francis, G. W. (1981). *Chem. Phys. Lipids*, **29**, 369.
43. Nichols, P. D., Guckert, J. B., and White, D. C. (1986). *J. Microbiol. Met.*, **5**, 49.
44. Vincenti, M., Guglielmetti, G., Cassani, G., and Tonini, C. (1987). *Anal. Chem.*, **59**, 694.
45. Barber, M., Merren, T. A., and Kelly, W. (1964). *Tetrahedron Lett.*, **18**, 1063.
46. Murata, T. (1977). *Anal. Chem.*, **49**, 2209.
47. Evershed, R. P., Prescott, M. C., and Goad, L. J. (1990). *Rapid Commun. Mass Spectrom.*, **4**, 345.
48. Marai, L., Myher, J. J., and Kuksis, A. (1983). *Can. J. Biochem. Cell Biol.*, **61**, 840.
49. Hasegawa, K. and Suzuki, T. (1973). *Lipids*, **8**, 631.
50. Hasegawa, K. and Suzuki, T. (1975). *Lipids*, **10**, 667.
51. Curstedt, T. (1974). *Biochim. Biophys. Acta*, **360**, 12.

52. Horning, M. G., Casparrini, G., and Horning, E. C. (1969). *J. Chromat. Sci.*, **7**, 267.
53. Barber, M., Chapman, J. R., and Wolstenholme, W. A. (1968). *J. Spectrom. Ion Phys.*, **1**, 98.
54. Myher, J. J., Kuksis, A., and Yeung, S. K. F. (1978). *Anal. Chem.*, **50**, 557.
55. Satouchi, K. and Saito, K. (1979). *Biomed. Mass Spectrom.*, **6**, 396.
56. Johnson, C. B. and Holman, R. T. (1966). *Lipids*, **1**, 371.
57. Wakeham, S. G. and Frew, N. M. (1982). *Lipids*, **17**, 831.
58. Vetter, W. and Meister, W. (1981). *Org. Mass Spectrom.*, **16**, 118.
59. Harvey, D. J. (1990). *Spectrosc. Int. J.*, **8**, 211.
60. Villani, F. J. and King, M. S. (1963). *Org. Synth., Coll. Vol.*, **4**, 88.
61. Harvey, D. J. and Tiffany, J. M. (1984). *Biomed. Mass Spectrom.*, **11**, 353.
62. Budzikiewicz, M., Djerassi, C., and Williams, D. H. (1967). *Mass spectrometry of organic compounds*. Holden-Day, San Francisco.
63. Morris, W. R. (1969). *Biochem. Biophys. Acta*, **176**, 537.
64. McDaniel, C. A. and Howard, R. W. (1985). *J. Chem. Ecol.*, **11**, 303.
65. Nes, W. R. (1985). In *Methods in enzymology*, Vol. 111 (ed. J. H. Law and H. C. Rilling), p. 3. Academic Press, Orlando, Florida.
66. Rahnier, A. and Benveniste, P. (1989). In *Analysis of sterols and other biologically significant steroids* (ed. W. D. Nes and E. J. Parish), pp. 223–49. Academic Press, New York.
67. Goad, L. J. (1991). In *Methods in plant biochemistry*, Vol. 7 (ed. B. V. Charlwood and D. V. Barnthorpe), pp. 369–434. Academic Press, New York.
68. Fenwick, R. G., Eagles, J., and Self, R. (1983). *Biomed. Mass Spectrom.*, **10**, 382.
69. Pramanik, B. N., Zechman, J. M., Das, P. R. and Bartner, P. L. (1990). *Biomed. Environ. Mass Spectrom.*, **19**, 164–70.
70. Munster, H., Stein, J., and Budzikiewcz, H. (1986). *Biomed. Environ. Mass Spectrom.*, **13**, 423–7.
71. Jensen, N. J., Tomer, K. B., and Gross, M. L. (1987). *Lipids*, **22**, 480.
72. Kim, H. Y. and Salem, N. (1986). *Anal. Chem.*, **58**, 9.
73. Kim, H. Y. and Salem, N. (1987). *Anal. Chem.*, **59**, 722.
74. Lawson, A. M., Gaskell, S. J., and Hjelm, M. (1985). *J. Clin. Chem. Clin. Biochem.*, **23**, 433.
75. Gould, V. J., Turkes, A. O., and Gaskell, S. J. (1986). *J. Steroid Biochem.*, **24**, 563.
76. Veares, M. P., Evershed, R. P., Prescott, M. C., and Goad, L. J. (1990). *Biomed. Environ. Mass Spectrom.*, **19**, 583.
77. Evershed, R. P. and Prescott, M. C. (1989). *Biomed. Environ. Mass Spectrom.*, **18**, 503.

Appendix
Suppliers of specialist items

Alltech Associates Inc., 2051 Waukegan Road, Deerfield, IL 60015, USA.

Alltech Applied Science Ltd., Units 6–7, Kellet Road Industrial Estate, Carnforth, Lancs LA5 9XP, UK.

Amersham International plc, Lincoln Place, Green End, Aylesbury, Buckinghamshire HP20 2TP, UK.

Anachem, Charles Street, Luton, Bedfordshire LU2 0EB, UK.

Analtech Inc., 75 Blue Hen Drive, PO Box 7558, Newark, Delaware 19711, USA.

Bast of Copenhagen, Ingerslevsgade 44 DK-1705, Copenhagen V, Denmark.

BDH Ltd., Apparatus Division, PO Box 8, Dagenham, Essex RM8 1RY, UK.

Beckman Nuclear Systems Operations, 2500 Harbor Boulevard, Fullerton, CA 92634, USA.

Camag, Sonnenmattstrasse 11, CH-413 Huttenz, Switzerland.

Camag Scientific Inc., PO Box 563, Wrightsville Beach, NC 284480, USA.

Camlab Ltd., Nuffield Road, Cambridge CB4 1TH, UK.

Drummond Scientific Co., 500 Parkway, Broomall, PA 19008, USA.

Dyson Instruments Ltd., Hetton Lyons Industrial Estate, Hetton, Houton-le-Spring, Tyne and Wear DH5 0RH, UK.

Fluka Chemika-BioChemika, Industriestrasse 25, CH 9470, Bucks, Switzerland.

Hamilton Bonaduz AG, PO Box 26, Ch-7402 Bonaduz, Switzerland.

Hamilton Company, PO Box 10030, Reno, Nevada 89520, USA.

V. A. Howe Ltd., 12–14 St Ann's Crescent, London SW18 2LS, UK.

Iatron Laboratories Inc., 1–11–4 Higashi-Kanda, Chiyoda-ku, Tokyo 101, Japan.

Kodak Clinical Diagnostics Ltd., Mandeville House, 62 The Broadway, Amersham, Bucks HP7 OH7.

Macherey-Nagel & Co., D-5160 Duren, Werkstrasse 6–8, Germany.

E. Merck, Frankfurterstrasse 250, D6100, Darmstadt 1, Germany.

Molecular Dynamics Ltd., 4 Chaucer Business Park, Kemsing, Sevenoaks, Kent TN15 6PL, UK.

New England Nuclear, Du Pont (UK) Ltd., Medical Products Department,

Biotechnology Systems Division, Wedgwood Way, Stevenage, Hertfordshire SG1 6YH, UK.

Newman-Howells Associates Ltd., Wolvesey Palace, Winchester, Hants SO23 9NB, UK.

Nu Chek Prep. Inc., PO Box 295, Elysian, MN 56028, USA.

Pharmacia LKB Biotechnology AB, Bjorkgatan 30, S-751 82 Uppsala, Sweden.

Pharmacia LKB Biotechnology, 800 Centennial Avenue, PO Box 1327, Piscataway, NJ 08855–1327, USA.

Packard Instrument Company, 2200 Warrenville Road, Downers Grove, IL 60515, USA.

Shimadzu Corporation, International Marketing Division, Shinjuku Mitsui Bldg, 1–1, Nishi-Shinjuku 2-chome, Shinjuku-ku, Tokyo 163, Japan.

Sigma Chemical Company, PO Box 14508, St Louis, MO 63178, USA.

Sigma Chemical Company Ltd., Fancy Road, Poole, Dorset BH17 7TG, UK.

Supelchem, Shire Hill, Daffron Walden, Essex CB11 3AZ, UK.

Supelco Inc., Supelco Park, Bellefonte, PA 16823–0048, USA.

Trivector Scientific Ltd., Sunderland Road, Sandy, Bedfordshire SG19 1RB, UK.

Waters Associates, Milford, Massachusetts, USA.

Whatman Laboratories Ltd., Unit 1, Coldred Road, Parkwood, Maidstone, Kent, UK.

Index